Praise for **BEHEMOTH**

"You may have no detailed knowledge of factories except that they can be converted into cool lofts. In that case, you'll learn much from historian Joshua Freeman." —Jonathan Rose, *Wall Street Journal*

"[An] immersive, trivia-packed history." —Henry Grabar, *Slate*

"An effortless and engaging guide." —David Sessions, *New Republic*

"Freeman has written a superb account.... Almost every page contains a memorable fact or an intriguing thought." —Ian Jack, *Guardian*

"Excellent." —*National Book Review*

"Freeman uses the history of the factory as a way to re-examine how workers are treated worldwide." —Bradley Babendir, *Pacific Standard*

"A global tour of three centuries, from English textile mills to Detroit steel plants to Chinese iPhone factories." —Tom Beer, *Newsday*

"Fascinating." —Richard N. Cooper, *Foreign Affairs*

"[A] compulsively readable cultural history of the birth and development of factories and their impact on society."
—Henry L. Carrigan Jr., *BookPage*

"A tour de force. . . . Taking us on a whirlwind tour of giant factories from Lancashire to Detroit, from Wolfsburg to Nova Huta, from Mahalla al-Kubra to Shenzhen, Freeman charts the rise and spread of this peculiar form of organ[...] *B[eh]emoth* is an extraordinary book that offers cou[...] [...]ry of the past two centuries." [...] Beckert

"*Behemoth* is a wonderful book—capacious, beautifully written, and filled with brilliant historical insights that help shed light on why the future of manufacturing is of such tremendous political importance today."

—Kim Phillips-Fein

"A bracing, morally insightful account of how modernist dreams have so often turned into the nightmares that entrap millions of workers and debase whole societies. Joshua B. Freeman writes history of the boldest and most provocative sort."

—Nelson Lichtenstein

BEHEMOTH

BEHEMOTH

A HISTORY OF THE FACTORY
AND THE MAKING OF THE
MODERN WORLD

Joshua B. Freeman

W. W. NORTON & COMPANY

Independent Publishers Since 1923

NEW YORK LONDON

For information about permission to reproduce selections from this book,
write to Permissions, W. W. Norton & Company, Inc.,
500 Fifth Avenue, New York, NY 10110

For information about special discounts for bulk purchases, please contact
W. W. Norton Special Sales at specialsales@wwnorton.com or 800-233-4830

Manufacturing by LSC Communications Harrisonburg
Book design by Lovedog Studio
Production manager: Anna Oler

Library of Congress Cataloging-in-Publication Data

Names: Freeman, Joshua Benjamin, author.
Title: Behemoth : a history of the factory and the making of the modern world /
Joshua B. Freeman.
Description: First Edition. | New York : W. W. Norton & Company, [2018] |
Includes bibliographical references and index.
Identifiers: LCCN 2017051960 | ISBN 9780393246315 (hardcover)
Subjects: LCSH: Factory system—History. | Industrial revolution—History.
Classification: LCC HD2351 .F68 2018 | DDC 338.6/4409—dc23
LC record available at https://lccn.loc.gov/2017051960

ISBN 978-0-393-35662-5 pbk.

W. W. Norton & Company, Inc., 500 Fifth Avenue, New York, N.Y. 10110
www.wwnorton.com

W. W. Norton & Company Ltd., 15 Carlisle Street, London W1D 3BS

1 2 3 4 5 6 7 8 9 0

As always,
for Debbie, Julia, and Lena

Rereading your book has made me regretfully aware of our increasing age. How freshly and passionately, with what bold anticipations, and without learned and systematic, scholarly doubts, is the thing still dealt with here! And the very illusion that the result will leap into the daylight of history tomorrow or the day after gives the whole thing a warmth and vivacious humour—compared with which the later "gray in gray" makes a damned unpleasant contrast.

—Karl Marx, in an 1863 letter to Friedrich Engels
about *The Condition of the Working Class in England*

At sea, the sailors . . . manufacture a clumsy sort of twine, called *spun-yarn*. . . . For material, they use odds and ends of old rigging called "*junk*," the yarn of which are picked to pieces, and then twisted into new combinations, something as most books are manufactured.

—Herman Melville,
Redburn: His First Voyage (1849)

Contents

INTRODUCTION

WE LIVE IN A FACTORY-MADE WORLD, OR AT LEAST most of us do. Almost everything in the room I am writing in came from a factory: the furniture, the lamp, the computer, the books, the pencils and pens, the water glass. So did my clothes, shoes, wristwatch, and cell phone. Much of the room itself was factory made: the sheetrock walls, the windows and window frames, the air conditioner, the parquet floor. Factories produce the food we eat, the medicines we take, the cars we drive, the caskets we are buried in. Most of us would find it extremely difficult to survive, even for a brief time, without factory-made goods.

Yet in most countries, except for factory workers themselves, people pay little attention to the industrial facilities on which they depend. Most consumers of factory products have never been in a factory, nor do they know much about what goes on inside one. In the United States, it is the absence of factories rather than their presence that gets publicized. The loss of roughly five million manufacturing jobs between 2000 and 2016[1] led to sharp critiques, from the right and the left, of the international trade agreements blamed for their disappearance. Factory jobs are deemed "good jobs," with little examination of what they actually entail. Only occasionally do factories themselves become a big story, as when in 2010 the mistreatment of Chinese workers who assembled iPhones and other electronic gear briefly became subject to international scrutiny.

Things weren't always this way. Factories, especially the largest and most technically advanced, were once objects of great wonder. Writers, from Daniel Defoe and Frances Trollope to Herman Melville and Maxim Gorky, marveled at them, or were horrified. Tourists, ordinary and celebrated—Alexis de Tocqueville, Charles Dickens, Charlie Chaplin, Kwame Nkrumah—visited them. In the twentieth century, they became a favorite subject of painters, photographers, and filmmakers, leading artists like Charles Sheeler, Diego Rivera, and Dziga Vertov. Political thinkers, from Alexander Hamilton to Mao Zedong, debated their significance.

From eighteenth-century England on, observers recognized the revolutionary nature of the factory. Factories visibly ushered in a new world. Their novel machinery, workforces of unprecedented size, and outflow of uniform products all commanded attention. So did the physical, social, and cultural arrangements invented to accommodate them. Producing vast quantities of consumer and producer goods, giant industrial enterprises brought a radical break from the past, in material life and intellectual horizons. The large factory became an incandescent symbol of human ambition and achievement, but also of suffering. Time and again, it served as a measuring rod for attitudes toward work, consumption, and power, a physical embodiment of dreams and nightmares about the future.

In our time, the ubiquity of factory-made products and the lack of novelty in the existence of the factory has dulled appreciation of the extraordinary human experience associated with it. At least in the developed world, we have come to take factory-made modernity for granted as a natural condition of life. Yet it is anything but. Only a brief flash in the history of humankind, the age of the factory does not go as far back as Voltaire's first play or the whaling ships of Nantucket. The creation of the factory required exceptional ingenuity, obsession, and misery. We have inherited its miraculous productive power and long history of exploitation without giving it much thought.

But we should. The factory still defines our world. For nearly half a century, scholars and journalists in the United States have been announcing the end of the industrial age, seeing the country as transforming into a "postindustrial society." Today, only 8 percent of American workers are in manufacturing, down from 24 percent in 1960. The factory and its workers have lost the cultural purchase they once had. But worldwide, we are in a heyday of manufacturing. According to data compiled by the International Labor Organization, in 2010 nearly 29 percent of the global workforce labored in "industry," down only a bit from a 2006 prerecession high of 30 percent and considerably above the 1994 figure of 22 percent. In China, the world's largest manufacturer, in 2015, 43 percent of the workforce was employed in industry.[2]

The biggest factories in history are operating right now, making products like smartphones, laptops, and brand-name sneakers that for billions of people around the world define what it means to be modern. These factories are staggeringly large, with 100,000, 200,000, or more workers. But they are not without precedent. Outsized factories have been a feature of industrial life for more than two centuries. In each era since the factory arrived on the stage of history, there have been industrial complexes that have stood out on the social and cultural landscape by dint of their size, their machinery and methods, the struggles of their workers, and the products they produced. Their very names—Lowell or Magnitogorsk or now Foxconn City—have broadly evoked sets of images and associations.

This book tells the story of these landmark factories as industrial giantism migrated from England in the eighteenth century to the American textile and steel industries in the nineteenth century, the automobile industry in the early twentieth century, the Soviet Union in the 1930s, and the new socialist states after World War II, culminating in the Asian behemoths of our time. In part, it is an exploration of the logic of production that led at some times and places to the intense concentration of manufacturing in massive, high-profile facilities and at other times and places to its dispersion and social invisibility. Equally, it is a study of how and why giant factories became carriers of dreams and nightmares associated with industrialization and social change.

The factory led a revolution that transformed human life and the global environment. For most of human history, up to the initial stirrings of the Industrial Revolution and the creation of the first factories in the early eighteenth century, the vast majority of the world population was rural and poor, living precarious existences plagued by hunger and disease. In England, in the mid-eighteenth century, life expectancy did not reach forty, while in parts of France only half of all children lived to see their twentieth birthday. Average annual per capita growth of global economic output during the period between the birth of Jesus and the first factory was essentially zero. But in the eighteenth century it began nudging up and between 1820 and 1913 approached 1 percent. In the years since it has been higher, with a peak, between 1950 and 1970, of nearly 3 percent. The cumulative effect of the increased production of goods and services has been utterly transformative, measured most basically in life expectancy, now over eighty in the United Kingdom, a bit higher in France, and nearly sixty-nine globally. Steady supplies of food, clean water, and decent sanitation have become the norm in much of the world, no longer restricted to tiny pockets of the wealthy in the most advanced areas. Meanwhile, the surface of the earth, the composition of the oceans, and the temperature of the air have been profoundly altered, to the extent of threatening the species itself. Not all of this was strictly the result of the Industrial Revolution, let alone the giant factory, but much of it was.[3]

In both capitalist and socialist countries, the giant factory was promoted as a way to achieve a new and better way of life through increased efficiency and output from advanced technology and economies of scale. More than simply a means to boost profits or reserves, large-scale industrial projects were seen as instruments for achieving broad social betterment. As factories came to embody the idea of modernity, their physical structures and processes were hailed by writers and artists for their symbolic and aesthetic characteristics. But even as giant factories inspired utopian dreams and reveries of machine worship, they also brought on fears about the future. For many workers, social critics, and artists, the big factory meant proletarian misery, social conflict, and ecological degradation.

Understanding the history of giant factories can help us think about what kind of future we want. The outsized factory has been a marvel at reducing unit costs and pouring out massive quantities of goods. Yet these testaments to human ingenuity and labor often proved short-lived. Most of the facilities discussed in this book no longer exist or function at much reduced scales of operation. In Europe, the Americas, and most recently Asia, the abandoned factory has become a distressing, all-too-common sight. The concentration of production in a few massive complexes again and again created vulnerabilities, as pools of available workers dried up and employees began asserting claims to proper compensation, humane treatment, and democratic voice (demands manufacturers in many countries are confronting today). Heavy capital investment reduced flexibility when new products and production techniques emerged. Industrial wastes and heavy energy consumption led to ecological despoilment. What has kept the model of industrial giantism alive has not been its sustainability in any one locale but its reemergence, over and over, in new places, with new workforces, natural resources, and conditions of backwardness to be exploited. Today, as we may well be witnessing the historic apogee of the giant factory, economic and ecological conditions suggest that we need to rethink the meaning of modernity and whether or not it should continue to be equated with ever more material production in vast, hierarchically organized industrial facilities of the kind that were the bane and the glory of the past.

As once landmark factories in Europe and the United States closed, leaving behind physical ruins and social misery, a nostalgia for the factory and its world has grown up, particularly in blue-collar communities. Websites lovingly document factories long shuttered, what some scholars have dubbed "smokestack nostalgia" or, more cuttingly, "ruin porn." There are literary versions, too. In an essay about Philip Roth, Marshall Berman noted his novel *American Pastoral*'s theme of "the tragic ruin of America's industrial cities." Roth "writes vividly about the decay, but his writing really takes off when he tries to imagine the city as a Utopia of industry. The voice he develops to tell this story could be called Industrial Pastoral. The common feeling here is that life was far more 'real' and more 'authentic' yesterday, when men in boots made things, than it is today, when it is a lot harder to say what it is we do all day." Berman reminds us, "One important quality of pastoral vision is that it leaves out dirty work."[4]

Some of the power of factory nostalgia comes from the association of the factory with the idea of progress. Out of the Enlightenment emerged the notion that through human effort and rationality the world could be transformed toward greater abundance, well-being, and moral order, a central belief of both the entrepreneurs who led the Industrial Revolution and the socialists who were their harshest critics. The factory was repeatedly portrayed as an instrument of progress, an almost magical means to achieve modernity, part of a larger Promethean project that also brought us the great dams, power plants, railways, and canals that have transformed the surface of our planet.

Today, for many people, the very idea of progress seems quaint, even murderous, an artifact of the Victorian era that could not survive world war, genocide, and abundance. The modern appears old-fashioned in a declared-to-be postmodern world. For others, the notion of progress retains a powerful grip on their imaginations and a deep moral significance, informing a yearning for a return to—or arrival at—a world of large-scale industry.

Understanding the giant factory requires coming to grips with the ideas of progress and modernity. Rather than a narrow exercise in the study of architecture, technology, or industrial relations, a full history of the giant factory takes us beyond factory walls to changing moral, political, and aesthetic sensibilities and the role of the factory in producing them.

Modernity, with which the factory has been linked, is a slippery term. It can simply denote the quality of being modern, something contemporary, existing at the current moment. But it often has served as more than a neutral categorization. Until the nineteenth century, the modern usually was unfavorably compared to the past. Then, in the age of the factory, modern increasingly came to connote improved, desirable, the best that can be. Modern entailed a disavowal of the past, a rejection of the old-fashioned for the most up-to-date, an embrace of progress. One dictionary defines modernity as "characterized by departure from or repudiation of traditional ideas, doctrines, and cultural values in favour of contemporary or radical values and beliefs."

Modernism in the arts and literature, arising in the nineteenth century, took modernity as its battle cry, in what Jürgen Habermas called "the cult of the New," even as it sometimes critiqued or mocked it. Novelty became its own virtue, a weapon in an assault on conventional values and ruling authorities. The factory system and the dizzying rate of change it made possible were its precondition. Not surprisingly, the factory itself became a favored subject for modernist artists.[5]

This study focuses on very large factories, the largest of their time measured by the number of workers they employed, not all factories.[6] Giant factories served as templates for the future, setting the terms of technological, political, and cultural discussion. They were not typical. Most factories were much smaller and less sophisticated. Very frequently, they had worse conditions for their workers. But giant factories monopolized public attention. Debates about the meaning of the factory tended to focus on the industrial behemoths of the day.

There have been few studies of the factory, let alone the giant factory, that cut across time and space. Rarely has it been considered as an institution in its own right, with a distinctive history, aesthetic, social characteristics, political salience, and ecological impact.[7] But much has been written about particular factories. That is especially true of the factories discussed in the pages that follow, for they were selected in part because they were so celebrated or condemned in their time. Without the work of other scholars, as well as the wealth of journalistic accounts, government reports, visual representations, fictional portrayals, and first-person descriptions, this study would not be possible. The work of my predecessors is particularly impressive because while some factories have been proudly shown off by their creators, many others, from the earliest English textile mills to the giant factories of today, have been carefully shrouded in an effort to protect trade secrets and hide abusive practices.

To many inhabitants of the modern world, the factory may seem distant from their everyday routines and concerns. It is not. Without it, their lives could not exist as they are. Except in some very isolated places, we are all part of the factory system. Given the great costs as well as great benefits of the giant factory, we owe it to ourselves to understand how it came to be.

BEHEMOTH

CHAPTER 1

"LIKE MINERVA FROM THE BRAIN OF JUPITER"

The Invention of the Factory

IN 1721, A STONE'S THROW FROM ALL SAINTS' CHURCH (now Cathedral) in Derby, England, the first successful example of a factory, as we use the term today, was built on an island in the River Derwent. Unlike many older types of buildings—the church, mosque, palace, or fortress, the theater, bathhouse, dormitory or lecture hall, the courtroom, prison, or city hall—the factory is strictly a creature of the modern world, a world it helped create. As far back as the ancient world, there were episodic large assemblages of workers to make war or build structures such as pyramids, roads, fortifications, and aqueducts. But until the nineteenth century, manufacturing generally took place on a far more modest scale, engaged in by craftsmen and their helpers working alone or in small groups or by family members making goods for home consumption. In the United States, as late as 1850, manufacturing establishments on average employed fewer than eight workers.[1]

With John and Thomas Lombe's Derby Silk Mill, the factory seemingly popped into existence fully developed, without infancy.[2] A picture of the mill is immediately recognizable to our eyes as a factory. A five-story, rectangular brick building, its façade punctured by a grid of large windows, in outward appearance it closely resembles thousands of the factories which were to come, including many still operating. Inside

Figure 1.1 Sir Thomas Lombe's Derby Silk Mill in 1835.

it had all the main characteristics of a modern factory: a large work-force engaged in coordinated production using powered machinery, in its case driven by a twenty-three-foot-high waterwheel. The combination of externally powered equipment and numerous people working together in one space might not seem like much today, but at the time it represented the beginning of a new world.[3]

The first factories were built not out of grand social visions but to take advantage of mundane commercial opportunities. The Lombes put up their factory to profit from a shortage of organzine, a kind of silk yarn used for warp. To make cloth, yarn, called the weft, is woven over and under a crossing set of yarns, called the warp. Because alternate strands of warp are repeatedly pulled up to allow the weft to be pushed through, they need to be stronger. To make organzine, long threads produced by silkworms were wound into skeins. These had to be put onto bobbins, twisted, "doubled" with other threads, and then twisted again to make yarn, a process known as silk throwing. While on the continent machines were being used to throw silk, in England it was done using spinning wheels, a process too slow to meet the demand from weavers.

In 1704, a Derby barrister built a three-storied, water-powered mill to house imported Dutch silk-throwing machines, but he proved unable to produce quality yarn. Thomas Lombe, a local textile dealer, tried next, sending his half brother John to northern Italy to study the methods used there. Defying laws banning the disclosure of information about the construction of silk-throwing machinery, he returned with several Italian workers and enough information for the Lombes, working with a local engineer, to build and equip their factory. Children apparently did much of the work inside.

Thomas Lombe claimed that his mill was never a great success, in part because of his difficulty in getting raw silk from Italy. This may have been a strategy to discourage competitors and convince Parliament to extend the patent he took out on his machines. Instead, in 1732 the British government, to promote industrial development, gave Lombe a large cash payment in return for making public a model of his machinery.[4]

The factory system spread slowly. In 1765 there were just seven mills producing organzine, though one, near Manchester, by the end of the century had two thousand workers, a gigantic enterprise by contemporary standards. More common were smaller mills using power-driven machinery to produce tram, a weaker type of silk thread used for weft.[5]

While entrepreneurs, driven by practical calculations, moved cautiously in following the Lombes' footsteps, observers almost immediately recognized the novelty and importance of the Derby mill. Daniel Defoe visited the factory—"a vast Bulk"—in the 1720s, declaring it "a Curiosity of a very extraordinary Nature." Like Charles Dickens's Thomas Gradgrind, fictional archetype of the early industrial age, Defoe, in the face of this modern marvel, fell back upon "Fact, fact, fact!" "nothing but Facts!" Anticipating the gee-whiz wonder of so many future descriptions of large factories, he recounted how the Lombe machinery "contains 26,586 Wheels and 97,746 Movements, which work 73,726 Yards of Silk-thread, every time the Water-wheel goes round, which is three times in one Minute, and 318,504,960 Yards in One Day and Night."[6] James Boswell, who visited the same mill a half century later, in the stream of tourists who came to see this new thing under the sun, more tersely described the machinery as "an agreeable surprize."[7]

Alone, the Derby mill might have remained "a Curiosity of a very extraordinary Nature." But it turned out to be the opening of the factory age. In its wake came ever more factories, which would radically transform the British economy and ultimately world society. The large factory would prove to be the leading edge and the leading symbol of a broader Industrial Revolution that created the world we live in.

Cotton

The lasting importance of the Lombes' factory was not as a template for silk mills but as a template for cotton mills. Limited demand, foreign competition, and difficulty obtaining suitable raw material restricted British silk production. But cotton was a different story, becoming the driving force for the Industrial Revolution and the creation of the factory system we still use today.

Cotton cloth, used for clothing and decoration, long predated the first British cotton mills. By the sixteenth century, textiles produced in India by spinners and weavers working at home with simple, hand-powered equipment were being exported to Europe, West Africa, and the Americas. A century later, they had become a truly global commodity.[8]

Until the late seventeenth century, it would have been rare to have seen someone in Europe wearing cotton clothes; imported cotton textiles were used largely for household decoration. Most clothing was made out of other fibers: wool, flax, hemp, or silk.[9] But the quality and variety of cotton cloth soon made it a favorite for European garments. With increasing population and rising income pushing up demand, local merchants tried to take over at least some of the processes for making cotton textiles from foreign producers, an early example of what would later be called import substitution.[10] Instead of importing calicos—cotton cloth with printed patterns—European traders began buying plain white Indian cloth, which they had decorated by local artisans. By the mid-eighteenth century, large-scale calico printing shops, some with hundreds of workers, were operating in various parts of Europe.[11] English merchants also began weaving imported cotton yarn with flax to produce fustians.[12]

In 1774, Britain ended restrictions on producing and decorating all-cotton textiles, earlier put into place to protect the silk and wool industries. Deregulation, along with fustian production, contributed to ballooning demand for cotton yarn.[13] Merchants, artisans, and entrepreneurs set out to capture the market with locally produced product. But the obstacles they faced were considerable.

Simply getting enough raw cotton was the first problem. Indian producers used Indian-grown cotton, but the European climate was unsuitable for its cultivation. In the late eighteenth century, Britain imported cotton from all over the world, including Asia and various parts of the Ottoman Empire. Supply lagged behind demand, leading to the increasing cultivation of cotton in the Americas using slave labor, first in the West Indies and South America and then, after the introduction of Eli Whitney's cotton gin (patented in 1794), in the southern United States. By the early nineteenth century, over 90 percent of the cotton used in Britain was grown by slaves in the Americas. As British textile production exploded, cotton growers in the United States moved westward into the Mississippi River valley, where a brutal empire arose on the labor of enslaved Africans ("food for the cotton-field," Frederick Douglass called them). Thus, the rise of the factory system, with its association with modernity, was utterly dependent on the spread of slave labor. "Without slavery you have no cotton; without cotton you have no modern industry," wrote Karl Marx—an overstatement, but one with much truth.[14]

The technical demands of turning raw cotton into weft and warp presented a second challenge. As Edward Baines wrote in his 1835 *His-*

tory of the Cotton Manufacture in Great Britain, whereas "silk needs only that the threads spun by the worm should be twisted together, to give them the requisite strength," "[c]otton, flax, and wool, having short and slender filaments, require to be spun into a thread before they can be woven into cloth." The raw cotton used in Britain had individual fibers generally less than an inch long. To convert it into yarn it had to be "carded," combed to pull apart the fibers and line them up in parallel to create a "sliver." Slivers were then drawn out to a prescribed thickness ("roving") and twisted to gain strength. Both the last step and all the processes together were called "spinning."[15]

Until the 1760s, spinning was a domestic industry, with men doing the heavy work of carding, while women used spinning wheels to create finished yarn and children helped out in various ways. As Blaines noted, "the machines used . . . were nearly as simple as those of India." However, it cost more to produce cotton yarn in Britain than in India and the quality was lower, too fragile to use as warp. And there was not enough of it; it took at least three spinners along with a few ancillary workers to keep one weaver (generally male) in yarn, meaning weavers often had to go beyond their own household for supplies, a problem exacerbated by the introduction of the flying shuttle in the 1730s, which greatly increased weaving productivity.[16]

Conditions were ripe for a radical change. Expanding fustian, hosiery, and cotton textile production ensured inventors and investors a payoff if they could increase the output, improve the quality, and lower the cost of cotton yarn. Merchant entrepreneurs already had experience with large-scale production through their organization of extensive networks of domestic spinners and weavers, who were given raw materials by a central agent to make specific types of yarn or cloth and paid by the piece. Though the banking system in the textile districts had limited financial and technical capacity, manufacturers, merchants, and landed gentry had capital resources to back new enterprises. A large, underemployed agricultural workforce constituted a potential labor pool for large-scale industry.[17]

In the last decades of the eighteenth century, English inventors, artisans, and merchant manufacturers developed a series of machines to boost the quality and quantity of locally produced cotton yarn. James Hargreaves developed the first mechanical spinning device in 1764, the jenny. It proved of limited use, since it could only produce weft and required a skilled worker to operate. Richard Arkwright was more successful. A tinkerer, who had had his ups and downs as a barber, wig maker, and public house owner, Arkwright applied for a patent on a spinning machine in 1768 and seven years later for carding equipment. With partners, he first built a mill in Nottingham that used horses to power spinning machines. He soon switched to water power, long used for sawmills, grain mills, mineral-crushing mills, and paper mills, building a factory in Cromford, an isolated spot sixteen miles up the River Derwent from where the Lombes had built their mill. Once he perfected his carding and spinning machinery, Arkwright and various partners built additional factories along the Derwent and then elsewhere. Arkwright's profits from his mills and royalties from his patents made him a very rich man.[18]

In part to circumvent Arkwright's patents, other carding and spinning machines were developed, including Samuel Crompton's spinning mule, giving those seeking to go into cotton yarn manufacturing a choice of equipment, some better suited for warp and some for weft. The boosts in productivity were startling: the earliest jennies increased output per worker sixfold or more, while Arkwright's equipment, once perfected, proved several hundredfold more efficient. In the late eighteenth century, the first power looms for weaving were introduced, mechanizing the next step in textile production. The early looms had many problems and could produce only low-quality fabric. As a result, hand-weaving remained dominant in cotton production until the 1820s and even later in worsted and wool. But with incremental improvements, power looms gradually became the norm in virtually all forms of weaving.[19]

Arkwright's Nottingham mill employed three hundred workers,

Figure 1.2
English inventor and entrepreneur Sir Richard Arkwright in 1835.

about the same number as the Lombes'. His first mill in Cromford was smaller, with about two hundred employees, mostly children. A second mill he put up in Cromford had eight hundred workers. Jedidiah Strutt, a hosiery manufacturer and early partner of Arkwright's, erected a mill complex in Belper, seven miles south of Cromford, that employed 1,200 to 1,300 workers by 1792, 1,500 in 1815, and 2,000 by 1833. The complex of mills in New Lanark, Scotland, which Arkwright helped build but Robert Owen and his partners took over, had 1,600 to 1,700 workers in 1816. By then, steam-powered cotton mills were being erected in urban areas, with several factories in Manchester employing over a thousand workers. The giant factory had arrived.[20]

Why the Giant Factory?

Why did cotton manufacturers adopt factory production? And why did their factories grow so large? This was a subject of considerable discussion both at the time that the first big mills were built and among scholars in more recent times. Popular accounts of the birth of the factory often present it as a technologically driven imperative, the outgrowth of a series of paradigm-shifting inventions, like Arkwright's spinning machines. But as many scholars have shown, there was no simple relationship between mechanical innovation, social organization, and production scale.

Early mechanized spinning equipment did not require a factory environment. The first models of Arkwright's machines were small and could be powered by hand in a cottage setting. That also was true of the early jennies and mules. Arkwright apparently promoted centralized factory production not because of technical considerations but to protect his ability to collect patent royalties. Reasoning that if his machines were widely used in domestic production they inevitably would be copied without his receiving payments, he only licensed his equipment to be used in units of a thousand spindles or more, practical only in large, water-powered mills of the sort he himself constructed (hence his spinning machines were dubbed "water frames"). Even then, Arkwright tried hard to keep information about his equipment secret; in 1772, he wrote to Strutt, "I am Determind for the feuter [future] to Let no persons in to Look at the wor[k]s."[21]

Even as large factories became a familiar sight in the early nineteenth century, they were not the most common mode of production in the British textile industry. Nonfactory production, far from disappearing, continued and even grew in various sectors of the industry. As late as the mid-nineteenth century, many textile manufacturers had both factories for spinning and weaving and networks of domestic handweavers.[22] Furthermore, well into the nineteenth century, the typical British textile mill was small. In 1838, the average cotton mill had 132 workers, the average woolen mill just 39. In Lancashire, the most important textile region, in 1841, only 85 of 1,105 mills employed more than 500 workers.[23]

Factories did not necessarily follow the Lombe/Arkwright model of a single manufacturer operating an entire, powered plant. Some factories housed large numbers of workers using hand-powered equipment. Also, until the 1820s, it was common for mills to rent space and power to multiple small employers. In 1815, two-thirds of Manchester cotton firms occupied only part of a factory. One Stockport mill housed twenty-seven master artisans, who collectively employed 250 workers, a system not unlike that common in metalworking factories, where artisans rented individual work spaces and access to steam power. In the woolen industry, into the mid-nineteenth century, one historian wrote, "multiple tenancy of mills and the subletting of room and power were common features." There were even some "cooperative" mills used by subscribing small producers. In the silk industry, when steam-powered looms began being used in the 1840s and 1850s, the technology was adapted to domestic production. Steam engines were erected at the end of rows of cottages occupied by weavers, each with a few looms, with power transmitted through shafts into the small buildings.[24]

Myriad arrangements, then, of technology, scale of production, and business organization could be found for nearly a century after the first large, water-powered cotton mills were constructed. Only in the mid-nineteenth century did steam- or water-powered equipment located in factories owned and operated by single entities become the dominant model in all the major subdivisions of the British textile industry. And even then, what by the standards of the day could be considered very large factories—mills employing over a thousand employees—were the exception, not the rule, in both urban and rural settings.[25] But the very large mills received a disproportionate amount of attention, both at the time and since, because they were seen as the cutting edge of not only industry and technology but also of social arrangements.[26]

Why did the owners of these facilities choose to go big, to adopt the large, centralized factory model? Charles Babbage, the great English mathematician and inventor, devoted a whole chapter "On the Causes and Consequences of Large Factories" in his influential 1832 book, *On the Economy of Machinery and Manufacturers*. Babbage began with the obvious, that the introduction of machinery tended to lead to greater production volume, resulting in "the establishment of large factories." A leading student of the division of labor, he contended that efficient production units had to be multiples of the number of workers needed for the most efficient division of labor in a particular production process. He also noted various economies of scale. These included the cost of maintenance and repair workers and accounting staff, who would be underutilized in too small a factory. Additionally, centralizing various stages of production in one building reduced transportation costs and made one entity responsible for quality control, making lapses less likely.[27]

But what exactly was large? Babbage elucidated the factors that set a floor on efficient size, but not how to determine optimal size. In the cotton industry, only a few workers were needed to operate each spinning or weaving machine. In practice, during the first decades of the nineteenth century, it seemed as if there were few production economies achieved by the giant cotton factories that midsize or even smaller enterprises did not share. In the late nineteenth century, the pathbreaking economic theoretician Alfred Marshall noted that "There are . . . some trades in which the advantages which a large factory derives from the economy of machinery almost vanish as soon as a moderate size has been reached. . . . [I]n cotton spinning, and calico weaving, a comparatively small factory will hold its own and give constant employment to the best known machines for every process: so that a large factory is only several parallel smaller factories under one roof."[28]

Writing just after Babbage, Leeds journalist Edward Baines echoed some of his explanations for the adoption of the factory model, while adding a few that pointed in a different direction. Centralization, he argued, allowed greater supervision of every stage of production by a skilled overseer. It also lessened the risk of waste and theft of materials. Finally, it facilitated the coordination of various stages of the production process, preventing "the extreme inconvenience which would have resulted from the failure of one class of workmen to perform their part, when several other classes of workmen were dependent upon them."[29]

In sum, centralizing gave manufacturers the ability to better supervise and coordinate labor, the work of many individuals who under the putting-out system would be supervising their own labor (and that of family members) in far-flung domestic settings.

Scholars trying to explain the rise of the factory system have elaborated Baines's arguments. Until the 1970s, historians of industrialization stressed technology as the driving force for change. David Landes began his long chapter on "The Industrial Revolution in Britain" in his 1969 classic, *The Unbound Prometheus,* by stating, "In the eighteenth century, a series of inventions transformed the manufacture of cotton in England and gave rise to a new mode of production—the factory system." New machines opened the possibilities for increased productivity and profits, sparking a series of organizational and social shifts, often quite sudden, including the rise of the large factory and the industrial "revolution" that accompanied it.

The academic neo-Marxist revival that began just as Landes was finishing his tome led to a reconsideration of the story, pointing toward advantages in labor supervision rather than technical superiority in the rise of the factory system. Concentrated workers could be made to work longer and harder than dispersed workers, while creating more consistent products and limiting endemic theft and embezzlement. Thus the example of early factories that brought workers under one roof without introducing power machinery or changes in production methods. But other scholars then challenged the idea that the reorganization of labor accounted for the savings gained by factory production, pointing instead to some of the advantages that Babbage, Baines, and Marshall noted in moving multiple processes, which in the past had been conducted by external agents, into a single location and within a single firm: inventories could be reduced, transportation costs lowered, and production more closely aligned with shifts in demand.[30]

Concurrent with these debates over the reasons for the adoption of the factory model was a growing literature that rejected the idea that the industrialization entailed a radical break with past practices. Rather, economic historians argued, a less visible process of "protoindustrialization" laid the basis for later, more dramatic and widely noted changes that came to be labeled the Industrial Revolution. By the early eighteenth century, in England and elsewhere in Europe, merchants and entrepreneurs were organizing increasingly large networks of home-based producers, selling to broadening markets, and accumulating capital. In the process, urban-based manufacturing migrated to the countryside, where excess and off-season agricultural workers provided a ready labor source. Thus large-scale, rural-based manufacturing already had emerged before the invention of power-driven machinery and large mills, making the seemingly revolutionary leap not quite so great.[31]

Even with the old and new explanations for the rise of the factory system, it remains unclear why cotton mills so quickly reached a very large size, in the 1,000- to 1,500-worker range, but thereafter stopped growing, with new mills tending to be smaller. In the early days of the factory, the economics of water power may have made large-sized plants attractive, given the relative scarcity of sites and the capital investment needed to construct dams and channels to deliver a steady flow to waterwheels. At New Lanark, the largest mill complex in Great Britain, workers had to carve out a hundred-yard rock tunnel to get water to the mill wheels. Steam power provided greater flexibility. While some steam-powered mills also were large, perhaps as a way to quickly grab market share, historian V. A. C. Gatrell suggested that, after the first wave of cotton mill construction, new entrants saw few economies and greater risks in matching the size of the pioneer plants, recognizing that managerial constraints could make larger plants *less* efficient.[32]

Perhaps plant size did not simply reflect economic calculus. At a moment when most wealth in Britain took the form of land ownership or government bonds, large factories provided a way to establish social status. Arkwright built a castle, Willersley, near his Cromford mills, having bought most of the surrounding land. The ex-barber was soon acting as a paternal grandee, building a chapel and school (with compulsory attendance) for the children who made up much of his workforce and sponsoring festivals for his workers. In an extravagant gesture that symbolized the social elevation made possible by his inventions and mills, Arkwright lent the Duchess of Devonshire five thousand pounds to cover her gambling debts. His son, while continuing to operate the family mills, invested heavily in land and government bonds and provided mortgages to the gentry and even the nobility, becoming the richest commoner in Britain. The Strutt family, though better established than Arkwright, followed a similar trajectory. Frances Trollope portrayed the use of a large factory to transform social status in her 1840 novel, *The Life and Adventures of Michael Armstrong the Factory Boy*, with Sir Matthew Dowley building a mansion on an estate from which the "grim-looking chimney cones" of his factory could be seen.[33] As it would do over and over again, the giant factory brought into existence not only a new mode of production but also a new class of wealthy industrialists who sought to join the ruling elite.

Creating the Factory World

Cotton mills were on an entirely different scale than the small commercial and residential buildings in the river valleys and towns where they first appeared. England had big buildings, buildings bigger than the biggest new cotton mills. The great cathedrals were much larger. And in the seventeenth and eighteenth centuries, new types of large urban buildings sprung up: hospitals, barracks, citadels, prisons, colleges, warehouses, and dockyards. But cathedrals and other big buildings had interior spaces organized for very different activities than manufacturing.[34] To accommodate large-scale production, power-driven machinery, and masses of workers, new architectural designs and improved building techniques and materials were needed. Innovations to meet the specific needs of the cotton industry soon spread beyond it, shaping the built environment in England and elsewhere for the next two centuries.

Arkwright apparently modelled his first Cromford mill on the

Lombes', also five stories high. Its "long, narrow proportions, height, range of windows ... and large areas of relatively unbroken interior space," wrote historian R. S. Fitton, "became the basic design in industrial architecture for the remainder of the eighteenth and through the nineteenth centuries." Arkwright's second mill at Cromford was seven stories high and 120 feet long, and a third mill he erected nearby was 150 feet long and topped by a cupola.[35]

Arkwright used timber post and beam construction for his mill interiors, leaving them vulnerable to the ever-present danger of fire, with so much flammable thread and cloth lying around and cotton dust in the air. In the early 1790s, William Strutt (Jedidiah's son) erected a mill with cast-iron columns, ironclad wood beams, and brick-arch floor supports to reduce the fire danger. Soon after, Charles Bage, a friend of Strutt, designed a five-story flax mill that was the world's first completely iron-framed building, forerunner of all the iron- and steel-structured buildings that were to come, including the skyscrapers that steel framing made possible. Improvements to iron beams quickly followed; replacing wood timbers with iron not only reduced the danger of fire but increased the distances which could be spanned, allowing wider floors to accommodate the large, self-acting spinning mules that were introduced in the 1820s. To heat their multistory mills (which reduced thread breakage), Arkwright and the Strutts followed the example of the Lombe mill, designing complex systems to circulate warm air.[36]

Power looms, which became increasingly common in the second and third decades of the nineteenth century, did not easily fit into existing mills, because their operation created such strong vibrations that they could not safely be situated above the ground floor. Instead, it became the common practice to build single-story weaving sheds, often abutting spinning mills or in their yards. To light these extensive structures, their roofs had rows of pitched ridges, with windows on one side of each ridge to bring in indirect sunlight. The "sawtooth roof" soon topped all sorts of industrial buildings and can still be seen on both sides of the Atlantic.[37]

In early textile factories, complex arrangements of shafts and gears distributed power from waterwheels to individual machines. Water power was cheap and efficient, as long as there was a steady flow of water. That meant that mills had to be sited on rivers with substantial, steady flows, like the Derwent. Even then, sometimes there was not enough water, leading some mill owners, including Arkwright, to experiment with using steam engines—recently perfected to drain mines—to lift water into reservoirs, which could steadily supply water to a waterwheel.

The thin supply of labor in the often isolated areas with good mill sites presented a bigger problem. (Arkwright chose Cromford for his mills in part because it was near a lead mine, hoping to hire miners' wives and children.)[38] Using steam power to directly drive spinning and weaving equipment, though more expensive, allowed mills to be built in urban areas, giving access to larger labor pools and obviating the need of mill owners to supply housing.

Technically, only minor modifications were needed to adapt mill design from water to steam power, but the change had huge effects. Steam engines required coal-fired boilers, leading to a vast expansion of the coal industry, which became another driving force of the Industrial Revolution. Steam-powered mills contributed mightily to environmental degradation, both from coal mining and the volumes of soot and black smoke emitted from their boilers. In *Hard Times*, Dickens described the "rattling" and "trembling" of factory steam engines, pistons going up and down "like the head of an elephant in a state of melancholy madness," and boilers spewing out "monstrous serpents of smoke." Black smoke and polluted air came to emblemize Manchester and other urban centers of textile production and the Industrial Revolution itself.[39]

Another innovation, first seen in cotton mills, was the elevator, a clever solution to the challenge of rapidly moving people and material in and out of multistory buildings. Primitive water-powered hoists were installed in several Strutt mills around the turn of the nineteenth century. A large 1834 Stockport mill, designed by William Fairbairn, included a steam-powered elevator in each wing, a device so new that a contemporary description had no language for it, calling the shafts "upright tunnels."

Fairbairn was a key figure in the diffusion of design innovations. His company could provide a complete, fully equipped factory to specification; "The capitalist has merely to state the extent of his resources, the nature of his manufacture, its intended site, and facilities of position in reference to water or coal, when he will be furnished with designs, estimates, and offers." Fairbairn's firm built plants around the world, including a wool factory near Istanbul for the Sultan of Turkey and a giant spinning and weaving complex in Bombay.[40]

Nothing better captured the sense of invention in the textile districts of Britain than the "Round Mill" built at the Strutts' factory complex in Belper. The three-story, circular stone building, divided into eight segments, apparently derived from Samuel and Jeremy Bentham's panopticon. In its center stood an inspection station from which a supervisor could observe activity in the entire building, realizing the ideal of constant surveillance the Benthams championed. The Strutts may have adopted the Benthams' design to minimize the risk from fire, as the central overlooker could shut off any of the building segments by closing doors, isolating the flames and protecting the rest of the structure.[41] Though the Round Mill found few direct imitators, the idea of continual surveillance would become ever more part of the factory regime, never more so than in our own times.

Change rippled outward from the large factory, beyond its own walls. Mill owners had to develop physical, social, and psychological infrastructure to make factory production possible. Simply getting people and material to and from rural mills required extraordinary effort. When Arkwright arrived at Cromford, the nearest road suitable for wheeled vehicles lay miles away; small bales of raw cotton had to be carried over the moors by packhorses until 1820, when mill owners built a new road alongside the Derwent. Even with workers walking as much as four or five miles to work each day, not enough people lived near rural mills to staff them fully. So many early mill owners built housing for their workers, sometimes even what amounted to whole new villages, with churches, schools, inns, and markets.[42]

Feeding so many people clustered near isolated mills also presented a challenge. Some manufacturers set up their own farms to supply their workers with foodstuffs. Very commonly workers received only a small part of their wages in cash with the rest in the form of rent payment for company-owned houses and "truck," goods or credit at company stores ("tommy shops") that sold food, coal, and other supplies, often above market prices and of low quality, a source of smoldering resentment among workers.[43]

Truck helped solve another problem mill owners faced, a shortage of currency with which to pay their employees. Small-denomination coins did not circulate in sufficient quantities to meet big payrolls, which were extremely unusual before the mills. Hoarding exacerbated the problem. Mill owners had to improvise, paying workers with tokens or foreign currency overstamped with new denominations or issuing their own notes, which they hoped local merchants would accept.[44]

Difficult as these challenges were, they paled before the problem of discipline. For Andrew Ure, a leading booster of the emerging factory system, the greatest hurdle mill operators faced was "training human beings to renounce their desultory habits of work, and to identify themselves with the unvarying regularity of the complex automation." Of course, hand-powered, domestic manufacturing—like all work—required discipline, too, but it was a different kind of discipline, with the pace of work keyed to the completion of particular tasks. Much as in farming, intense activity alternated with slack times. Domestic producers interspersed carding, spinning, and weaving with household chores, farming, other kinds of labor, and leisure. Famously, in many trades workers used "Saint Monday" (and sometimes "Saint Tuesday," too) to take care of personal business, recover from hangovers or bring on new ones, socialize, or simply laze about, putting in few hours of productive labor. Recalling the days of hand spinning, one witness testified before an 1819 parliamentary commission that "it was generally the practice to drink the first day or two of the week and attempt to make it up by working very long hours towards the close of the week."

The sometimes romanticized autonomy of domestic labor extended only to heads of households, generally men. Wives, children, apprentices, and employed journeymen did not have the same control over their time; they were subject to external discipline regulating not only their periods and pace of work but all aspects of the production process. Discipline was familial, embedded in their general subservience to the head of their household. It could be harsh but it still was task-oriented, with production for the market intermixed with production for the home, domestic chores, and, if they were lucky, recreation.

By contrast, factory production required coordinated activity by dozens or hundreds of workers, who were expected to start and stop work at the same time, day after day. Companies developed elaborate sets of rules and systems of fines and punishments for their violation. Overseers monitored when workers came and went and what they did inside the mill. Some workers had their activities regulated by the demands of the machines they worked on, having to do a particular task at a particular point in the cyclical operation of the apparatus. Ure pooh-poohed the strain of such machine-paced work; the "piecers" in fine spinning, children charged with retying broken threads, had "at least three-fourths" of each minute off, making it, in his view, an easy job. Friedrich Engels, writing a decade later, saw it differently: "To tend machinery—for example, to be continually tying broken threads—is an activity demanding the full attention of the workers. It is, however, at the same time a type of work which does not allow his mind to be occupied with anything else. . . . [It] gives the operative no opportunity of physical exercise or muscular activity. . . . It is nothing less than torture of the severest kind . . . in the service of a machine which never stops." "In handicrafts and manufacture," Marx wrote in *Capital*, "the workman makes use of a tool, in the factory, the machine makes use of him."[45]

If, as David Landes put it, "The factory was a new kind of prison; the clock a new kind of jailer," that in turn created another problem, how to be punctual in a world in which workers did not own clocks. In the past, workers never needed to be punctual or to key their work to particular moments of time. To enforce the new time discipline, some factories rang morning bells to awaken their workers. In urban districts, workers hired a "knocker-up" who used a long pole to knock on their upstairs window each morning to make sure they arose in time for work. Eventually, the knocker-up became a stock figure on Lancashire music-hall stages, adding to the original meaning of the term a second one that it retains to this day.[46]

Factory Tourism

Although recent scholarship has debunked the idea that the factory system arose from the genius of a few inventors and entrepreneurs who changed everything, drawing a subtler picture of economic and social changes that began long before the Industrial Revolution took off, the Industrial Revolution nonetheless *was* a revolution, and seen as such at the time. Contemporary observers had no doubt that the cotton mill and the changes it wrought represented a technical, economic, and social break from the past. From the late eighteenth century on, factories, factory villages, and manufacturing cities drew tourists, journalists, and philanthropists from continental Europe and North America as well as Great Britain itself.[47] Part of the attraction was their novelty. W. Cooke Taylor, the son of an Irish manufacturer who toured the industrial districts of Lancaster in the early 1840s, wrote that "The steam-engine had no precedent, the spinning-jenny is without ancestry, the mule and the power-loom entered on no prepared heritage: they sprung into sudden existence like Minerva from the brain of Jupiter."[48]

The scale and setting of mill buildings, whether in rural river valleys or crowded industrial cities, startled visitors. British poet laureate Robert Southey wrote that the approach to the New Lanark mills reminded him "of the descent upon the baths of Monchique," built by the Romans in southern Portugal. Like many other observers, Southey searched for precedents to understand the novelty he confronted. The view, he wrote, surprised him because there was "too a regular appearance" of the buildings, which "at a distance might be mistaken for convents, if in a Catholic country." Alexis de Tocqueville, who visited Manchester in 1835, likened mills to "huge palaces," a common comparison in a world with few secular structures of such scale. One German visitor to northern England wrote that he "might have arrived in Egypt since so many factory chimneys ... stretch upwards towards the sky like great obelisks." "Just when they seem engaged in revolutionizing themselves and things, in creating something that has never yet existed," Marx wrote three decades after Southey visited New Lanark, "precisely in such period of revolutionary crisis," people "anxiously conjure up the spirits of the past to their service and borrow from them names, battle cries, and costumes in order to present the new scene of world history in this time-honored disguise and this borrowed language."[49]

Even more than mill buildings themselves, the machinery they contained mesmerized visitors. In *Michael Armstrong*, Trollope wrote of visitors being given a mill tour: "It is the vast, the beautiful, the elaborate machinery by which they were surrounded that called forth all their attention, and all their wonder. The uniform ceaseless movement, sublime in its sturdy strength and unrelented activity, drew every eye, and rapt the observer's mind in boundless admiration of the marvelous power of science!" Trollope bemoaned the visitors' inattention to the child laborers nearby: "Strangers do not visit factories to look at them; it is the triumphant perfection of British mechanism which they come to see." French socialist and feminist Flora Tristan wrote of a steam engine she saw in England: "In the presence of the monster, you have eyes and ears for nothing else."[50]

The modernity of the mills dazzled observers. To lengthen hours of operation, in the early nineteenth century mill owners began installing gaslights, a spectacle that drew visitors from near and far. In *Hard Times*, Dickens described morning in "Coketown" as "The Fairy palaces burst into illumination." The size of the mills and accompanying warehouses even made possible new types of entertainment. Sam Scott drew an immense Manchester crowd in 1837 when he leapt off the roof of a five-story warehouse into the River Irwell, surviving to repeat the stunt in Bolton. Another daredevil, James Duncan Wright, attracted even larger crowds in the 1850s with his act of using a pulley to slide down ropes attached to mill chimneys, which he claimed made him the fastest man alive.[51]

Debating the Factory System

For all the wonder of the buildings and the machines, though, it was the broader social innovation—what came to be called the "factory system"—on which discussion, debate, and conflict centered during the first half of the nineteenth century. An imprecise term, the "factory system" generally referred to the whole new mode of production that came with the factory, including the workforce that had to be assembled, the conditions of labor and of life for those workers, and the impact of the factory on economic and social arrangements. Cooke Taylor, allied with the new manufacturers, recognized that because England was "already crowded with institutions," the rapid development of mechanized factory production "dislocated all the existing machinery of society." "A giant forcing his way into a densely-wedged crowd," he wrote, "extends pain and disturbance to the remotest extremity: the individuals he pushes aside push others in their turn . . . and thus also the Factory system causes its presence to be felt in districts where no manufactures are established: all classes are pressed to make room for the stranger."[52]

For many of its critics, and even some of its supporters, the exploitation of labor, particularly of child labor, became their focus in judging the new system. Though underutilized agricultural workers were a draw for manufacturers, the scale of the factories made recruiting and retaining a workforce a challenge, especially in the countryside. Many local men proved reluctant to take mill jobs, unwilling to submit to the unaccustomed close supervision and discipline that came with them. In any case, mill owners did not want adult men for most positions, preferring women and children whom they could pay less and who did not have the sense of pride and craft that came from apprenticeship training. Mechanical power eliminated the need for most heavy labor, especially in spinning. Instead, the new yarn-making equipment largely required constant monitoring to look for broken threads, full bobbins, and other problems that needed to be quickly addressed, work that necessitated nimble fingers and alert minds but not strength. So mill owners recruited a workforce that was young and primarily female. In 1835 Ure estimated that a third of cotton mill workers in England were under twenty-one; a half in Scotland.[53] Many were very young; at Cromford, some employees were only seven years old (though the firm preferred to hire workers starting between ages ten and twelve). In some spinning mills, virtually the only adults present were overseers. Today, in the United States, factories are associated with masculinity, but in their early days they were spaces largely occupied by women and children.[54]

Conditions for mill workers were harsh. Entering a factory for the first time could be a terrifying experience: the noise and motion of the machinery; the stifling air, full of cotton dust, in many mills kept oppressively warm to reduce breakage; the pervasive stench from the whale oil and animal grease used to lubricate the machinery (before petroleum products were available) and from the sweat of hundreds of laboring people; the pale countenances and sickly bodies of the workers; the fierce demeanor of the overseers, some of whom carried belts or whips to enforce their discipline. In weaving rooms, the deafening clatter of scores of looms, each with a shuttle being batted back and forth some sixty times a minute, made it impossible for workers to hear one another.

Figure 1.3 *Carding, Drawing, and Roving*, a somewhat idealized 1835 illustration of English factory life.

In the early decades, mill owners generally ran their factories day and night, with two twelve- or thirteen-hour shifts (including an hour break for dinner), following the schedule pioneered by the Derby Silk Mill. Children worked both shifts. With Sunday the only day off, work-weeks of over seventy hours were normal. To keep exhausted children awake and working, supervisors and adult workers hit them with straps, hands, and even wooden poles (though there was much debate about how common such abuse was).[55]

Perhaps not surprisingly, early mill owners often found themselves unable to fully staff their mills with willing workers. So some turned to unwilling workers. Workhouses—the prisonlike residences of last resort for orphans and the destitute—were tapped for child workers, whom parish officials apprenticed to mill owners, giving them full legal authority over their charges and making it a criminal act for the children to run away. In Yorkshire, it was not uncommon for 70 percent or more of a mill's workforce to be parish apprentices. At New Lanark, before Robert Owen took over management, some of the apprentices were as young as five years old. Ordinary apprentices, signed up by their parents, also could be jailed for running away. So could workers who signed fixed-term contracts if they quit before their termination date. Further, an 1823 law made any worker who left his or her job without notice liable to three months imprisonment. Thus the power of the state helped assemble and keep in place a workforce for the new factory system. What's more, it was not uncommon for the state and an employer to effectively be one and the same, since mill owners sometimes served as magistrates who judged cases of desertion involving their own workers.[56] Legally unfree labor, not only in the growing of cotton but in the mills themselves, played an essential role in the early decades of the factory system.

Today, in popular discourse and mainstream ideology, the Industrial Revolution is often associated with individual liberty and what is called the free market.[57] But in the early years of the factory system, it was as likely to be dubbed a new form of slavery as a new form of freedom. Joseph Livesey, a well-known journal publisher and temperance campaigner, himself the son of a mill owner, wrote of the apprenticed children he saw in mills during his childhood, "They were apprenticed to a system to which nothing but West Indian slavery can bear any analogy."[58] In *The Life and Adventures of Michael Armstrong*, Trollope wrote that apprenticed paupers suffered "miserable lives, in labour and destitution, *incomparably more severe*, than any ever produced by negro slavery." In its structure, *Michael Armstrong* is a version of the slave rescue narrative, recounting the frustrated efforts of the heroine, the rich daughter of a factory owner, to liberate Armstrong from his villainous apprenticeship at an isolated factory and his ultimate escape.[59]

The metaphor of slavery for factory labor no doubt reflected the intense debate over slavery itself during the early decades of the nineteenth century, leading up to emancipation in the British Empire in 1834. Still, it was a measure of how horrifying factory labor was seen to be that so many observers equated it with chattel slavery. One self-described "Journeyman Cotton Spinner" wrote of the terrible heat in spinning rooms, where workers had no breaks: "The negro slave in the

West Indies, if he works under a scorching sun, has probably a little breeze of air sometimes to fan him; he has a space of ground, and time allowed to cultivate it. The English spinner slave has not enjoyment of the open atmosphere and breezes of heaven." Engels, writing of English textile workers just a few years after Trollope, believed "Their slavery is more abject than that of the negroes in America because they are more strictly supervised." He also bemoaned that, as in slavery, the wives and daughters of workers were forced to gratify the "base desires" of manufacturers. Elsewhere, Engels compared workers under the factory system to "the Saxon serf under the whip of the Norman baron." Similarly, in Benjamin Disraeli's novel *Sybil, or the Two Nations*, one character pronounces that "There are great bodies of the working classes of this country nearer the condition of brutes than they have been at any time since the Conquest." Richard Oastler entitled his 1830 letter in the *Leeds Mercury*, which launched the Ten Hours Movement to reduce factory working hours, "Yorkshire Slavery."[60]

For Robert Southey, the association of slavery with the factory system did not stem from particular abuses but from the nature of the system itself. Calling the New Lanark mills under Owen, who even before his radical turn was known for his humane treatment of workers, "perfect of their kind," he nonetheless felt that "Owen in reality deceives himself. He is part-owner and sole Director of a large establishment, differing more in accidents than in essence from a plantation: the persons under him happen to be white, and are at liberty by law to quit his service, but while they remain in it they are as much under his absolute management as so many negro-slaves." The factory system, Southey believed, even at its best, tended "to destroy individuality of character and domesticity." At its worst it was outright devilish; after visiting a Manchester cotton factory, he wrote "that if Dante had peopled one of his hells with children, here was a scene worthy to have supplied him with new images of torment."[61]

Some critics of the factory system—and some defenders of slavery—questioned the very distinction between free labor and slavery, given the circumstances in which mill workers lived. British laborers were "slaves of *necessity,*" wrote Samuel Martin in 1773, unable to "mitigate their labours" or "increase their wages." Owen asked of factory operatives, "Are they, in anything but appearance, really free labourers? . . . What alternative have they or what freedom is there in this case, but the liberty of starving?"[62] Here lay a critique that went to the very heart of the spread of market relations, part and parcel of the Industrial Revolution.

Besides the ill-treatment of labor, environmental despoilment figured heavily in critiques of the factory system. Over and over again, accounts of Manchester and other industrial centers noted the darkness and foul air. Scottish geologist Hugh Miller wrote of Manchester in 1845: "One receives one's first intimation of its existence from the lurid gloom of the atmosphere that overhangs it." Similarly, Cooke Taylor wrote, "I well remember the effect produced on me . . . when I looked upon the town . . . and saw the forest of chimneys pouring forth volumes of steam and smoke, forming an inky canopy which seemed to embrace and involve the entire place." The air was so polluted, Taylor observed, that everyone who could live outside Manchester proper did so.[63] Major General Sir Charles James Napier, appointed in 1839 to command the northern district of England, which included Manchester, described the city as "the entrance to hell realized," with its rich and poor, immorality, and pervasive pollution; the whole city, was "a chimney."[64]

Water pollution was as severe as air pollution. Hugh Miller recounted the befouling of the River Irwell from cloth dyes, sewage, and other waste, so it resembled "considerably less a river than a flood of liquid manure, in which all life dies."[65] Maybe the most impressive aspect of Sam Scott's leap was not the five-story drop but his surviving the toxic brew into which he plunged.

The environmental damage of cotton manufacturing extended far beyond mill sites themselves. Cotton growing required deforestation and it rapidly depleted soil, one reason why in the United States it migrated (along with its slave labor force) from the eastern seaboard

Figure 1.4 *Cotton Factories, Union Street, Manchester*, an 1835 engraving showing the proliferation of factories in England and resulting pollution.

to the Mississippi Valley. Coal mining polluted rivers and scarred the landscape.[66]

Perhaps the most famous critique of the factory system—at least the most remembered in our era—captured its despoilment of nature in just a few words, William Blake's decrying of the "dark Satanic Mills" that blotted England's "mountains green" and "pleasant pastures," in an 1804 verse that formed part of the preface to his long visionary poem *Milton*. Set to music in 1916 under the title "Jerusalem," Blake's words today are sung throughout the English-speaking world, in churches and soccer stadiums alike. At least in part, Blake seemed to be reacting directly to the smoke-blackened sky that was becoming a feature of urban English life. Near his home in London, a large, steam-powered grinding mill operated until consumed by fire in 1791 (by some reports as a result of arson by angry workers). Yet for Blake, it was not just smoke that made mills "Satanic." For the great mystical poet, the mill symbolized a spiritual descent from a preindustrial England on which God had smiled, a metaphor for a whole way of life Blake was determined to overcome in order to build a new Jerusalem "In England's green & pleasant Land."[67]

Urban poverty was often portrayed as another form of despoilment, another fall from grace. The mechanization of the cotton industry brought an enormous increase in the population of the districts in which factories were located. The population of Lancashire almost doubled, from 163,310 in 1801 to 313,957 in 1851. "What was once an obscure, poorly-cultivated bog," Engels wrote in 1845, "is now a thickly-populated industrial district." Factory towns, like Manchester, Glasgow, Bolton, and Rochdale, "experienced a mushroom growth." Manchester and adjacent Salford more than tripled in population, from 95,000 in 1800 to more than 310,000 in 1841. In Lancashire alone, in 1830 there were more than 100,000 cotton mill workers.[68] Rural migrants from elsewhere in England made up much of the new industrial workforce, as did newcomers from Scotland and Ireland, where rural poverty pushed thousands upon thousands to emigrate.[69]

The densely packed working-class neighborhoods that sprung up near mills were in their own way as novel and disturbing as the mills themselves. The congregation of so many workers in one place was unprecedented. Taylor wrote "The most striking phenomenon of the Factory system is, the amount of population which it has suddenly accumulated on certain points." "[H]ad our ancestors witnessed the assemblage of such a multitude as is poured forth every evening from the mills of Union Street [in Manchester], magistrates would have assembled, special constables would have been sworn, the riot act read, the military called out, and most probably some fatal collision would have taken place." What was so frightening to Taylor were not just the sheer numbers but the fact that the factory workers were new creatures, an unknown and uncontrolled breed, "new in its habits of thought and action, which have been formed by the circumstances of its condition, with little instruction and less guidance, from external sources."[70]

Writing at almost the same time, Engels, in *The Condition of the Working Class in England*, provided some of the most graphic descriptions we have of the miserable living conditions of English factory workers: their destitution, the meagerness and filth of their dwellings, their tattered clothing, the awful smell of their homes and the streets they lived on. (Manchester's assistant poor-law commissioner described streets "so covered with refuse and excrementitious matter as to be almost impassable from depth of mud, and intolerable from stench.") Like Blake, Engels compared life under the factory system to an idealized vision of preindustrial life, the "idyllic" world of cottage textile workers, who "vegetated happily," self-sufficient if outside the realm of intellectual or political consciousness. For Engels, it was not just the poverty of the new working class that appalled him, but also the work itself, the machine-paced production, the "iron discipline" demanded by overseers, the "endless boredom." "No worse fate can befall a man than to have to work every day from morning to night against his will at a job he abhors."[71]

In the end, though, for Engels, like Taylor, the most significant aspect of the concentration of large numbers of workers in mills and factory neighborhoods was the creation of a new social formation, a "proletariat . . . called into existence by the introduction of machinery." Urbanization, wrote Engels, "helps weld the proletariat into a compact group with its own way of life and thought and its own outlook on society." Historian E. P. Thompson summed up prevailing sentiment in nineteenth-century England: "However different their judgments of value, conservative, radical, and socialist observers suggested the same equation: steam power and the cotton-mill=new working class." And that class, for Engels and many others, meant the coming of a new stage of history.[72]

Of course, the factory system had its defenders, in the national debate it provoked and more specifically around efforts, beginning at the start of the nineteenth century, to protect child and female workers, primarily by limiting their hours of work.[73] A few factory defenders claimed there were no problems, or at least not any that were the responsibility of mill owners. Andrew Ure—who Marx dubbed "the Pindar of the automatic factory"—argued that the beating of children in woolen factories working on "slubbing machines" (which prepared yarn for spinning) was strictly the fault of the adult "slubbers." Slubbing machines were hand powered, allowing their operators, Ure claimed, to slack off, leading them to beat their assistants in their efforts to catch up. Powered equipment, by setting the pace of labor, would eliminate the abuse of children. After acknowledging such problems also existed in cotton spinning mills using steam or water power, Ure retreated to simple denial, writing that in his visits to factories in Manchester and surrounding districts he "never saw a single instance of corporal chastisement inflicted on a child, nor indeed did I ever see children in ill-humour. . . . The work of these lively elves seemed to resemble a sport, in which habit gave them a pleasing dexterity."[74]

W. Cooke Taylor acknowledged poverty among mill workers and granted "juvenile labour to be a grievance." He blamed neither the factory system nor the mill owners but depressed economic conditions stemming from Britain's extended conflict with France and restrictions on trade, a view echoed by Charlotte Brontë in her novel *Shirley* (set at the time of the Napoleonic Wars). For Taylor, there was one thing worse than juvenile labor, "juvenile starvation." "I would rather see boys and girls earning the means of support in the mill than starving by the roadside, shivering on the pavement, or even conveyed in an omnibus to Bridewell." As a propagandist against the Corn Laws, which put a tariff on imported grain, for Taylor the solution to the ills of the factory lay in free trade, which would expand markets abroad and cheapen food at home.[75] Thomas Carlyle shared Taylor's view that the ills of the factory system were not intrinsic to it: "Cotton-spinning is the clothing of the naked in its results; the triumph of man over matter in its means. Soot and despair are not the essence of it; they are divisible from it." This faith, that the Promethean triumph of the factory fundamentally represents human progress and can be cleansed of its abuses, has remained a core liberal belief ever since.[76]

While reformers defended the factory system in spite of its faults, others opposed all efforts to regulate mills. In the debate over an 1833 bill to limit the working hours of mill children, the Chancellor of the Exchequer, Lord Althrop, feared that new rules would diminish Britain's competitiveness and reduce international demand for British textiles, hurting those meant to be protected. Some factory defenders opposed regulation on the grounds that property rights were absolute.[77]

A potentially powerful argument in the defense of the factory system—that if conditions were bad, they were no worse than elsewhere—gained little purchase, even though in many respects it was true. Cooke Taylor took a jab at the rural gentry—supporters of the Corn Laws—in claiming that conditions for agricultural workers were worse than for factory workers. Ure argued that the lot of handcraft workers was worse than "those much-lamented labourers who tend the power-driven machines of a factory," while children working in coal mines were worse off than in textile factories. Engels did not fundamentally disagree. His study of the condition of the English working class documented the miserable circumstances of miners, domestic workers, pottery workers, and agricultural workers, as well as mill workers. In his view, the "most oppressed workers" were not factory employees but "those who have to compete against a new machine which is in the process of replacing hand labor."[78]

Historian John Gray, in a study of the debate over factory regulation, showed how the mills came to symbolize the broad changes caused by industrialization and became the focus of efforts to ameliorate the often dreadful condition of workers, especially women and children. Nonfactory workers—some laboring for less money under harsher conditions—were all but ignored. The novelty of the factory system drew attention to the exploitation of its workforce, while the long-standing exploitation of agricultural workers, domestic producers, servants (encompassing nearly twice as many women as in the textile industry), and others went largely unnoted by politicians, journalists, and writers, who generally had little interest in the lower classes.[79]

The Factory Acts passed by Parliament in 1802, 1819, 1825, 1829, and 1831 regulated labor only in cotton mills and only the labor of children, doing nothing for the vast majority of British workers.[80] They had only very modest effect on actual conditions, lacking effective enforcement mechanisms. During the debate over the 1833 act—which did bring substantial changes, ending the employment of children under nine and limiting the hours and banning night work for older children—a Royal Commission endorsed the regulation of factories not because they were necessarily the site of the most onerous child labor but because regulation was more feasible in "buildings of peculiar construction, which cannot be mistaken for private dwellings" and where timekeeping was subject to "the regularity of military discipline" than at other worksites. Precisely because textile manufacturing had become so concentrated in large, well-known mills, it was more susceptible to regulation and improvement than dispersed employment. In the voluminous official inquiries and extended parliamentary debates about textile mill labor, Gray notes, "The identification of problems requiring intervention was dissociated from any systematic critique of industrial capitalism, and indeed became linked to a vision of the well-regulated factory as the site of social and moral improvement, as well as the symbol of economic progress." Thus the large factory became the vehicle for not only visions of ever-greater productivity and material bounty but also for the notion that a more humane version of the economic system soon to be dubbed capitalism was possible.[81]

Not everyone agreed. Engels said of the 1833 law, "By this Act the brutal greed of the middle classes has been hypocritically camouflaged by a mask of decency." Admitting that the law checked the "worst excesses of the manufacturers," he pointed to the ineffectiveness of some of its provision, like the requirement for two hours daily schooling for child mill workers, which Engels charged owners met by hiring unqualified retired workers as teachers. More profoundly, Engels, like Marx, believed that the exploitation of labor was an inherent characteristic of capitalism, of which the textile mills were the leading edge. For Marx and Engels, misery was not divisible from the factory system; for workers, it was its very essence.[82]

Marx's opus, *Capital*, is an often abstract analysis of the entire system of the creation, circulation, and reproduction of capital and attendant social processes. Today, to the extent it is studied, it usually is as a universal description and critique of capitalism as an economic system. Yet *Capital* is a book deeply rooted in a specific time and place, in an England when the textile industry reigned supreme. Cotton is everywhere in *Capital*: in Marx's explanation of key ideas, such as surplus value; in his account of broad historical developments, such as the transition from manufacture in the old sense of hand production to power-driven machine production; in his examination of a new set of class relations; and in his outrage at the exploitation of workers. The centrality Marx gave to the struggle over the working day in *Capital*, "a struggle between collective capital, *i.e.*, the class of capitalists, and collective labour, *i.e.*, the working-class," which he saw as the main battleground over the degree of exploitation of workers, mirrors the centrality of the hours issue in the national debate over regulating English cotton mills, which both Marx and Engels wrote about in great detail.[83]

Time after time, when in *Capital* Marx uses an example to illustrate his theories, he turns to the cotton mill. In a typical passage, in which he is explicating his method to calculate the "rate of surplus-value," Marx tries to explain to his readers "the novel principles underlying it" with examples: "First we will take the case of a spinning mill containing 10,000 mule spindles, spinning No. 32 yarn from American cotton, and producing 1 lb. of yarn weekly per spindle. We assume the waste to be 6%; under these circumstances 10,600 lbs. of cotton are consumed weekly, of which 600 lbs. go to waste. The price of the cotton in April, 1871, was 7 3/4 d. per. lb.; the raw material therefore costs in round numbers £342. The 10,000 spindles, including preparation-machinery, and motive power, cost, we will assume, £1 per spindle . . ." and on and on he goes for another half page of detailed calculations. There is nothing abstract here; Marx is talking about the ins and outs of the daily business of making cotton yarn, drawing much of his information from Engels, who spent nearly twenty years helping manage a Manchester cotton mill his family partly owned.[84] Thus, the cotton mill figured very large in the emergence of industrial capitalism and in the thinking of its most important critics, who gave a privileged place in their understanding of the capitalist system to a particular form of production and a particular group of workers who were seen as representing the future shape of society, even though at the time they still constituted a modest fraction of economic activity and of the working class.

Worker Protest

Journalists, critics, government investigating committees, novelists, even poets, almost all from the middle or upper classes, poured out a flood of words about the factory system during the first half of the nineteenth century. By contrast, we have only a tiny corpus of appraisals from workers themselves, most of whom, if not illiterate, had little occasion or capacity to record their thoughts in forms that would receive much attention or survive through the years.[85] To the extent we can reconstruct the attitude of workers toward the factory system, we have to do so largely by looking at their actions, not their words.

One relevant word, though, was brought into the English language by workers, "Luddite." Today "Luddite" is widely used as a catchword for technophobes, opponents of machine-based advancement, stripped from its original context.[86] The word came from the bands of workers and their supporters who in 1811 and 1812 and again from 1814 to 1817 attacked textile machinery, mills, and mill owners in the Midlands and in northern England, claiming they were acting under the command of General (or sometimes Captain or King) Ned Ludd.

Britain had a long history of machine-breaking as a form of protest and pressure, which predated the Luddites and continued after them. In the textile industry alone, incidents of machine wrecking occurred as early as 1675, with an attack on silk-weaving machines, and continued through the 1820s with periodic assaults on cotton equipment. Both Hargreaves and Arkwright had early installations of their machines destroyed by mobs, leading Arkwright to design his Cromford complex to be easily defended, with building placements, walls, and gates restricting access.[87] But the Luddites represented a more extensive, threatening, and enthralling episode of machine breaking than anything before or after.

Luddite attacks generally were preceded by letters threatening the destruction of machines and buildings and even murder unless employers met specified demands. An 1811 letter, apparently sent to a hosier named Edward Hollingsworth, read (as transcribed from the damaged original) "Sir if you do not pull don the Frames or stop pay [in] Goods onely for work or m[ake] Full fason my Company will [vi]sit y^r machines for execution agai[nst] [y]ou...," signed "Ned Lu[d]."[88]

The framework knitters, who made stockings, lace, and other woven goods on looms they sometimes owned but often rented from merchant-hosiers, were the first group of Luddites to go into action. To cut labor costs, merchants increased the rent and introduced wide looms, on which, instead of making a single item, large pieces of knitted material could be produced and then cut and sewed to make cheap goods, including stockings. Also, many merchants began paying in truck rather than cash. Faced with declining income and what they saw as the debasement of their trade, the frameworkers rallied under the banner of the mythical General Ludd, targeting wide frames and merchants who were cutting wages. Over the course of a year, an estimated one thousand knitting frames in Nottinghamshire, Leicestershire, and Derbyshire were destroyed. It took the passage of a law making frame-breaking a capital crime to halt the attacks.

The "croppers" in West Riding, Yorkshire, formed a second battalion in King Ludd's army. Croppers did the final, highly skilled finishing work on woven wool, raising the nap and using large, heavy, hand shears to cut and even the surface. The introduction of gig mills, to raise the nap, and shearing frames, to trim it, threatened to eliminate cropping as a skilled, well-paid craft. After trying unsuccessfully to use lawsuits and parliamentary lobbying to check the advance of the new machines, the croppers took to armed attacks on mills housing the machinery, including a successful assault by some three hundred Luddites on a mill near Leeds and an armed battle at a mill in Rawfolds that left two Luddites dead (and provided the plot for *Shirley*). Soon after, a particularly hated mill owner was assassinated. To restore order, four thousand troops were sent to occupy West Riding.[89]

In Lancashire, a third eruption of worker violence broke out, including food riots and assaults on mills using steam-powered weaving equipment. The mill attacks—including one by a crowd of over a hundred, marching behind a straw effigy of General Ludd, which burned down a mill owner's house before being fired on by a military unit, killing at least seven protesters—reflected the impact of mechanization on the hand-loom weavers. Initially, the factory system led to boom times for handweavers, as spinning machinery produced a bountiful supply of cheap yarn and a growing demand for weavers. The hand-loom workforce probably exceeded a half million between 1820 and 1840, outnumbering all factory textile workers. But the weavers' "golden age," as E. P. Thompson called it, was short-lived. The entrepreneurs who supplied the weavers with yarn and bought their products pressed down wages, even before power mills began providing substantial competition. Once they did, the downward pressure on wages and living standards became horrific, as mass impoverishment—sometimes literal starvation—descended on the weavers and their families. Looking back from not long after power weaving finally all but eliminated hand work, Marx wrote that "History discloses no tragedy more horrible than the gradual extinction of the English hand-loom weavers." And it was not just in England that the incorporation of weaving into the factory system took its toll; governor-general of India William Bentinck reported in 1834–35 "The bones of the cotton-weavers are bleaching the plains of India."[90]

Luddism, though the focus for much of the debate about industrialization, for the most part was only indirectly connected, when connected at all, to the giant factories that had popped up since the late eighteenth century. Hosiery knitting generally occurred in modest-sized workshops. Wool finishing likewise generally did not take place in massive mills. Only the attacks on power looms occurred on the terrain of the factory behemoth.

Luddites generally were more concerned with particular grievances against particular employers than with abstract opposition to technology. Some machine wrecking was part of a tradition of what Eric Hobsbawm called "collective bargaining by riot," using the destruction of property to pressure employers to raise wages and make other concessions. Many of the Luddites themselves operated machinery, albeit hand-powered, and most depended on factory-produced yarn for their livelihoods.[91]

Rather than as an expression of opposition to machinery or the mill system, Luddism is better understood as one of many forms of protest against the miseries workers—in factories, competing with them, and not engaged with them at all—experienced during the helter-skelter industrialization of the first half of the nineteenth century. Worker action took the forms that it did in part because other forms of collective activity were blocked. The concentration of workers in factories and urban neighborhoods created a critical mass for political discussion and labor organization, the context in which "The working-class made itself," as Thompson famously wrote.[92] But the outlets for action were limited.

Workers were shut out of direct participation in governance through most of the nineteenth century, with women and working-class men excluded from voting during the decades when the factory emerged as a key social institution. Workers did seek redress from Parliament, proposing laws, gathering signatures on petitions, testifying at commission hearings, and sending delegations to lobby members, but generally with scant results. The demand of the Chartists, who led massive popular mobilizations in the 1830s and 1840s, for universal male suffrage and the democratization of Parliament, fell on deaf ears.[93]

The government also severely limited the ability of workers to join together to pressure employers to improve wages and conditions. In reaction to the late eighteenth-century growth of proto-trade unions (among nonfactory workers) and the fear among British rulers, brought on by the French Revolution, of any type of radicalism or popular action, Parliament passed a series of laws—most importantly the 1800 Combination Act—against worker organization. Between 1792 and 1815, the government built 155 military barracks in industrial areas.[94]

In spite of legal prohibitions, workers formed open and secret organizations, held strikes, and joined in marches and mass demonstrations. The 1810s saw the first substantial walkouts by factory workers, some involving thousands of cotton spinners. The government reaction was heavy-handed, arresting, imprisoning, and transporting to the colonies leading activists and, in the case of some Luddites, hanging them. When in 1819 some sixty thousand protestors gathered in Manchester to demand democratic reforms, a military unit made up of local manufacturers, merchants, and storeowners charged the peaceful crowd, killing eleven people and leaving hundreds wounded in the so-called Peterloo Massacre. The government response was to pass still more repressive legislation, among other things banning meetings with more than fifty people present.

The 1820s brought yet more strikes, machine breaking, and reform campaigns, followed in the 1830s by a massive push to win legislation limiting factory working hours. In 1842 came a widespread strike among mill workers and miners, called the Plug Riots because strikers removed the plugs from steam engines, rendering them inoperable. By the 1850s, larger and more stable (though still mostly local) unions began to form among textile workers. Some launched large, prolonged, though generally unsuccessful strikes. Over a half century after the first factory giants had been erected, in spite of repeated, episodically large-scale efforts, the workers within them still lacked any effective political or organizational method for improving their lot or shaping the society in which they lived.[95]

Late eighteenth- and early nineteenth-century Great Britain often has been portrayed as a freer society than continental Europe. Some scholars, like Landes, suggest that this was one reason why the Industrial Revolution took off there first.[96] But for workers, especially factory workers, Britain was far from a free society. Factories grew up under an autocratic political regime, at least as far as it concerned working people. Workers did not have the right to vote, they did not have the right to assemble, they did not have the right to join together to bargain collectively with their employers, they did not have the right to quit their jobs whenever they wanted to, they did not have the right to say whatever they thought. Nothing better symbolized the support the state gave the emerging industrial system than the hanging of workers for the crime of not attacking persons but inanimate objects, breaking machines. Later to be extolled as the triumph of a new kind of freedom, the factory system was nurtured by severe restrictions on the rights of those whose labor made it possible. It took—and continued to take— the repressive power of the state to enable the giant factory to take root in unbroken soil.[97]

Becoming Ordinary

In the second half of the nineteenth century, cotton mills became less central to discussions and struggles about the structure of British society and the shape of its future. For one thing, they were no longer novel. By then, generations had grown up with large mills as a part of the world they lived in. Other, newer marvels had taken the lead as symbols of modernity, most importantly the railroad, which drew extraordinary attention from writers, artists, and the general public. In 1829 some ten to fifteen thousand people assembled in Lancashire to watch a competitive test of newly designed locomotives. The next year, when the first modern railway line opened, linking Liverpool to Manchester, dignitaries filled the first train and huge crowds lined the tracks. Trains became, as Tony Judt put it, "modern life incarnate."[98]

Textile mills no longer held first place, either, in sheer size, as other types of worksites came to rival or exceed them. The railroad system had a huge workforce, with some shops that built and maintained equipment employing as many workers as large textile mills. Other industries, especially metalworking, also built very large plants. By the late 1840s, the Dowlais iron works in Wales employed some seven thousand men in a complex that included eighteen blast furnaces, puddling ovens, rolling mills, and mines, dwarfing even the largest textile mill."[99]

Changed economic and political circumstances also muted attention to the textile factory. In the mid-nineteenth century, the British economy began to significantly improve, with growing international markets for English textiles contributing to increased revenue and improved conditions for workers. Legislation also began easing the lot of mill employees, especially the 1831 Truck Act, which required workers to be paid in cash, the 1833 act regulating child labor, and an 1847 law that limited the working day for children and women in mills to ten hours, realizing a long-time goal of working-class reformers. When Engels returned to Manchester in 1849, just seven years after he began research for what became *The Condition of the Working Class in England*, he found a very different city, more prosperous and peaceful. "The English proletariat," he complained, "is actually becoming more and more bourgeois."[100]

The transformation was as much political as economic. The failure of the Chartists to win their demands, in spite of their huge success in mobilizing support, took much of the wind out of the sails of the radical movements. At the same time, Chartism, with its emphasis on male suffrage, shifted attention away from female and child mill workers to adult men: artisans, construction workers, and other nonfactory laborers. The campaign against the Corn Laws, which began in 1838 and triumphed eight years later, rearranged the political terrain, too, in effect bringing workers and mill owners into alliance against the landed gentry, at least on this one, much debated issue. Further easing tensions, more mill owners began adopting paternalist practices, which had been prevalent among some of the earliest textile manufacturers, like Arkwright and Strutt, but rejected by many others.[101]

Textile workers continued to protest conditions they faced in the mills, but their struggles were no more prominent than those of miners and other groups acting through unions. After the mid-nineteenth century, the attention of middle-class reformers and observers shifted away from the mill, even as conditions for mill workers, though improved, remained often oppressive and child labor, albeit slightly older, continued to be widely used into the twentieth century. The questions surrounding the large textile mill and the factory system it brought into being devolved into part of a more general and less apocalyptic debate about the rights and standards of labor. By the time Charlotte Brontë published *Shirley* in 1849, she viewed the great dramatic struggles over the factory system as something from the past, with the large cotton mill having been, all in all, a source of social betterment.[102]

By then, the giant cotton factory had led to new ways of organizing production, new sets of social relations, and new ways of thinking about the world. All but its most adamant defenders recognized that in the short run the big factory had brought with it massive human suffering, both among the workers in the mills and those displaced by them. Yet for many, the mill held forth the promise of a better world. In an unpublished article which would become the basis of *The Communist Manifesto*, Friedrich Engels wrote: "Precisely that quality of large-scale industry which in present society produces all misery and all trade crises is the very quality which under a different social organization will destroy that same misery and these disastrous fluctuations."[103] For better and for worse, the extraordinary social invention that first appeared with the Lombes' mill and the early cotton spinning mills, the factory behemoth, represented a giant leap toward a new world, our modernity.

CHAPTER 2

"THE LIVING LIGHT"

New England Textiles
and Visions of Utopia

D URING AN 1842 TOUR OF THE UNITED STATES, Charles Dickens spent a day visiting Lowell, Massachusetts, the largest cotton manufacturing center in the country. Founded just twenty years earlier, the midsize city, set in the countryside, had become a bustling conglomeration of mills, boardinghouses, and churches, its streets lined with trees and flowers and filled with lively young women. If he were to make a comparison between Lowell and the factories of England, Dickens wrote, "The contrast would be a strong one, for it would be between the Good and Evil, the living light and deepest shadow." Dickens was far from alone among European travelers in seeing Lowell as a different order of society than the manufacturing centers of Britain. Englishman John Dix wrote in 1845 that "a more striking contrast than that afforded by Manchester . . . and Lowell, can scarcely be imagined." Michael Chevalier, a French political economist, described manufacturing as "the canker of England," which at least "temporarily involves the most disastrous consequences." By contrast, he found Lowell "neat, decent, peaceable and sage." Novelist Anthony Trollope, son of Frances Trollope (who wrote *Michael Armstrong*), dubbed Lowell "a commercial Utopia."[1]

European writers visiting Lowell—a regular stop on the circuit of

New World wonders—were particularly taken by its pastoral setting and young, female workforce. "COTTON MILLS! In England the very words are synonymous with misery, disease, destitution, squalor, profligacy, and crime!," wrote Dix. "How different from the neighborhood in which we now are, where the only sound which is heard above the whirling of spindles, and the clatter of machinery, is the chirp of the locust or the song of the robin." Chevalier found the sight of Lowell "new and fresh like an opera scene." Witnessing "neatly dressed" young women working "amidst the flowers and shrubs, which they cultivate, I said to myself, this, then, is not like Manchester." Dix, too, was impressed by the "healthy, good-humored, pretty faces, and honestly-earned habiliments" of the Lowell workers, who, he wrote, "belonged to another race of beings" compared to their Manchester counterparts.[2]

If, in the Old World, cotton mills came to be seen as dystopian, in the New World, they were repeatedly hailed as beacons of a bright future. As it turned out, many of the characteristics of New England textile manufacturing that won such praise—the bucolic surroundings of the factories, the neat mill towns, and the attractive young female workers—lasted but a few decades. But other aspects of the Lowell system of manufacturing, which commanded less notice from casual visitors, endured, anticipating what nearly a century later would come to be called "mass production." By promoting a vision of the mill town as a morally uplifting and culturally enlightening community, and developing a system of cheap, standardized manufacturing, Lowell spread the idea that both economic and social betterment could be achieved through technically advanced industry. Lowell reduced fears of industrialization while equating progress with the efficient production of consumer goods. Doing so made the New England textile industry an important episode in not only the history of the giant factory, but also in the development of our modern world.[3]

Beginnings

Lowell was not the first attempt to establish cotton manufacturing in the United States. Earlier, the industry had begun to develop along the same lines as in England. In the late eighteenth century, a few efforts were made to build spinning and carding machines, including one that, like Arkwright's early mill, used horses as a source of power.[4] But success only came when, in an echo of the Lombes' theft of Italian technology, textile machinist Samuel Slater evaded a British ban on the emigration of skilled manufacturing workers, in place until 1825. Like the Italians, the British hoped to maintain through the force of law a monopoly on advanced technology—textile machinery could not be exported until 1843—but the effort proved futile.

Slater, born in Belper, amid the world's first successful cotton mills, had apprenticed with Jedidiah Strutt, living with the Strutt family and working in one of its mills, where he became familiar with Arkwright's equipment. In 1789, he slipped out of England, telling no one of his plans. Arriving in America, he quickly hooked up with Moses Brown, a partner in the Rhode Island merchant company Almy and Brown, who hired him to build and equip a water-powered mill in Pawtucket, Rhode Island. Compared to the brick or stone English mills, the Almy and Brown factory was very modest, a two-and-a-half story wood structure, housing machinery made almost entirely out of wood. Starting up slowly, it initially did carding and spinning with a workforce of nine local children. By 1801 it had over a hundred children at work.[5]

The Almy and Brown factory soon spawned new operations, as Slater and other mechanics who worked there launched their own enterprises, often in partnership with merchants. The Slater-style mills remained small, as the rivers they were built on generally could power only modest operations. Also, there were not enough nearby children to sustain large plants, with no poorhouses, as there were in England, to turn to for forced recruits. Mills advertised for large families to hire, with men to work as skilled mechanics and children minding machines. But with labor scarce in the thinly populated United States, recruiting workers proved difficult. So production grew not through increased mill size but by replication, with factories moving further into the back country, where untapped pools of labor could be found. By 1809, at least twenty-seven mills were operating in Rhode Island, eastern Connecticut, and southern Massachusetts.[6]

The American mills mimicked English practices, most obviously in the extensive use of child labor, including children as young as four. In another carry-over, they generally paid their employees, except for skilled mechanics, with credit at a company store rather than cash, reflecting, as in Britain, a shortage of small currency, as well as limited working capital. To conserve cash and retain workers, mills usually paid wages only once a quarter, or even less frequently, and delayed for weeks giving final payouts when workers quit.

At first, total factory output remained modest. For one thing, the demand for cotton yarn was limited. Most Americans wore flax or wool clothing. Those who preferred cotton could buy British exports. For another thing, raw cotton was hard to obtain. Little cotton was being grown in the United States when Slater got started, so at first he used cotton imported from Cayenne and Surinam, only later adding Southern-grown cotton to the mix.[7]

But output soared in the second decade of the nineteenth century. The Napoleonic Wars, the Embargo Act (in effect from 1807 to 1809), and the War of 1812 disrupted English imports just when a growing taste for cotton clothing and an increasing market for cotton cloth from settlements west of the Alleghenies boosted demand. Looking to cash in, merchants and mechanics launched a wave of spinning mill construction across the Northern states. Weaving remained strictly hand done. In Pennsylvania, full-time skilled artisans produced fine-quality cloth. In New England, some mills set up networks of outworkers to weave, but rarely as a full-time occupation. Frustrated by the difficulty in getting outwork returned in a timely fashion, Almy and Brown hired weavers to work in the company's factory.[8]

It was in this context that Francis Cabot Lowell conceived of a different way to produce cotton cloth. A wealthy Boston merchant, Lowell, during an extended sojourn in Britain, decided that big profits could be made through the large-scale integrated production of textiles, using powered equipment for all phases of the operation within a single factory. At the time, few British firms spun and wove in the same plant and no power loom had ever been used in the United States, because of Britain's technology embargo. On returning home, Lowell hired a skilled mechanic, Paul Moody, to help him build machinery modeled after what he had seen in England. By 1814, they had a power loom successfully operating and a dressing machine to prepare the warp.[9]

Meanwhile, Lowell formed a joint-stock company, the Boston Manufacturing Company, with other Boston merchants to build and operate a mill. The investors realized that with the full-scale resumption of British trade after the War of 1812, their opportunities for profits in international commerce would be reduced. Manufacturing promised to be a rewarding alternative, even as they continued to be active in trade and real estate speculation.

Creating the company was a radical innovation. In the early nineteenth century, stockholder corporations were rare, with each needing a separate enabling state law. Generally, they were used only for enterprises considered public utilities, like building a canal. The corporate form had great advantages; it allowed aggregation of capital on a scale few individuals could afford and shared risk among multiple parties, a practice well known to merchants, who often formed partnerships to finance ship journeys. Joint stock corporations also facilitated enterprise continuity when investors chose to withdraw their funds and eased the process of inheritance, important for the rich, largely passive stockholders who would be drawn to the textile industry. (Corporations gained an additional advantage when they were granted limited liability in most New England states during the 1830s and 1840s.) Within five years, Boston Manufacturing raised $400,000 in capital (soon raised to $600,000). By contrast, as late as 1831 the average capitalization for 119 mills in Rhode Island fell below $45,000.[10]

To begin operations, Boston Manufacturing bought a mill site in Waltham, up the Charles River from Boston, where a paper mill already was using water power. There the company built a four-story brick mill, forty feet wide and ninety feet long, topped by a cupola housing a bell to call employees to work. Though not much larger than the largest existing U.S. cotton mills, the Waltham factory fundamentally differed in that it housed weaving as well as spinning equipment, so that within a single structure bales of raw cotton were turned into finished cloth. Also, Boston Manufacturing recruited a different kind of workforce than earlier mills, hiring, in addition to a few skilled male workers, local young women to operate both the spinning and weaving equipment.[11]

The Boston Manufacturing looms were crude, requiring course yarn—much coarser than what was being used in England—to avoid excessive breakage. As a result, the mill could only produce basic, heavy cloth. Initially the company turned out yard-wide white sheeting, of the sort then being imported from India, a product popular in the growing Western settlements, where home spinning and weaving were less common than in New England and durability was valued. Some of the cloth was sold in the South to make clothing for slaves. The company distributed all of its output through a single agent, paid on commission, rather than by the consignment system other mills used. Lowell cleverly protected his market by lobbying to have the 1816 Tariff Act place a higher duty on cheap imported textiles than on higher-priced goods of the sort that the Rhode Island mills were producing, effectively locking out foreign competition.[12]

The Waltham mill, completed in late 1814, proved almost immediately profitable. In 1817 Boston Manufacturing paid out its first dividend, 12½ percent. By 1822, the company had fully repaid its initial investors, with a cumulative dividend of 104½ percent. In 1816, the company built a second mill, close to the first, somewhat larger at 40 feet by 150 feet. A small separate building was erected for picking, the breaking up of the raw cotton bales that produced highly flammable cotton dust. Like the first mill, the second had towers on the outside of the main structure to hold stairs and toilets (which dumped their waste into the Charles River).[13]

With the completion of the second Waltham mill, a template for the northern New England textile industry was in place. Somewhat as had occurred a century earlier at the Lombe mill, a new model of production came together rapidly, followed by a long period of replication and incremental improvements, but no radical shifts. Nathan Appleton, an original Waltham investor, noted in 1858 "how few changes have since been made from the arrangements established . . . in the first mill built at Waltham."[14]

What made the Waltham system different and important? First, the integration of production within a single space and a single firm. Raw materials went into a mill and finished products came out. All the problems and costs associated with coordinating and transporting materials in various stages of production to and from different factories or outworkers and ensuring their quality were eliminated. Having all processes under one roof allowed productivity gains, such as spinning weft directly onto bobbins used in subsequent weaving.

Second, the Waltham model mills concentrated on making standardized products at high speed. Most Waltham-style mills produced only a single type of cloth or at most a few and ran their machinery at higher speeds than equivalent equipment in England. Innovations introduced by Lowell and Moody traded off flexibility for speed. Their "double speeder" roving frames, for example, were costly to reconfigure for different types of yarn, encouraging long production runs of the same product. Moody later introduced other changes to speed up equipment, including using leather belts rather than shafts to transmit power to individual machines and making main shafts out of wrought iron rather than wood. But the high-speed equipment could only produce relatively simple fabrics, not complex weaves, like ginghams, that had colored patterns, or other "fancy goods."

Third, the Waltham system automated as many processes as possible to reduce the need for skilled labor. Many Waltham-style machines had "stop-motion" features, which halted the equipment if a thread broke or another problem developed, reducing the needed skill of operatives and increasing the number of machines they could monitor.[15]

Fourth, the Boston group, in pioneering the use of the corporate form for manufacturing, linked big capital to goods production. The corporation would not become the norm for manufacturing outside of textiles for decades to come, but the advantages it brought eventually made it standard for large-scale industrial enterprises. With heavy capital investment in plant and equipment and large reserves, Boston Manufacturing and the companies modeled on it could build larger, more efficient factories and were better able to withstand the vicissitudes of the economy than smaller companies modeled on the Slater mills.

Fifth, the use of a single selling agent rather than multiple jobbers created a close identification between particular products and particular companies, a step toward what would later be called branding. Sometimes it was the selling agent rather than the mill that decided what products should be made, much like how, nearly two centuries later, brand-name companies and giant chain stores would tell clothing, shoe, and electronics manufacturers precisely what to produce. The sales agent, rather than the manufacturer, felt the pulse of the market.[16]

Finally, the Waltham model mills developed primarily as domestic, not international, enterprises. Much of the recent literature on the cotton industry stresses its global character. This certainly was the case in Britain, which imported raw cotton and exported cotton goods, a hub of world commerce. But the Waltham-Lowell mills used cotton grown in the United States and sold their products primarily within the nation's borders. In 1840, exports accounted for less than 8 percent of U.S. cotton cloth production, in 1860 still less than 10 percent, a nice source of profit and a safety valve for excess output, but not central to the industry.[17] The rich array of natural resources in the United States and its large, growing domestic market meant that American industry would develop primarily as a domestic enterprise, engaged with international markets but not dependent on them.

Lowell

Waltham established the model, but it was Lowell that became famous. The Boston Manufacturing Company founded the city to expand its capacity. After erecting a third mill in Waltham, the company directors decided to build a new complex to produce calicoes. Without enough water power in Waltham for additional mills, company leaders found a site twenty-three miles north of Boston, in what was then East Chelmsford, Massachusetts, where at Pawtucket Falls the Merrimack River dropped thirty feet, unleashing enormous energy.

Years earlier, a company called the Proprietors of Locks and Canals on the Merrimack had built a canal around the falls to permit navigation. Quietly, Boston Manufacturing bought up the stock of the older company and land along the river. To launch the new enterprise, in 1822 it created the Merrimack Manufacturing Company, offering shares to its investors. Using Irish laborers, the new company widened and deepened the existing canal and rebuilt the locks to create mill sites with adequate power. At a time preceding power equipment and dynamite, the infrastructure work, along with building and equipping new mills, proved extremely expensive. Only an assemblage of some of the richest men in New England could have financed industrial development on this scale.[18]

The mills built at the new site—and others later modeled on them—were much larger and more substantial than the early Rhode Island factories. Handsome, durable brick structures, without much ornament, they bore at least a superficial resemblance to the Lombes' mill, by then already a century old.[19] Technical considerations dictated their size and shape. The wooden shafts used to convey power from waterwheels could be extended only so long before breaking, no more than one hundred feet. Even after builders began centering mills over their waterwheels, allowing horizontal shafting on both sides, building length was limited. The need to bring in light from perimeter windows restricted mill width. So floor plates could not be very big, in the case of the Merrimack mills, 156 feet by 44½ feet. To create more space and fully utilize the power of the waterwheels, mills were built up, in the Merrimack model five stories high, including an attic and a basement. For greater capacity, Merrimack and other textile companies built multiple mills in clusters, sometimes arrayed around a central yard.

New England textile firms did not make much use of iron structural elements until the 1840s. Cast iron was expensive in the United States, while large wooden beams were readily available, familiar to local construction workers, and capable of supporting heavy weights and absorbing vibrations. Like the British, the Americans worried about the danger of fire, but they adopted a different approach to minimizing it, not attempting fireproof construction by replacing wood with iron and brick but instead seeking to retard the spread of flames by using very heavy timbers, not only for beams but also for flooring, which would be slow to catch fire and capable of continuing to support weight even if charred.

By 1825, Merrimack had completed five virtually identical mill structures and additional buildings for bleaching and calico printing. Each mill was self-contained, with both spinning and weaving equipment, capable of turning raw cotton into woven cloth.[20] As in Waltham, the new mills proved quickly profitable; within two years of commencing production, Merrimack paid its first dividend. To further expand, its directors came up with a strategy of creating additional firms, each of which would have its own stockholders and directors, with heavy overlap in ownership from company to company. The structure facilitated raising capital from new investors, while allowing existing stockholders to withdraw money from older companies to invest in new ones.

To advance the corporate metastasis, Merrimack transferred the land and water power it did not need to a reconstituted Locks and Canals

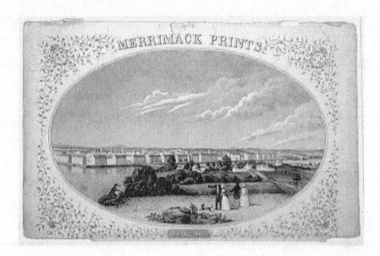

Figure 2.1 An engraving of Lowell, Massachusetts, in the 1850s, featuring a bucolic setting in the foreground.

company, which also took over the machine-shop operation of Boston Manufacturing. Like the William Fairbairn's company in England, Locks and Canals could provide what today would be called a turnkey facility. When new companies were formed—starting with Hamilton Manufacturing in 1824, followed by Lowell Manufacturing, Appleton Company, Lawrence Manufacturing, Boott Mills, Suffolk Manufacturing, and Tremont Mills—Locks and Canals sold them mill sites and machinery and provided water power (usually for a per spindle fee).

The owners carefully orchestrated the proliferation of companies. Rather than have firms compete with one another, each new company specialized in a different product: Merrimack, calicoes; Hamilton, twilled and fancy goods; Lowell, carpets as well as cotton cloth; and so on. Many of the companies shared the same selling agent and routinely exchanged cost information. Eventually there were ten major firms in Lowell operating a total of thirty-two mills.[21]

Merrimack and its progeny built more than mills; they built a whole city in what had been nothing but a thinly populated farming area. At the initiative of Merrimack, the mill sites and surrounding land were spun off from Chelmsford as a separate town, named for Francis Cabot Lowell, who died in 1817. With nowhere near a large enough local population to staff the factories that were rapidly going up, the first priority was to build housing for workers to be recruited from afar.

The feature outside observers usually focused on when they wrote about Lowell, the company boardinghouses full of lively young women, did not come from Waltham. Boston Manufacturing owned some housing in Waltham, but apparently rented it largely to male workers. Unmarried female workers either lived with their own families, if they were local, or with families not connected to the company. The boardinghouse model developed elsewhere. Shortly after Boston Manufacturing had its first mills up and running, it began selling machinery and patent rights to others setting up textile factories, who generally used the second Waltham mill as a template for their buildings. In New Hampshire, the Dover Manufacturing Company built two mills housing Boston Manufacturing machines and a new town, complete with street grid, company store, bank, commercial buildings, and boardinghouses for its female employees. The company rented the boardinghouses to housekeepers to manage, detailing rules for the residents. A similar complex in Great Falls, New Hampshire, likewise included boardinghouses for female workers. Apparently it was from these complexes that the builders of Lowell adopted the boardinghouse model.[22]

The Lowell boardinghouses were not uniform in design. The early structures, made out of wood, generally rose two stories high; later units, made of brick, three stories. By 1830, Merrimack owned, in addition to its production facilities, twenty-five wood tenements, four brick tenements, twenty-five cottages, a house for its agent, a church and parsonage, storage buildings, and a "Fire Department," along with a store and two warehouses in Boston. As Lowell grew, the textile companies helped finance a library, reading room, and lecture hall. By 1840, Lowell housed eight thousand textile workers, with a total population exceeding twenty thousand, making it the eighteenth-largest city in the United States.[23]

Figure 2.2 *Merrimack Mills and Boarding Houses,* an 1848 engraving by O. Pelton depicting boardinghouses at Lowell lined up and leading to a mill at the end of the street.

Scaling Up

Even as the core group of textile investors—what economic historian Vera Shlakman dubbed the "Boston Associates"—expanded their production in Lowell by forming multiple corporations, they expanded beyond Lowell by founding new mill towns across northern New England. In Chicopee Falls, on the edge of Springfield, Massachusetts, they helped launch four textile companies, mimicking the Lowell pattern of having an additional company to control land and water power and manufacture machinery. Other complexes arose in Taunton and Holyoke, Massachusetts; Nashua and Manchester, New Hampshire; and Saco and Biddeford, Maine. In the mid-1840s, when Lowell itself ran out of mill sites, a group of Boston investors developed a new town, Lawrence, nearby on the Merrimack River, which became a major wool and cotton center. In a few instances, the Boston group took over mills others had founded, like the complex in Dover.[24]

The Boston Associates companies were genuinely Boston companies. Their owners consisted largely of Boston residents who had made their fortunes before their textile investments. Most rarely visited their mills. Even companies with distant factories were run by a treasurer who lived in Boston, operating through an on-site agent. Selling and banking were done in Boston as well. The combination of absentee ownership and workers largely recruited from afar meant that the mills and mill towns often had few local roots. Industrial capitalism—which in the United States, as in England, had the textile industry at its lead—did not develop organically out of existing communities but was implanted, fully formed, by outside merchant capital.[25]

The textile complexes built by the Boston group dwarfed contemporary factories. An 1832 federal survey found that of the thirty-six manufacturing enterprises reporting more than 250 employees, thirty-one were textile companies. On the eve of the Civil War, manufacturing establishments in the United States employed on average only 9.34 workers. By contrast, Merrimack, the largest Lowell company, in 1857 had 2,400 workers, while six other companies in the city had over 1,000.[26]

Continuing growth, however, did not mean continuing innovation. After an initial burst of inventiveness, the Boston-based mill owners and managers proved a conservative lot, not introducing major technological changes for decades. Until the mid-1840s, individual mill buildings rarely exceeded by much the dimensions of the second Waltham mill, each housing 250 to 300 workers. The companies increased production by speeding up existing equipment and building new mills using their well-established template. Able to make a good return on their money by doing more of the same, the Boston investors felt little need for novelty.[27]

The question of power provides a good illustration. With plentiful water power, and coal farther away and more expensive than it was for British mills, New England mill owners did not widely adopt steam power until after the Civil War, long after it had become common in England. As a result, New England mill towns had none of the black smoke and soot so characteristic of British industry. When the growth of Lowell and the planning of Lawrence presented the possibility that companies on the Merrimack would run out of water power, instead of installing steam engines the mill owners bought real estate and water privileges at the outlet of Lake Winnipesaukee in New Hampshire, over sixty miles away, to direct more water into the river (outraging Ralph Waldo Emerson for what he saw as arrogance).[28]

The corporate arrangements adopted by the Boston textile investors allowed expansion on a scale unprecedented for manufacturing. In 1850, the mills they controlled accounted for about a fifth of all the cotton spinning in the United States. In Lowell alone, in 1857 the ten mill companies, the Lowell Bleachery, and the Lowell Machine Shop (spun off from Canals and Locks) together employed over thirteen thousand workers.[29]

But the Lowell model did not take full advantage of potential efficiencies that came with size. Within firms, running each mill building as a self-contained production unit meant that while there were some shared functions that no doubt lowered costs—most importantly buying raw cotton and selling finished goods—in other respects each building operated as a separate, modest-sized enterprise. The idea of a fully integrated, rationalized, multisite company still lay in the future. The Lowell mills did not begin to even calculate unit costs until the 1850s, so they had no way to know the advantages and disadvantages of different arrangements, sticking by habit to the system Lowell had introduced in the first Waltham mill. Even after the companies began connecting once-freestanding mill buildings to one another and completely ringing mill yards with buildings—to the dismay of workers who could no longer look out at town and country scenes—they continued to treat each mill as a separate entity. And because each cluster of four or five mills was organized as a separate corporation, other savings that might have accrued in purchasing, sales, and management were not realized.[30]

Amoskeag Manufacturing Company was the exception that suggested there might be greater efficiencies in a different organizational structure. Set up in the late 1830s to develop a new textile center on the Merrimack River in New Hampshire, along with a town grandiosely named Manchester, the company at first replicated the Lowell pattern, expanding through the creation of new corporate entities. But unlike in Lowell, eventually the separate companies began to consolidate under one management, until all the mills in Manchester were controlled by Amoskeag. The consolidated corporate structure facilitated expansion. At its peak in the early twentieth century, Amoskeag had 17,000 workers in thirty mills and many associated buildings, bordering the river for over a mile on one side and a half mile on the other. Its size allowed the company to be almost completely self-sufficient, using its own workers for even major construction projects and building most of its own machinery.[31]

The model of expansion through replication—many separate mill buildings, controlled by many separate companies—proved something of a dead end. When other companies began to approach and then exceed the size of the Boston Associates network, like the Pennsylvania Railroad, Standard Oil, and U.S. Steel, some experimented with interlocking directorates, but most quickly moved to consolidate corporate control and financial supervision, even with far-flung facilities.[32] Organizationally idiosyncratic, nonetheless it was the Waltham-Lowell system that first brought large-scale factories to the United States, and it was that system that until the Civil War represented industrialism in political and cultural discourse, a pole for criticism and, more often, praise of a new type of society.

Factory Girls

"The American factory girl," declared an 1844 article about Lowell in the *New-York Daily Tribune*, "is generally the daughter of a farmer, has had a common education at the district school, and has gone into the factory for a few seasons to acquire a little something for a start in life.

She spends some weeks or months of every year under her father's roof, and generally marries and settles in its vicinity. Many attend Lectures and evening schools after the day's work is over, and of the six thousand more than half regularly occupy and pay for seats in the numerous Churches of Lowell. . . . [H]ardly any where is Temperance more general or are violations of the law less frequent." The newspaper perhaps painted an overly rosy picture, but its description was basically accurate. It was the character of the Lowell "girls" and their life in the mill town that so impressed visitors from home and abroad and led them to sharply contrast American mills to British ones.[33]

Francis Lowell and his partners turned to farm girls as a workforce largely out of a lack of alternatives. The Lowell group sought to avoid the social disapproval that accompanied the wholesale employment of children and, in any case, their power looms required considerable strength to operate, necessitating adult operators. Unlike Britain, the United States had neither a surplus of urban male workers nor an overpopulated countryside to draw on. Perhaps in an earlier era slaves might have been used; in the much smaller Southern textile industry, they *were* used; by one estimate, more than five thousand slaves labored in Southern cotton and woolen mills by 1860. But by the time Lowell built the Waltham mills, slavery was all but over in the North.

Instead, the Waltham-Lowell–style factories found a brilliant solution in the recruitment of young women from rural New England. Unmarried, in their teens and twenties, they provided a well-educated workforce, accustomed to seeing and doing hard work and being subservient to male authority, but not so vital to their families that their withdrawal would create an economic or social crisis. And, to the mill owners' liking, they were a revolving labor force. When they became unhappy or the mills lacked work, they could return to their families rather than staying nearby and making trouble, avoiding the discontent and disorder that came in England with the creation of a permanent proletariat.[34]

For these workers, the mills represented an opportunity before marrying to expose themselves to a wider world, while economically helping themselves and their families. Few came from destitute homes, desperate for additional income, as was so commonly the case in Great Britain. Rather, they typically came from middling families, daughters of farmers or rural artisans. But money did play a big part in why they came. Typically, they kept their earnings, using them to buy clothes, accumulate a dowry, save money for normal school, or to set themselves up independently from their families. Many also sent money home, to help pay off a farm mortgage or family debts, to support a widowed mother, or to pay for a brother's education. A big attraction of the Waltham-Lowell–style mills was that they paid cash, not credit at a company store, like many of the Rhode Island–style mills. At the time, women had few other ways to make money, except domestic service (which many New Englanders rejected as subservient), schoolteaching (more seasonal than factory labor), or seamstressing.

But money was not the whole story. The mills also provided an escape from families, rural life, boredom, and isolation, a chance to experience a new, more cosmopolitan world of independent living, consumer goods, and intense sociability. Earning their own living gave women a sense of independence and relieved their parents of a burden. Ironically, the mills themselves made redundant one of the main contributions young women had made to the family economy, spinning yarn and weaving cloth at home for family use or for the market.[35]

There were other components of the mill workforce besides young women. Especially in the early days, there was a strict sexual division of labor. Women held almost all the jobs operating machinery, except for picking and carding. Men did all the construction, maintenance, and repair work and held all the supervisory positions. In addition, the mills recruited skilled male workers from England and Scotland for specialized jobs for which there was no pool of qualified native workers, including calico printing and producing woolens. A small number of children worked in the mills, too (though the Lowell mills generally did not hire anyone under age fifteen), as did a few older, married women.

The Hamilton Manufacturing Company was probably typical in 1836, with women making up 85 percent of its workforce. Over time, the percentage of female workers dropped, at least modestly. In 1857, excluding the all-male Lowell Machine Shop, the Lowell textile workforce as a whole was a bit over 70 percent female.[36]

The Lowell-style mills rarely had to advertise for workers. Young women—a sample of Hamilton workers found their average age on hiring just under twenty—came on their own after hearing about the mills, often joining or sending for sisters, cousins, or friends. The *Lowell Offering*, a magazine of poetry and fiction by mill workers, not only received extensive praise from visitors, it also served as a form of job advertising for the companies (which quietly subsidized it). When nearby hinterlands became tapped out of workers, the mills sent recruiters to scour the more distant countryside, bringing back their finds on wagons before railroads eased transportation.[37] Female mill workers typically had a relatively short tenure. Most estimates agree that women stayed on average something like four or five years, commonly returning home for stretches while employed.[38]

From the start, mill owners calculated that parents would allow their daughters to live on their own and work in the mills only if they were assured of their safety and well-being. For the mills "To obtain their constant importation of female hands from the country," wrote the *Burlington* [Vermont] *Free Press* in 1845, "it is necessary to secure *the moral protection of their characters while they are resident in Lowell.*" To that end, the companies established what the paper termed a system of "moral police." Elaborate company rules regulated workers off the job as well as on. The Middlesex Company declared that it would "not employ any one who is habitually absent from public worship on the Sabbath, or whose habits are not regular and correct." Workers were forbidden from smoking or using any kind of "ardent spirit" in the mills and were generally required to live in company-owned boardinghouses unless they had family living nearby. The boardinghouses, in turn, had their own sets of rules, including a ten o'clock curfew and, in at least one case, the requirement that all residents be vaccinated for smallpox (which the company agreed to pay for). The matrons who ran the boardinghouses had to report rules violators, who could be fired. Companies required workers to sign one-year contracts and give two weeks' notice before quitting. They circulated among themselves lists of workers who had been discharged or who had quit before the end of their contracts, whom they agreed not to hire, and imposed fines for lateness and poor-quality work.[39]

Company paternalism was not simply regulatory or punitive; especially in the early years the companies tried to make the mills attractive places to work and the mill towns attractive places to live. Lowell was carefully laid out, with trees lining its broad streets and an orderly placement of the mills, boardinghouses, and commercial structures. Companies planted trees and put in flower beds around their buildings and in their mill yards and allowed workers to grow plants and flowers on windowsills inside the factories. One newly arrived worker in Manchester, impressed by the brick houses and "very handsome streets," wrote her sister that she thought it "a beautiful place." The sociability of the mill towns, especially Lowell, with its lectures and literary societies, was widely praised, though also somewhat exaggerated, since, given the very long hours of work, workers had limited time for other activities. Still, cities like Lowell and Manchester looked and felt very different than the crowded, filthy, impoverished English textile centers like Wigan, Bolton, and the namesake Manchester.[40]

The experience of working in a factory and living in a factory town transformed the women who flocked to Lowell, Manchester, Chicopee, and the like. Augusta Worthen, two of whose sisters had worked in Lowell, later recalled that the young women from her town, Sutton, New Hampshire (population 1,424 in 1830), who traveled to take jobs in Lowell or Nashua had "a chance to behold other towns and places, and see more of the world than most of the generation had ever been able to see. They went in their plain, country-made clothes, and after working several months, would come for a visit, or perhaps to be married, in their tasteful city dresses, and with more money in their pockets than they had ever owned before." For one group in particular, mill work could be utterly altering, widows and older unmarried women, dependent on their relatives for support. Mill worker Harriet Robinson later remembered them "depressed, modest, mincing, hardly daring to look one in the face.... But after the first pay-day came, and they felt the jingle of silver in their pocket and had begun to feel its mercurial influence, their bowed heads were lifted, their necks seemed braced with steel, they looked you in the face, sang blithely among their looms or frames, and walked with elastic step to and from their work."

Many mill workers returned to their hometowns to marry, some-times settling down to farm lives much like those of their parents. But a detailed study, by historian Thomas Dublin, of women who had worked for Hamilton Manufacturing found that they typically married at a somewhat later age than women from their hometowns who had not gone to a mill, were far less likely to marry a farmer, and were more likely to settle down in a city, with quite a few staying in Lowell after marrying. Although the New England countryside itself was changing, with improvements in transportation and the spread of commercial relations, for young workers the mill experience accelerated the tran-sition out of a world of semi-subsistence agriculture into an emerging commercial society. Even those women who settled back home were never quite the same as those who never left.[41]

Unlike British textile workers, the young women who flocked to the New England mills left behind a veritable flood of words. Almost all literate, they kept diaries, wrote letters back home and to one another, contributed to *The Lowell Offering,* its successor, *The New England Offering*, and labor papers like *The Voice of Industry*, and, in a few cases, wrote memoirs or autobiographies. In their letters, money is discussed frequently: wage rates, how much could be earned in alternative types of employment, expenses, and so on. Work itself does not figure as strongly as activities outside of work, family news, or religion. There are very occasional comments on the pace of work, but surprisingly little description of the mills. Social life and saving money—the reasons why so many workers left their homes—remain at the forefront, while work tasks and the factories in which they occurred seemed to have been taken for granted.[42]

Perhaps one reason was that, at least in the first decades, mill workers generally did not consider their labor especially arduous. "Many of the girls who come to Lowell, from the country," an 1843 editorial in *The Lowell Offering* noted, "have been taught by their good mothers that industry is the first of virtues." Responding to claims about the unhealthful effects of factory labor, the editorial declared mill work "light—were it not so there would not be so many hurrying from their country homes to get rid of milking cows, washing floors, and other such healthy employments."

Just as in England, often new hires walking into a mill for the first time found the noise and motion of the machinery overwhelming, the experience of sharing a huge work space with scores of others disorienting, and the tasks tiring. But acclimation usually followed. Though the intensity of jobs varied considerably, at least in the early years, when the companies were still perfecting machinery and operations and profits were high, many jobs were not especially taxing. In the spinning and weaving rooms, workers often had stretches of free time while they monitored equipment, waiting for a thread to break or a bobbin to need replacing, in some case defying rules to read or socialize.[43]

But the work was work. In a review of Dickens's *American Notes*, *The Lowell Offering* quoted approvingly his comment about the "Lowell operatives" that "It is their station to work. And they do work. . . . upon an average, twelve hours a day; which is unquestionably work, and pretty tight work too." Repetitive actions over the long days brought boredom and fatigue. The air in the mills was often foul, especially during the winter when candles and lamps were needed for light, and the noise could become oppressive. Often it was too hot or too cold. And many workers resented the tight regulation of their lives, what some came to call "factory tyranny."[44]

Mill town life also had its downside. Some newcomers found being surrounded by so many other people, after having spent their lives on isolated farms or in small villages, disconcerting. The boardinghouses were crowded, with four to six women sharing each bedroom (two to a bed), affording little privacy (though that was nothing new for those who had grown up in large New England farm families, crammed into close quarters). But the opportunities for richer social, intellectual, and religious life than possible in their hometowns—and to make money—seemed to outweigh the challenges of urbanity for most of the newcomers.[45]

Conditions, however, were not static; they deteriorated over time. An extended burst of mill building—both of the Slater and Lowell types—began narrowing the gap between supply and demand for cloth. By 1832, some five hundred cotton mills operated in New England alone. To keep up dividends in the face of growing competition and falling prices, the Boston-based corporations sought to cut costs. Payroll was not necessarily their biggest expense. In some years, companies paid more for raw cotton than for the labor to convert it into cloth. But it was an expense over which they had control.[46]

Companies reduced labor costs in multiple ways. Sometimes they simply lowered wage rates, which for many workers were piece rates. In March 1840, for example, the directors of Merrimack Manufacturing voted "That in consequence of the depression of the times a reduction of the wages of the operatives is indispensable," authorizing the company treasurer to cut wages "to the point that they may be deemed expedient & practicable." The companies also began running machinery at higher speeds, taking advantage of technical improvements in shafting and equipment. And they began assigning spinners and weavers more machines to monitor. Whereas once a weaver might have been assigned one or two looms, by the 1850s it was common to assign three or four. As output—and the strain of work—went up, piece rates were reduced, so that wages rose at most modestly. A study of four Lowell-style mills in northern New England found that between 1836 and

1850 productivity increased by almost a half, while wages went up only 4 percent.[47]

In the 1830s, in response to wage cuts, a few dramatic if brief flashes of protest occurred. They came at a moment of increasing labor organization nationally, as a language and politics of worker mobilization emerged. An announcement by the Lowell mills in early 1834 of a forthcoming 12½ percent wage cut set off a wave of meetings, petitions, and agitation, seeking to reverse the decision. When a mill agent fired a leader of the protest, other workers walked out with her, parading the streets and visiting other mills, calling for their employees to walk out, too. Some eight hundred women joined the "turnout." But it was short-lived and unsuccessful. Within less than a week the strikers had either returned to their jobs or quit them, and the wage reduction went through as planned.

Two years later, 1,500 to 2,000 workers took part in a much better organized turnout, protesting a hike in the price of room and board in the company boardinghouses, effectively another wage cut. At some mills, the walkout lasted for weeks, with at least one company having to shut down a mill, consolidating its nonstriking workers in its others to keep production going. A newly formed Factory Girls' Association, with a reported 2,500 members, coordinated the strike. Though the exact outcome remains unclear, at least some mills partially or fully rescinded the increase.[48]

These were not the first mill worker strikes; there had been earlier, brief walkouts in Pawtucket, Rhode Island, and Waltham and Dover, Massachusetts. But the Lowell walkouts were bigger and carried more symbolic weight because they took place in the most celebrated factory town in the nation. Also, though the organized labor movement in the United States had been developing in fits and starts since soon after the Revolution, walkouts by women and factory workers were still a novelty.

In other ways, though, the Lowell strikes fit a national pattern, in which the language of republicanism and the spirit of the Revolution were invoked to mobilize workers against what was seen as an emerging tyranny of economic power. "We circulate this paper," read one petition circulated during the 1834 strike, "wishing to obtain the names of all who imbibe the spirit of our Patriotic Ancestors, who preferred privation to bondage. . . . The oppressing hand of avarice would enslave. . . . [A]s we are free, we would remain in possession of what kind Providence has bestowed upon us, and remain daughters of freemen still." Strikers saw wage reductions and the power to impose them as not just a menace to their economic well-being but also to their independence and respectability, threatening to reduce them to the opposite of freemen—or daughters of freemen—slaves. Just as in England, workers feared that the mill might not be a source of freedom but of its opposite. During the 1836 walkout, strikers walking in procession down the Lowell streets sang:

> *Oh! isn't it a pity, such a pretty girl as I—*
> *Should be sent to the factory to pine away and die?*
> *Oh! I cannot be a slave,*
> *I will not be a slave,*
> *For I'm so fond of liberty*
> *That I cannot be a slave.*

There was something light-hearted in this—the verse parodied the song "I won't be a nun," which went "I'm so fond of pleasure that I cannot be a nun"—but something serious, too.[49]

In Lowell, the walkout movement proved short-lived. But worker criticism of the factory system, if anything, became more common during the 1840s. As in England, reformers focused on the long hours of labor. "The great master evil in operation in Lowell, and too generally in American factories," the *New-York Daily Tribune* wrote, "is that of Excessive Hours of Labor." New England factories rarely operated around the clock, but the workday was very long. In Lowell, in the mid-1840s it generally lasted between 11½ and 13½ hours on weekdays, with somewhat shorter hours on Saturdays, with Sundays off. [50]

Following the same path as in England, New England textile workers sought legislative restriction of working hours—to ten hours—first for children and then for workers generally. Mill workers petitioned legislatures, formed organizations, including Female Labor Reform Associations in Lowell and Manchester, held picnics and parades, and published appeals in an effort to reduce the hours of work. Massachusetts and Connecticut did pass laws limiting working time for children, but unlike in Britain, American mill workers did not win meaningful legislation covering adult workers. Some Lowell mills reduced working time slightly, but despite an impressive organizational effort, the ten-hour movement effectively failed. [51]

Paradise or *Paradise Lost?*

The dissatisfaction of mill workers with their jobs, employers, and what they perceived as an unrepublican disparity in wealth and power made little impression on the stream of visitors who came to see the mills. [52] Davy Crockett, then a Whig congressman from Tennessee, visited Lowell just months after the 1834 strike (less than two years before his death at the Alamo). Crockett wrote that he "wanted to see the power of machinery . . . . [and] how it was that these northerners could buy our cotton, and carry it home, manufacture it, bring it back, and sell it for half nothing; and, in the mean time, be well to live; and make money besides." Like so many others, he was fascinated by the manufacturing processes and charmed by the "girls" who "looked as if they were coming from a quilting frolic." "Not one," he reported, "expressed herself as tired of her employment or oppressed with work," not surprising given that Crockett was accompanied by Abbott Lawrence, one of the most prominent mill owners. "I could not help reflecting," Crockett continued, "on the difference of condition between these females, thus employed, and those of other populous countries, where female character is degraded to abject slavery."

Though a bitter opponent of Andrew Jackson, Crockett's view of Lowell resembled that of the president, who had visited a year earlier. (Jackson was not the first president to visit a textile mill; James Monroe toured Waltham in 1817.) Leading Lowell investors hoped to charm Jackson at a moment of intense debate over tariffs, a matter in which they had great interest. They largely succeeded, organizing a procession of thousands of female workers in white dresses carrying parasols and wearing sashes reading "Protection to American Industry" and taking the president on a tour of the Merrimack mills.[53]

By the mid-1830s, it was not surprising for political opponents to be in agreement about Lowell-style manufacturing. In the era of the American Revolution, many leaders, like Thomas Jefferson, worried that manufacturing would threaten the agrarian nature of the country, on which, they believed, liberty, virtue, and republicanism rested. Industry, they feared, would bring the social ills and divisions it bred in Britain. But by the War of 1812 a broad consensus jelled that the United States needed its own manufacturing industries to ensure its strength and independence. Furthermore, even many critics of industrial development came to believe that the physical and political setting of the United States would shape a system of manufacturing shorn of the evils that accompanied it in Europe. Using water power rather than steam meant that American mills were dispersed in towns and small cities, avoiding the congestion and urban ills of Manchester and other British mill cities. Using young, country women as short-term workers avoided the creation of a debased proletariat. What was wrong with Old World manufacturing, American political and intellectual leaders came to believe, was not manufacturing but the Old World. Lowell, many contended, demonstrated that manufacturing in the New World could coexist with democratic values, moral purity, and pastoral harmony.[54]

Not everyone was so sanguine. Poet and abolitionist John G. Whittier was often quoted for his smile-inducing 1846 description of "The

Factory Girls of Lowell": "Acres of girlhood—beauty reckoned by the square rod, or miles by long measure!—The young, the graceful, the gay—flowers gathered from a thousand hill-sides and green vallies of New England." Whittier praised the Lowell workers for their "hope-stimulated industry," teaching "the lessons of Free Labor," a sharp contrast to the "whip-driven labor" of the slave plantation. But later in the same article he chastised the "good many foolish essays written upon the beauty and divinity of labor by those who have never known what it really is to earn one's livelihood by the sweat of the brow—who have never, from year to year, bent over the bench or loom, shut out from the blue skies, the green grass, and the sweet waters, and felt the head reel, and the heart faint, and the limbs tremble with the exhaustion of unremitted toil." Whittier acknowledged "much that is wearisome and irksome in the life of the factory operative."[55]

Labor reformer Seth Luther sharply took to task politicians who praised the cotton mills based on whirlwind tours: "For an hour or more (not fourteen hours) he seems to be in the regions described in Oriental song, his feelings are overpowered.... His mind being filled with sensations, which from their novelty, are without a name, he explains, 'tis a paradise." But for Luther, "if a cotton mill is a 'paradise,' it is *Paradise Lost*,'" a site of unhealthy long hours, poorly paid workers, and tyrannical overseers.[56]

Critics of New England mill conditions, unlike in England, rarely claimed that factory conditions were as bad as or worse than slavery. Ralph Waldo Emerson was something of an exception when, in a bitter commentary on Lowell, he equated black slaves in the South with female mill worker "slaves" and criticized mill owners for wanting to live in luxury without working, "enjoyment without the sweat."[57] But critics still turned to slavery for metaphors of oppression. An 1844 letter in the *Manchester Operative*, for instance, likened the mill bell calling workers to their tasks to "a slave driver's whip," while for a New Hampshire worker the unrestrained power of overseers was equivalent to that of slave drivers. A few critics—though not many—acknowledged that while the mills were not a form of slavery themselves, they were deeply embedded in the slave system, dependent on slave labor to grow the cotton they used and producing textiles sold to slave owners to clothe their chattel.[58]

Though supporters and critics generally agreed that New England mills were not as bad as those in Britain, some argued that the difference might be temporary. Seth Luther declared that the "misery in horrid forms . . . in the manufacturing districts of England" was "directly produced by manufacturing operations" and that the United States was "following with a fearful rapidity the '*Splendid Example of England*.'" Luther highlighted the employment of children, very common in the Rhode Island–style mills, a harsh reality usually elided by the focus of contemporary observers and later historians on the Lowell-style mills. Luther decried the lack of education that inevitably resulted from toiling long hours in the "*palaces of the poor.*" To the cry of manufacturers that "it is not so bad as it is in England yet," he responded that one might as well "say the Cholera is not so severe in Boston yet as it has been in New York."[59]

Anthony Trollope came to similar conclusions. The superior conditions and paternalist institutions of Lowell, he suggested, were made possible by its comparatively small size by English standards. (On the eve of the Civil War, there were nearly four times as many cotton workers in Britain as in the United States.) Scaling up, Trollope envisioned, would require moving from water to steam power. If Lowell made the switch and "spread itself widely," he wrote, "it will lose its Utopian characteristics." John Robert Godley made a similar point in his 1844 *Letters from America*, questioning whether Lowell could be used to demonstrate that "the evils which have in Europe universally attended the manufacturing system are not inevitable in it." Lowell, he noted, was set up and developed "under eminently favorable circumstances." Over time, as the population of the United States grew, wages fell, and manufacturing increased in importance, he doubted that "the favourable contrast which the New England factories now present to those of England, France, and Germany, can possibly continue." A quarter century later, Edward Bellamy, a lifelong resident of Chicopee Falls and author of the blockbuster utopian novel *Looking Backward*, also saw European conditions of poverty and social division coming to the United States. He had "no difficulty," he wrote, "in recognizing in America, and even in my comparatively prosperous village, the same conditions in course of progressive development."[60]

Herman Melville at least implicitly suggested that the United States already had the same kind of class division that manufacturing had brought to England in his 1855 story, "The Paradise of Bachelors and the Tartarus of Maids." The first part of the story portrays a group of well-fed, self-indulgent London lawyers, while the latter part recounts a winter visit to a paper factory in an isolated New England valley, apparently based on Melville's visit to a paper mill in Dalton, Massachusetts (which is still operating). The narrator expresses his awe at the ingenuity and operation of the papermaking machine, "this inflexible iron animal," "a miracle of inscrutable intricacy." But he is horrified by the pale, unhealthy looking, silent "girls," the unmarried women who come from "far-off villages" who operate the machinery, "mere cogs to the wheels," a far cry from the way Lowell workers usually were portrayed. Rather than in a "commercial Utopia," Melville's young women were trapped in "Tartarus," a province of the underworld, while far away the wealthy barristers fed and liquored themselves.[61]

New England reformer Orestes Brownson was more explicit in seeing the nation divided into "two classes," laborers and capitalists. In a widely debated essay on "The Laboring Classes," Brownson used Lowell as an example in decrying the effect of factory labor on workers and the growing gap between industrialists and their employees, suggesting that only a radical recasting of society could re-create true community.[62] Seth Luther concurred: "The whole system of labor in New England, *more especially in cotton mills*, is a cruel system of exaction on the bodies and minds of the producing classes, destroying the energies of both, and for no other object than to enable the 'rich' to 'take care of themselves,' while 'the poor must work or starve.' "[63]

Alexis de Tocqueville, too, saw a growing class division in the United States, brought about by factory production. The efficiencies of large factory production, he predicted, would enrich manufacturers to the extent that they would become a new aristocracy, threatening democracy, while workers were physically and mentally disadvantaged by the narrow, repetitive nature of factory tasks. "Whereas the workman concentrates his faculties more and more upon the study of a single detail, the master surveys a more extensive whole, and the mind of the latter is enlarged in proportion as that of the former is narrowed." The industrial class divide, the concentration of workers, and the cyclical nature of the economy might well endanger "public tranquility," a problem, in Tocqueville's view, that would require more government regulation to avoid.[64]

Faded Visions

The debate over Lowell raised what already had become a recurring question: Was the factory system inherently oppressive to workers and threatening to social cohesion or did its nature change with its environment? Over time, the critical views of Brownson, Tocqueville, and Luther became more widely shared. In England, the cotton mill quickly brought a broad acceptance of the idea that it was creating a new type of class society. In the United States, there was an interregnum during which the large factory was associated with the idea that industry and republican community could coexist. But by the time of the Civil War, changes in the factory system itself, evident in Lowell and other cotton centers, faded visions of "commercial Utopia."

Above all, it was the transformation of the workforce that changed the public perception of the New England mills. By the late 1840s, fewer young New Englanders were coming to the mill towns as a result of growing displeasure with the pay, hours, and increased workload, evident in the strikes of the 1830s and the ten-hour movement. Also, for young women other alternatives to staying in rural homes opened up. Railroads made it easier to move to urban centers or out West. With the spread of public education, the number of jobs for teachers swelled and salaries improved.[65]

Fortunately for the mills, in the mid-1840s, just as the influx from the countryside diminished, a new labor pool materialized with mass migration from famine-gripped Ireland. Between 1846 and 1847 alone, immigration from Ireland more than doubled, and by 1851 it more than doubled again. There were always Irish workers in Lowell and other mill towns; Irish men dug the canals and helped build the factories. But before 1840, the textile companies generally spurned Irish women; in 1845, only 7 percent of the Lowell mill workforce was Irish. Necessity ended the discrimination; by the early 1850s, about half the textile workers in Lowell and other mill towns were Irish. At the Hamilton mill, by 1860 over 60 percent of the employees had been born abroad.[66]

The increasing number of immigrant workers brought other changes. More children began being hired in Lowell-style mills, especially boys, as whole families needed to work to support themselves, a reversion to the pattern in the early Slater-type mills. The gendered division of labor broke down as male immigrants accepted jobs once reserved for women, paid wages that in the past only women would take. At Hamilton, in 1860, 30 percent of the workforce consisted of adult men.

Immigrant family labor contributed to the decline of the boardinghouse system and company paternalism. Lowell firms put up mills at a faster pace than they built housing, and after 1848 they stopped building housing entirely. Institutional arrangements once needed to attract rural young women and reassure their parents became increasingly superfluous, as the companies acknowledged in the 1850s when they dropped requirements for church attendance and boardinghouse residence for single women. A growing proportion of the workforce—including more and more single women—lived in non-company-owned boardinghouses or in rented tenement apartments. The company boardinghouses lingered on—between 1888 and 1891 a quarter of the workers at the Boott mills were still living in company-

Figure 2.3 Winslow Homer's 1868 engraving of New England factory life, *Bell-Time.*

owned housing—but they declined in importance as the immigrant workforce grew.[67]

As the novelty of the mills wore off, the "acres of girlhood"—or at least of native-born girlhood—shriveled, and company paternalism diminished, travelers, politicians, and writers lost interest in Lowell. But even as public attention moved away, the mills continued to expand. The Civil War stimulated growth. With cotton all but unobtainable and raw cotton prices soaring, many Lowell mills sold off their cotton inventories for windfall profits, reducing or stopping their own operations. Some took advantage of the hiatus to expand and modernize. The Boott mill added two buildings and replaced much of its machinery. In the postwar years, it built yet another mill and began supplementing water power with steam. By 1890 it employed over 2,000 workers, large but nowhere nearly as large as the Merrimack mills, with over 3,000 workers, and the Lawrence mills, with over 4,500.

In nearby Lawrence, the economic downturn in 1857 drove three mills into bankruptcy, but the war brought a boom. Unlike in Lowell, the Lawrence mills generally held on to their cotton to continue production. Old mills expanded and new ones shot up and continued to grow after the war, on a scale surpassing anything in Lowell. To hedge against cotton goods' booms and busts, most Lawrence mills also produced woolens or worsteds. One Lawrence factory, the Wood Mill, controlled by the American Woolen Company, in the early twentieth century had more than seven thousand workers. Altogether, Massachusetts employment in the cotton textile industry soared from 135,000 in 1870 to 310,000 in 1905. In New Hampshire, Amoskeag Manufacturing Company expanded until it was the largest textile factory complex in the world.[68]

As the New England mills kept growing, Irish workers were joined and partially replaced by an influx of French Canadians. In the early twentieth century, other immigrant groups began working in the mills, too, largely Southern and Eastern Europeans but smaller groups, like Syrians, as well. For some of the newcomers, the experience of mill work felt not much different than it had for the early New Englanders. Cora Pellerin, a French Canadian who began working at Amoskeag in 1912 at age eleven, thought "It was paradise here because you got your money, and you did whatever you wanted to with it." But for many others, the experience of mill work and mill-town life was far less positive, as working conditions deteriorated and widespread poverty came to characterize the factory towns. "By 1910," according to historian Ardis Cameron, "readers of Charles Dickens would have found Lawrence's dull streets and cluttered alleys, its black canals and purple ill-smelling river, its vast piles of soot-covered brick buildings, its flimsy, damp privies whose waste oozed down open sewers and meandered through the city's shaded backyards a familiar landscape."[69]

After the 1850s, when New England mills appeared in the news, it usually was because of untoward developments. In January 1860, the seven-year-old Pemberton mill in Lawrence fell down, its poorly made cast iron columns unable to withstand the weight and vibrations of its machinery. In the collapse and subsequent fire that engulfed the rubble and those trapped in it, some one hundred people died and many more suffered serious injuries. It remains to this day one of the worst industrial disasters in U.S. history. Newspapers and magazines as far away as Hawaii reported on the catastrophe, recounting "heart-rending and appalling scenes" and featuring drawings of rescues and the charred remains of victims. Some papers—going beyond the conclusions of a coroner's inquest, which held the mill architect responsible—blamed the calamity on "the wealthy Boston philanthropists" who owned the mill and the "flagrant disregard" of company leaders for the "safety of their employees," tarnishing the reputation of the mill owners.[70]

Child labor also brought the mills unflattering public attention. Textile was among the industries targeted in an early twentieth-century campaign to keep children out of mines and mills. The photographs that Lewis Hine took in 1909 for the National Child Labor Committee of children working at Amoskeag became iconic.[71]

Labor strife further buried the notion that the New England mills would avoid the ills of European industry. After the Civil War, textile strikes became increasingly common. Some involved relatively small groups of skilled male workers, like mule spinners. In other cases, women workers or alliances cutting across skill and gender lines conducted the strikes. Lawrence workers staged small strikes in 1867, 1875, and 1881, and a long strike in 1882, which received national attention. Failed strikes took place in Lawrence in 1902 and in Lowell in 1903.[72]

The last time the country was captivated by a vision of the future emanating from the mills of New England came in 1912, when some fourteen thousand workers in Lawrence went on strike for two months to protest a pay cut, instituted in response to a state law reducing working hours. "The strike in Lawrence," declared socialist Congressman Victor Berger, "is a rebellion of the wage-working class against unbearable conditions." Led by fiery organizers from the Industrial Workers of the World (IWW), women and men from forty different ethnic groups banded together, creating multilingual committees to direct the struggle. The militancy and solidarity displayed by the Lawrence strikers—the type of semiskilled immigrant workers mainstream labor

Figure 2.4 Lewis Hine is famous for his startling portraits of child workers, including this one of a girl working in the Amoskeag textile factory in 1909.

leaders had written off as impossible to organize—inspired radicals and unionists across the country to imagine that a new labor movement and a transformed nation were coming.

Mill owners and government officials set out to crush the strike with a declaration of martial law, a ban on public meetings, the arrest of strike leaders on trumped-up charges, the mobilization of the National Guard, and physical attacks on the strikers and their supporters. When the strikers, running out of food and money, started sending their children out of town to live with supporters, the police and militia tried to stop them, clubbing adults and children alike at the railway station. The owners overplayed their hand, as a wave of national outrage contributed to their decision to grant a substantial pay increase, ending the strike in a workers' victory.[73]

After the strike, the IWW failed to consolidate its power. It took another two decades before New England mill workers finally created stable unions. By then, the end was near. Slow to modernize, and facing ever-growing competition from lower-cost Southern mills (some financed by New England mill owners), the mills put up by the Boston Associates began shutting down in the early twentieth century. Amoskeag closed in 1936, the rebuilt Pemberton mill in 1938, and the last of the original Lowell mills in the 1950s. Bits and pieces of textile production continued in Lawrence and elsewhere in New England, but the great experiment launched by Francis Cabot Lowell was over.[74]

Well before the Lowell factories began closing, the United States had surpassed Great Britain as the world's greatest industrial power. By the mid-1880s, more goods flowed out of American factories than British ones. By World War I, the manufacturing output of the United States topped that of Britain, France, and Germany combined. The meteoric growth of American manufacturing reflected, in part, the growing size of the country itself, which in 1890 approached a population of 63 million, far larger than Britain, with 33 million; France, with 38 million; and Germany, with 49 million, allowing the high output of cheap standardized goods for the home market.[75]

Lowell had helped usher in America's industrial age and its global industrial dominance. It had been born amid a blaze of positive publicity because it promised the fusion of mechanized manufacturing with republican values, creating a "commercial Utopia" that would confirm the United States as a land of new beginnings and infinite possibilities, free of the class divisions and inequalities of the Old World. The success of Lowell in creating a different social and cultural model for manufacturing helped ease long-standing national concerns about the impact of industrialization on what was still an agrarian republic, allowing a new consensus equating progress with increased productivity through mechanization and large-scale enterprise. By the time the Lowell mills receded from view, Americans had squarely embraced a vision of the future built on the bedrock of industry. Ironically, when the mills dominated national news for a final time, in 1912, before fading into oblivion and decay, it was because of the very kind of class warfare the promoters of Lowell had claimed its system would avoid.

CHAPTER 3

"THE PROGRESS OF CIVILIZATION"

Industrial Exhibitions, Steelmaking, and the Price of Prometheanism

ON MAY 10, 1876, THE INTERNATIONAL EXHIBITION of Arts, Manufactures, and Products of the Soil and Mine opened in Philadelphia, a celebration of the hundredth anniversary of the Declaration of Independence. One hundred thousand people heard speeches by assorted dignitaries, sixteen national anthems, the premiere of Richard Wagner's "Centennial Inauguration March," the "Hallelujah Chorus" sung by a thousand-voice choir, and a 100-gun salute. But for many visitors the highlight of the day came when President Ulysses S. Grant and Emperor Dom Pedro II of Brazil led the crowd into the immense Machinery Hall. There they climbed onto the platform of the forty-foot-high Corliss double walking beam steam engine. When each man turned a valve before him, the 56-ton, 1,400-horsepower engine came to life, turning twenty-three miles of shafting that powered hundreds of machines filling the wood and glass building.

The Centennial Exhibition, as it was commonly known, was an extravaganza occupying 285 acres, attended, during its six-month run, by nearly ten million visitors, equivalent to about a fifth the population of the United States. With exhibits from thirty-seven nations, its displays were encyclopedic, featuring everything from exotic plants

Figure 3.1
President Ulysses S. Grant and Emperor Dom Pedro II of Brazil start the Corliss engine at the Centennial Exposition in Philadelphia in 1876.

and prize cattle to fine art and historical artifacts. But machines and machine-made products overwhelmed all else.

The fourteen-acre Machinery Hall contained a dizzying array of industrial equipment, including a complete printing operation that produced a newspaper twice a day, a railroad engine, metalworking and woodworking machinery, brickmaking machines, and spinning and weaving equipment from Saco, Maine. In the section called, in the taxonomic terminology typical of the exhibition, "Machines, Apparatus, and Implements Used in Sewing and Making Clothing and Ornamental Objects," a visitor could watch a pair of suspenders mechanically made with his or her name woven into the fabric. Among the new inventions unveiled were the typewriter, the telephone, and a mechanical calculator. A huge variety of machine-made products could be found in the Main Exhibition Building. Smaller buildings, like the Singer Sewing Machine Company building and the Shoe and Leather Building, housed still more machines and machine-made products. Even the Agricultural Hall was full of machines, from reapers to windmill-driven pumps to chocolate-making equipment.[1]

It was a peculiar way to mark the one hundredth anniversary of the United States. There was plenty of patriotic imagery and patriotic kitsch. But the weight of the exhibition lay elsewhere, in the celebration of the technological marvels of the day, of the great productivity and inventiveness of the United States, of its progress as measured by its mastery of the mechanical realm. It took an ideological leap to see the connection between the American Revolution and the Corliss engine.

The concentration on mechanical marvels and industrial bounty measured how much views of national greatness and progress had changed during the half century since the Lowell mills opened. With little dissent, Americans had come to see machines and mechanical production as central to the meaning of the national experience, as integral to modernity. Americans had deep, sometimes violent disagreements about the structure and values of their society, as Reconstruction in the South came to a bitter end, workers suffering through a devastating economic depression launched the largest strikes the country had ever seen, and wars against Native Americans raged in the West. But about machinery and what it made possible, there was not much discord.[2]

Americans believed that machines were opening the door to a new age of unprecedented bounty, freedom, and national power. The steam engine took center stage. It seemed to defy the gods, as Prometheus had, capturing fire from them and putting it to work. Tench Coxe, a Philadelphia merchant who worked closely with Alexander Hamilton in the late eighteenth century promoting manufactures, even used the term "Fire" for the steam engine.

The miraculous power of steam fully revealed itself with the introduction of the first practical steamboats not long after the Revolution. John Fitch began operating a steam-driven ferry between Trenton and Philadelphia in 1790. In 1807, Robert Fulton's *North River*, equipped with a British-built steam engine, travelled up the Hudson River from New York to Albany. Four years later, his *New Orleans* introduced the steamboat to the Ohio-Mississippi river system, opening up the western frontier of the United States to commercial development. Two-way shipping on the Mississippi facilitated the spread of cotton culture and, with it, slavery.

But it was not just the effect of the steamboat that drew admiration; it was the boat itself, its speed, power, and unnatural beauty. Mounting a steam engine on a boat radically changed the experience of time, space, and distance, making once-epic journeys, like the trip from St. Louis to New Orleans, achievable in just a few days. Writer Edmund Flagg declared "There are few objects more truly grand—I almost said sublime—than a powerful steamer struggling with the rapids of the western waters." For Flagg and others, the contrast between the steamboat, the creation of mankind, and the wild, natural setting of the Mississippi contributed to making the scene so memorable, bordering on the sublime, which for the nineteenth-century observer meant not just awesome or beautiful but frightening, unsettling, and overwhelming, too.

Americans and Europeans traveling into newly settled lands often saw the steamboat to be a carrier of civilization itself, or at least their idea of civilization. But it did not take a wild Western setting to make the steamboat seem exalted. In 1848, Walt Whitman, echoing Flagg, wrote of the engine room of a Brooklyn ferry, "It is an almost sublime sight that one beholds there; for indeed there are few more magnificent pieces of handiwork than a powerful steam-engine swiftly at work." Three years later he said the United States had become a nation "of whom the steam engine is no bad symbol."[3]

The railroad soon eclipsed the steamboat as a symbol of modernity. Steam-driven trains were even more widely seen, more widely used, and more widely praised than steamboats. In the year of the Centennial, Whitman wrote in "To a Locomotive in Winter": "Type of the modern! emblem of motion and power! pulse of the continent!" By radically reducing the time, cost, and difficulty of moving people and things, the railroad tied the nation together, spreading commercial relations and disseminating ideas and sensibilities. With the railroad came new landscapes, a new sense of time, and a new cosmopolitanism.[4]

Exhibiting Modernity

Even standing still, the steam engine became a symbol of progress and national prowess, part of the broader celebration of machinery and manufactured goods, so evident at the Centennial Exhibition. Before the Philadelphia fair and continuing long after it, public exhibitions were built around the processes, symbols, and products of mechanical manufacturing, equating them with modernity. In 1839, for example, the Massachusetts Charitable Mechanic Association held its second exhibition at Boston's Quincy Market. Over the course of twelve days, seventy-thousand people attended. Among the exhibits were an operating miniature railroad, a small steam engine that powered other machinery, planning machines, a "cassimere shearing machine," printing presses, and knitting machines. Displayed goods included textiles from Lowell, looking glasses, cabinets, coaches, saddles, hosiery, hats, caps, furs, confectionery, soaps, perfumes, boots, cannons, rifles, swords, hardware, cutlery, locks, pumps, fire engines, and musical instruments. Defending against the belief that manufacturing was undermining republican virtue, James Trecothick Austin, in an address at the exhibition, tried to dismiss "the supposed conflicting interests of the various classes in American society." "Our splendid manufactures of silver," he said, "are worse than useless, if it is a sin against democracy to use a silver fork."[5]

The 1851 Crystal Palace Exhibition in London, officially the "Great Exhibition of the Works of Industry of All Nations," marked the beginning of the great international expositions and world's fairs, temples dedicated to progress and modernity as reflected in machines and machine-made objects. The building that contained the fair was at least as impressive as the exhibits within it. A huge iron and glass conservatory, the Crystal Palace was constructed entirely out of machine-made parts, so that after the exhibition closed it was easily disassembled and reconstructed on a different site. The British exhibit, by far the largest, had sections devoted to fine arts, "raw materials," "machinery," and "manufactures." The industrial tourism that the well-connected had indulged in with factory visits now was brought to the masses. Fifteen steam-driven machines for carding, spinning, and weaving took raw cotton and converted it to cloth while viewers stood nearby. The enormous display of manufactured goods educated attendees about the emerging consumer society, showing the myriad things that could be made and how they would make life better. "World exhibitions," Walter Benjamin would later write, were "sites of pilgrimage to the commodity fetish."[6]

The United States mounted a "Crystal Palace" exhibition of its own, the Exhibition of the Industry of All Nations, in 1853. The iron and glass exhibit hall, built in New York City on the site of what is now Bryant Park, was essentially a smaller version of the London building, with a dome added. It created a sensation; nothing like it had ever been seen in the New World. Like the London exhibition, it contained a hodgepodge of art, machinery, and manufactured products.[7]

Other countries, too, mounted international exhibitions. The French held a series of fairs in Paris, starting with the 1855 Exposition Universelle and its Palais de l'Industrie, intended by Napoleon III to top the London display. Succeeding exhibitions came in 1867, 1878, 1889, and 1900. Vienna put on an International Exhibition in 1873. Chicago created the large, well-attended, and widely celebrated 1893 Columbian Exposition. Other United States fairs followed in short order, including in Omaha (1899), Buffalo (1901), and St. Louis (1904).[8]

Even the 1895 Cotton States and International Exposition, held in Atlanta to highlight the economic recovery of the South under white rule and the continued reign of King Cotton, prominently featured a

Machinery Hall. One account called it "the heart" of the fair; "wheels, big and little, whirl in every quarter; dynamos generate untold volts of electricity; pumps and lathes, planes and drills are hard at work, all obediently responding to an unseen but irresistible force." "Southerners joined with millions of Yankee guests," wrote historian C. Vann Woodward about the Southern expositions held in the 1880s and 1890s, "to invoke the spirit of Progress and worship the machine."[9]

The Eiffel Tower, built for the 1889 Exposition Universelle, became the foremost icon of the international fairs. Gustave Eiffel, a successful French engineer, won a government competition for a centerpiece for the exposition celebrating the hundredth anniversary of the French Revolution. Made up of more than eighteen thousand wrought-iron members, fabricated at an off-site factory, the 312-meter tower soared to nearly twice the height of what had been the world's highest structure, the Washington Monument, completed just five years earlier. From the top, the tower offered vistas previously known only to a few balloonists, a preview of the bird's-eye view of the great metropolis that would become common only decades later, after the invention of the airplane.[10]

Before it was built, a group of prominent French artists, musicians, and writers protested what they called the "useless and monstrous Eiffel Tower," "the hateful column of bolted iron," which they declared would desecrate the beauty and honor of Paris.[11] But the tower almost immediately became celebrated as a symbol of modernity, portrayed as a new kind of beauty. Even before it was completed, George Seurat made it the subject of one of his best-known canvases. A flood of drawings, paintings, and lithographs followed, including works by Henri Rousseau, Diego Rivera, Marc Chagall, and, perhaps most delightfully, Robert Delauney, who returned to the subject over and over again. The tower proved an ideal subject for modernist approaches to representation, including pointillism and cubism. Pioneer filmmakers also engaged the tower, the subject of short films by Louis Lumière in 1897 and George Méliès in 1900.[12] So did writers. In Guillaume Apollinaire's poem "Zone," the tower herded the way to modernity:

At last you're tired of this elderly world

Shepherdess O Eiffel Tower this morning the bridges are bleating

You're fed up living with antiquity[13]

Blaise Cendrars concluded "Tower," his 1913 poem dedicated to Delauney,

You are all
Tower
Ancient god
Modern beast
Solar spectrum
Subject of my poem
Tower
World tower
Tower in movement[14]

The huge numbers of visitors to the expositions and the flood of positive publicity attested to the widespread admiration for the new industrialism—the steam engines, vast iron structures, and machinery on display. [15] Of course, not everyone was entranced. Guy de Maupassant declared, "I left Paris, and France, too, on account of the Eiffel Tower. It could not only be seen from everywhere, but it could be found everywhere, made of every kind of known material, exhibited in all windows, an ever-present and racking nightmare." The author tired of the crowds the 1889 fair attracted, among them "the people who toil and emit the odor of physical fatigue."[16]

Just how many working people actually attended the various fairs is difficult to say. The middle and upper classes apparently made up the bulk of the audience, better able to afford travel and entrance fees. The planners of the London Crystal Palace Exhibition paid considerable attention to attracting and controlling working-class visitors. Admission was cheaper on Mondays through Thursdays, facilitating visits by workers and their families, while allowing wealthier patrons to have Fridays and Saturdays largely to themselves. Many companies subsidized employee expeditions to the exhibition. The Philadelphia Centennial closed on Sundays, generally the only day workers had off, as a result of pressure from local clergymen, making it difficult for them to attend. But as in England, employers sponsored trips to the exposition for their workers.[17]

Working-class visitors generally seemed to have enjoyed the fairs—by some reports, they were more interested in machinery and less interested in fine art than their economic betters—but some leaders of workers' movements could not ignore what they saw as the exploitation that underlay the industrial bounty on display. Radical Chartist G. Julian Harney called the exhibits at the 1851 exposition "plunder, wrung from the people of all lands, by their conquerors, the men of blood, privilege, and capital." During the 1889 Paris fair, socialists from Europe and the United States gathered at congresses in the city. Friedrich Engels, who by then had retired from his Manchester cotton mill, stayed away. He wrote Laura LaFarge, Marx's daughter, "There are two things which I avoid visiting on principle, and only go to on compulsion: congresses and exhibitions." Paul LaFarge, Laura's husband, complained to Engels that "the capitalists have invited the rich and powerful to the *Exposition universelle* to observe and admire the product of the toil of workers forced to live in poverty in the midst of the greatest wealth human society has ever produced."[18]

Iron

The crystal palaces in London and New York, the great machinery halls, and the Eiffel Tower were possible because of advances in the iron industry. If the first half of the nineteenth century constituted the age of cotton, the decades after 1850 were the age of iron. By the time of the Centennial Exhibition, the largest manufacturing plants in Europe and the United States made iron and steel goods, not textiles. Iron mills and, later, steel mills supplanted textiles mills as symbols of modernity, as poles for debate about the nature of the society and what kind of future people sought.

Until the nineteenth century, iron was made only in small quantities for specialized products. Typically, in Europe and North America, the mining of ore, its conversion into iron, and the production of finished goods all took place at one site, by small groups of skilled workers. But by the middle of the nineteenth century, the rising demand for iron outstripped traditional production techniques, in which small furnaces, fueled by charcoal or coke, were used to remove oxygen and impurities from iron ore, producing metal which could be cast into finished goods or later reheated and converted into stronger, more malleable wrought iron.[19]

A huge boost in the demand for iron came from the spread of the railroad and the need for rails. In 1840, there were 4,500 miles of railway worldwide; by 1860, 66,300 miles; and by 1880, 228,400. At first, producing rails proved painfully difficult. Because not enough iron could be rolled at once to make a single rail, small bars had to be rolled into strips, which were layered, reheated, and rolled again. Quality was low; sometimes rails delaminated and, on heavily used lines, they wore out in as little as three months. American metallurgist Frederick Overman wrote in the early 1850s, "The application of science and machinery in the manufacture of iron does not exhibit so high a state of cultivation as we find in . . . the manufacture of calico prints and silks."[20]

That changed with a series of technical innovations that increased the quantity and quality of production. First came the blast furnace. Instead of forcing cold air through heated iron ore to remove the carbon in it, starting in 1828 in England and six years later in the United States, hot air, heated by the exhaust of the furnace itself, was used, greatly increasing the speed and efficiency of the process. Raising the temperature and pressure of the air yielded further gains. From a typical output in the 1850s of one to six tons of iron a day, by 1880 furnaces neared an output of one hundred tons a day.[21]

Iron produced by blast furnaces could be used to make some products by casting, like stoves and plows. But it was too brittle for many uses. Further reducing the carbon content to make wrought iron gave it greater strength and flexibility but required intensive labor, either repeated pounding at a forge or chemical transformation through a process known as puddling. Puddlers reheated cast-iron bars, so-called pig iron, along with scrap iron in special furnaces, stirring the mixture to oxidize the carbon and burn off impurities. Experience, skill, and physical strength were needed to control the process.

With a strong craft culture and a high level of unionization, puddlers forced iron manufacturers into what effectively was a partnership. The workers regulated all aspects of the puddling process, including how much iron to produce in each turn and their hours of work. They often paid helpers out of their own wages. In Pittsburgh, the most important iron center, a sliding scale linked puddlers' pay to their output and the selling price of iron, so that they shared any gains that resulted from higher productivity or improved market conditions. The men who operated rollers for shaping rails and other products also exerted near total control over the production process. In some mills, they negotiated a price per ton for an entire team of workers, which they decided among themselves how to divide.[22]

Early iron plants tended to be small, as puddling could make wrought iron in batches of only about six hundred pounds at a time. Soon, though, technical and financial considerations pushed up plant size. Rolling rails required expensive equipment; to be profitable, rail mills had to be operated around the clock, which necessitated a great deal of wrought iron. Some rail makers purchased iron from other firms, but the leading companies integrated backward, setting up their own blast and puddling operations. Switching fuel from charcoal to coke liberated them from the need to be near large tracts of forested land from

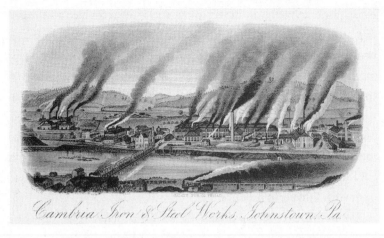

Figure 3.2 Cambria Iron and Steel Works in Johnstown, Pennsylvania, circa 1880.

which charcoal could be produced. Coal deposits and major rail lines made Pennsylvania particularly attractive for large-scale operations.

The Cambria Iron Works, near Johnstown, Pennsylvania, was for a time the most advanced mill in the United States, introducing a system of three-high rollers, which allowed iron to be moved backward and forward between shaping rollers, minimizing the need for reheating. Its rail mill extended over a thousand feet long and a hundred feet wide, far larger than the largest cotton mill. In 1860 it employed 1,948 workers, about as many as the largest Lowell mills. The Montour Iron Works in central Pennsylvania, another rail producer, had three thousand employees. Though like the first textile mills, iron mills often lay in rural areas or small towns near rivers, they proved far more disruptive, sprawling over large sites and spewing out dark smoke. One European traveler described iron-mill smoke in Pittsburgh as giving "a gloomy cast to the beautiful hills which surround it."[23]

The introduction of the Bessemer process led to a further leap in the scale of mill complexes. Puddling created a bottleneck in the production of iron goods, both because of its small-batch process and the strong control exerted by the puddlers. The Bessemer process, developed by Englishman Henry Bessemer in the mid-1850s, provided an alternative way of turning blast-furnace iron into a stronger, more malleable metal. As modified by subsequent inventors, it worked by forcing air into molten pig iron, allowing oxygen to combine with the carbon in the metal, thus removing it, with a manganese-based ore introduced to remove excess oxygen and sulphur. The end product fell somewhere between pig iron and wrought iron in its carbon content and proved more durable for rails than puddled metal. Its promoters dubbed it steel, appropriating the name of an older form of purified iron that had been very difficult to produce.

The Bessemer process worked best with iron made from ore with a low phosphorous content, more readily available in the United States than in Europe. So it was in the United States, starting right after the Civil War, that the Bessemer process was first widely adopted. Some products, like pipes, bars, and plates, continued to be made out of puddled iron even as the Bessemer and the later open-hearth processes for steelmaking spread. As late as the 1890s, one company, Jones and Loughlins, had 110 puddling furnaces. But thereafter iron production plummeted and the age of steel was firmly established.[24]

Even early on, Bessemer furnaces could convert five tons or more of iron into steel in one turn. To feed them, companies built bigger and bigger blast furnaces. Rather than making pig iron that later had to be reheated, they loaded Bessemer converters directly with molten metal. In Pittsburgh and Youngstown, they built bridges to allow trains with special ladle cars to carry liquid iron from blast furnaces on one side of a river to converters on the other. In the 1880s, some firms began taking ingots produced by converters directly to rolling mills, where, after adjusting their temperature in "soaking pits," workers rolled them without reheating. Thus heat and energy were conserved as the molten metal never was allowed to fully cool between its initial creation and the completion of finished products.[25]

Increased output, integration, and an ever-growing array of end products, requiring finishing mills for structural steel, wire, plate, and other goods, boosted iron and steel mills to unprecedented size. In Germany, the Krupp works in Essen, out of which came steel cannons that were crowd favorites at the Crystal Palace and other exhibitions, grew from seventy-two workers in 1848 to 12,000 workers in 1873. In France, the Schneider works in Le Creusot, an iron and steel producer which, like Krupp, came to specialize in armaments, had 12,500 workers in 1870.[26] In the United States, firms were quicker to mechanize and had fewer employees but also were growing. In 1880, the Cambria mill had the largest workforce in the industry, 4,200. Andrew Carnegie's Homestead plant, which displaced Cambria as the most technologically advanced mill in the United States, and which like Krupp and Schneider heavily engaged in producing armor, grew from 1,600 workers in 1889 to nearly 4,000 in 1892.

In 1900, of the 443 manufacturing establishments in the United States with over a thousand employees, 120 produced textiles, mostly cotton, and 103 iron or steel, so that half of all the large factories in the country were in these two industries. Among the very biggest plants, iron and steel dominated. Three of the four factories in the United States with over eight thousand workers made steel (Cambria, Homestead, and the Jones and Laughlins Pittsburgh plant), while the fourth made locomotives. Three more steel mills had between six thousand and eight thousand workers.[27]

The steel mill, as a production system, was far more complex than the cotton mill. Its products were less uniform. Rails made to standard specifications were produced in large quantities, but finishing mills also filled orders for myriad other goods, some in small numbers: structural steel in all shapes and sizes, steel sheets of varying dimensions, armor plate of different thicknesses and strengths, pipes, wire, bars, tinplate, and so on. Experienced workers and constant adjustment of machinery were needed to meet ever-changing specifications. Carnegie came to dominate the steel industry by running his business like the Lowell mills. "The surest way to continued leadership," he believed, was "to adopt policy of selling a few finished articles which require large tonnage." Bridges, he said, were "not so good because every order different." But beyond rails, which became relatively less important as the railroad system was built out and stronger rails required less frequent replacement, Carnegie's policy proved hard to imitate.[28]

A single worker operating a single machine could turn roving into thread or thread into fabric, but no one worker could produce a bar of pig iron or a steel rail. Instead, coordinated activity by teams of workers was needed. Even puddlers, the most autonomous metalworkers, worked in pairs, the heat and effort being so draining that they needed to spell each other. Each was assisted by a helper and sometimes a "boy." Larger groups of workers, some skilled and some laborers, operated blast furnaces, Bessemer and open-hearth converters, and rollers.

Unlike spinning and weaving, most iron and steel operations were not continuous. Blast furnaces were run nonstop, with raw materials poured in the top and iron tapped out at the bottom, until the linings burnt out or other problems developed, when they would be cooled and rebuilt. But most other processes were batch operations. Once a Bessemer converter was charged with molten iron, it took only eight to ten minutes before steel was poured out and the cycle restarted. Open-hearth converters took eight hours to complete their work— one reason why, though they produced higher-quality steel, companies were slow to adopt them. Unlike textile workers, many of whom did exactly the same thing all day long, ironworkers and steelworkers often took on varied tasks and alternated periods of intense labor with rest and recovery.[29]

In textile mills, many identical machines operated side by side, drawing power from a common source. Integrated iron and steel mills had many fewer machines (often with individual engines driving them), but they were linked in tighter sequential operation.

Some of those machines were gigantic. At Homestead, workers made armor from steel ingots that weighed as much as one hundred tons. After being rolled to the appropriate size, their ends were trimmed by a hydraulic press with a 2,500-ton capacity. They were then reheated to be tempered and cooled in a bath of 100,000 gallons of oil. Final machining was done with enormous equipment, like a planning machine weighing two hundred tons. The flywheel alone on one engine in the beam mill weighed one hundred tons. The Bethlehem Iron Company built an armor plant that had a 125-ton steam hammer, a massive, towering apparatus that dwarfed anyone standing nearby. Even equipment for handling raw materials grew to enormous size, like machines that could lift entire railcars full of ore or limestone and turn them upside down to load a blast furnace. Dignitaries at the 1890 opening of a steel mill in Sparrows Point, Maryland, rode in decorated gondola cars along the route iron ore would take, being pulled up to a charging platform over eight stories high.[30]

The Romance of Steel

"There is a glamor about the making of steel," John Fitch wrote at the beginning of his 1910 study of Pittsburgh steelworkers. "The very size of things—the immensity of the tools, the scale of production— grips the mind with an overwhelming sense of power. . . . majestic and illimitable." Fitch was only the latest in a long line of writers, artists, and journalists to be fascinated by the making of iron and steel. More than a half century earlier, Nathaniel Hawthorne was entranced by "exhibitions of mighty strength, both of men and machines" during a visit to an iron foundry in Liverpool, where he watched a twenty-three-ton cannon being made. "We saw lumps of iron, intensely white-hot, and all but in a melting state, passed beneath various rollers and . . . converted into long bars, which came curling and waving out of the rollers like great red ribbons." Hawthorne "found much delight in looking at the molten iron, boiling and bubbling in the furnace," with "numberless fires on all sides, blinding us with their intense glow."[31]

Fire was a big part of the allure of iron- and steelmaking, the intense heat, the white molten metal, the glowing red ingots. Heroic images of workers using fire to turn ore into metal were commonly featured in nineteenth-century journals, often depicted at night to heighten the effect of radiant metal in blast furnaces or Bessemer converters. Several of the drawings Joseph Stella made for the early twentieth-century Pittsburgh Survey showed men's faces lit by the glow of molten metal.

One of the most common allusions in writing about the Industrial Revolution was to Prometheus, for giving man powers of the gods. Fire was the greatest of his gifts, iron and steel the most Promethean industry. In seeking classical reference for an act of alchemy that seemed beyond the realm of ordinary mortals, the nineteenth century also looked to Vulcan, the Roman god of fire and metalworking. When the Pittsburgh-area puddlers organized a union in 1858, they called themselves the Sons of Vulcan. An 1890 account of a large steelworks in Newcastle, England, reported that in the foundry "modern Vulcans, in shirt-sleeves and with unbroken legs, are still casting thunderbolts." Artists commonly portrayed iron and steelworkers as intensely masculine, often bare-chested, with muscles rippling, a bit like ancient portrayals of Vulcan himself. The contrast was great to the typical representation of the English textile worker as a sickly child or the New England textile worker as a well-dressed young woman.[32]

But if for some iron- and steelmaking seemed the realm of the gods, for others it appeared the province of Satan, like the early English mills had been for Blake. Hawthorne described molten ribbons of iron as looking "like fiery serpents wriggling out of Tophet," the place in the Old Testament where worshippers burnt their children alive in sacrifices to Moloch and Baal, a hell on earth. Early in the twentieth century, the manager of an iron and steel mill in Pueblo, Colorado, wrote, "The steam, the fire, the fluid metal, the slag and the whir of the machine all ma[d]e it look like it was the Devil's Workshop." For Joseph Stella, Pittsburgh, "Often shrouded by fog and smoke, . . . ever pulsating, throbbing with the innumerable explosions of its steel mills—was like the stunning realization of some of the most stirring infernal regions

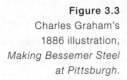

Figure 3.3
Charles Graham's
1886 illustration,
*Making Bessemer Steel
at Pittsburgh.*

sung by Dante." Similarly, Lincoln Steffens wrote, "I have never lost my first picture of Pittsburgh when I went there to write about it. It looked like hell, literally . . . with its fiery furnaces and the two rivers which pinched it in." [33]

Hellish though they might have been, iron and steel mills were often hailed as markers of national greatness and the advance of civilization. Their growth allowed the introduction of iron and steel implements on farms and in homes, the mechanization of other industries, a transformed landscape of railroads, bridges, and skyscrapers, and imperial power based on giant guns and steel warships. In 1876, George Thurston described the then-new Edgar Thomson steel mill in Braddock, Pennsylvania, as "a striking illustration of . . . the progress of civilization." "No grander monument to the growth of the nation . . . or the triumph of American manufactures and of American mechanics, could well be built." Mary Heaton Vorse, a left-wing journalist with a very different sensibility, nonetheless agreed in her 1920 book *Men and Steel*: "Our civilization is forged in the steel towns." And not just any civilization, but modernity: "Iron and Steel began the life of the moderns." Sociologist Sharon Zukin noted that "Steel has power because it has been the lifeline of industrial society. . . . Steel is linked upward to the national government by warfare and international trade, and downward to the local manufacturing community as an emblem of economic power." In the late 1940s, best-selling journalist John Gunther declared, "The basic power determinant of any country is its steel production."[34]

Class War

Fire, power, but one more thing, too, made iron and steel factories centers of public attention—labor strife. The English textile industry sparked a great debate about child labor and working conditions, if not much effective worker organization. The American textile industry was hailed, with considerable exaggeration, for harmonious relations between owners and workers. By contrast, labor conflict came to be strongly associated with the iron and steel industry, the site of some of the most dramatic episodes of what can only be called class war in the history of the United States.

In the decades after the Civil War, the growing power of industrial capital set off fierce economic and political struggles, at their broadest about what type of society the United States would be, and who would decide. Former slaves, farmers, women, and the unemployed mobilized, as a wide range of voices, far wider than we hear today—populists, monetary reformers, socialists, anarchists, social Darwinists, Christian reformers, feminists, and cooperativists—jumped into debates over social values and structures. Workers and the organizations they built composed the single most important force challenging the growing economic and political dominance of industrialists and financiers during what Mark Twain so aptly labeled the Gilded Age. Nowhere was labor conflict more intense than in the iron and steel industry.[35]

More than any other industry, iron and steel seemed to confirm the notion that the factory system was creating two, new, hostile classes. With much higher costs for starting up an iron or steel mill than a textile mill, capital tended to concentrate in a handful of powerful firms. For the men who controlled them, usually hands on, they were not one of many investments—as the textile mills were for the Boston Associates—but the source of all their wealth and power, the means to achieve some of the largest fortunes in the country. Their workers recognized what they were up against. The preamble to the constitution of the Amalgamated Association of Iron and Steel Workers declared "Year after year the capital of the country becomes more and more concentrated in the hands of the few . . . and the laboring classes are more or less impoverished. It therefore becomes us as men who have to *battle* with the stern realities of life, to look this matter fair in the face." Only admitting skilled workers, the Amalgamated's membership fluctuated with good times and bad, peaking in 1891 at more than 24,000 members. Its power rested on a sense of solidarity among its members and their skills, without which the mills could not operate.[36]

Or couldn't until they began mechanizing. New technology and the shift from iron to steel diminished the number of skilled workers and the level of skill needed in various phases of production. The move to steel also increased firm size and created an intensely competitive atmosphere, both of which worked against labor.

The market for rails, which drove the steel industry until late in the nineteenth century, fluctuated wildly, encouraging a ruthless management culture. During economic upswings, there were plenty of orders for everyone, but in downturns companies had to scramble and slash prices to keep their mills operating. Steel executives repeatedly made and broke deals with other companies to fix prices and divvy up markets, while pressing their subordinates to lower costs. Mechanization provided one route; reducing wages and extending hours another. But squeezing labor costs meant having to take on unions, leading to escalating battles in the 1880s and 1890s.[37]

The big companies took the lead in fighting the Amalgamated, equipped with the financial resources and multiple plants to win extended battles. Homestead saw some of the sharpest clashes. In 1882, the management of the plant (not yet owned by Carnegie) insisted that to keep their jobs employees had to sign an ironclad agreement not to join a union. Refusing, several hundred skilled workers struck for over two months, surviving repeated battles with private guards and the state militia until the general manager capitulated. Six years later, Carnegie used a four-month lockout and Pinkerton National Detective Agency guards to crush the union at his Edgar Thomson mill and go from a system of three shifts of eight hours to two shifts of twelve, a rout not only of unionism but of the standards it had defended.[38]

The next year, 1889, Carnegie tried to replicate his Thomson triumph at the Homestead mill, which he had purchased in 1883, plotting his moves while on a visit to Europe to see the great Paris exhibition. Once again, his company delivered a take-it-or-leave-it ultimatum to his workers, locked them out when they rejected it, and hired Pinkerton guards. But after two efforts to bring in scabs were repulsed by massive crowds of steelworkers and Homestead residents, the local manager gave in and negotiated a new agreement with the Amalgamated Association.[39]

When the contract expired at the end of June 1892, Carnegie sought to rid himself of the union once and for all. By then, nearly a decade of industrial strife had placed the issue of labor relations at the center of American life. Observers saw Homestead as a bellwether for the future of class relations. The high productivity of the Homestead plant and the Amalgamated's sliding scale made labor costs at the mill, according to Carnegie's calculations, above the norm, while allowing its skilled workers relative comfort, buying small houses in town (and electing one of their own to lead its government), purchasing some furnishings, living in decency.

The economic conflict had an ideological dimension. Carnegie and his partners, determined to drive down labor costs, wanted complete freedom to set wages and working conditions without union interference, to control what they saw as solely their property. By contrast, workers felt they had a moral claim on the company, having contributed to its success through their skill and toil. Many shared what at the time was a common democratic vision in which working people (or at least the white, English-speaking men among them) should—and in Homestead for a while did—have a say in both civic and industrial life.[40]

As the battle loomed, Carnegie again absented himself to Europe, leaving in charge his partner Henry Clay Frick. Again, the company prepared an offer it knew the union would reject. As it shut down operations and locked out its workers, Frick surrounded the mill with an eleven-foot-high fence, with gun ports and topped by barbed wire, and contracted with Pinkerton for three hundred guards.

All remained peaceful until the company tried to sneak the Pinkertons into Homestead on barges. In the middle of the night, union lookouts spotted them, alerting the town. As the *New York Herald* described it, "Like the trumpet of judgement blew the steam whistle of the electric light works at twenty minutes to three-o'clock this morning. It was the signal to battle, murder and sudden death, though not one of the thousands who heard and leaped from their beds to answer its signal dreamed of how much blood was to flow in response to its call." Workers and townspeople positioned on the steep banks of the Monongahela River kept the well-armed private army from disembarking, firing a cannon at them (which ended up killing a union backer by mistake), rolling flaming railcars toward the moored barges, raining down fireworks and sticks of dynamite, and pouring oil on the river and setting it on fire. Finally surrendering, the Pinkertons found themselves beaten, robbed, and humiliated by a gauntlet of strikers and local residents. Seven workers and three Pinkertons died in the fighting.

The union victory proved short-lived. Within a week, the governor of Pennsylvania sent 8,500 men—the state's entire National Guard—to occupy Homestead, where they remained until October. This massive

Figure 3.4 *An Awful Battle at Homestead, Pa.*, depicting the bloody clash between Carnegie's locked-out workers and the Pinkertons in the summer of 1892.

application of state power—accompanied by the indictment of well over a hundred workers on charges of murder, riot, and conspiracy—proved the key to the company's victory. With the troops in place, it began recruiting scab workers from around the country. On July 23, anarchist Alexander Berkman attempted to kill Frick—a rare American example of European-style propaganda by the deed—but the Carnegie executive proved a tough bird, surviving bullet and knife wounds, even helping to tackle his assailant. When a National Guardsman shouted, "Three cheers for the man who shot Frick," he was court-martialed and hung from his thumbs. In November, the union formally gave up.[41]

The fight at the Carnegie plant was closely followed across the country and abroad. A hundred reporters and sketch artists from major magazines, the press syndicates, and newspapers in Pittsburgh, New York, St. Louis, Chicago, Philadelphia, Baltimore, and London assembled in Homestead to cover the conflict. With special telegraph lines installed, news from the front lines immediately spread. Photographers documented the clash as well. Several companies sold stereoscopic images for home viewing, providing three-dimensional portrayals of the industrial war.[42]

The workers' defeat reverberated far and wide. Having been pushed out of the most advanced steel mill in the country, the Amalgamated saw its hold in the industry rapidly deteriorate. Within a year, more than thirty of the sixty-four mills in southwestern Pennsylvania rid themselves of the Amalgamated. In the iron industry, the union maintained strength among puddlers and sheet and tin plate workers, but even there company resistance and inept leadership gradually diminished its power. By 1914 it was down to just 6,500 members.[43]

Like the early British textile industry, the American iron and steel industry grew in an atmosphere of denied political rights—free speech, free assembly, rule of law. With unions weakened or eliminated, the steel companies came to exert near-autocratic control over not only the mills but the communities in which they were located. The town of Homestead sank into a dark era of suspicion and demoralization. In mid-1894, Hamlin Garland wrote in *McClure's Magazine*, "The town was as squalid and unlovely as could well be imagined, and the people were mainly of the discouraged and sullen type to be found everywhere labor passes into the brutalizing stage of severity." Theodore Dreiser, who lived in Homestead for six months that same year, found "a sense of defeat and sullen despair which was over all." More than a dozen years later, when John Fitch came to town, residents shied away from talking to him, fearful of company spies and retribution. Steel company influence over Homestead was so great that no halls could be found for any sort of union meeting. As late as 1933, four decades after the lockout, the only place in Homestead Secretary of Labor Frances Perkins could find to address a crowd of workers was inside a post office, an island of federal authority.[44]

In 1919, radical critic Floyd Dell called Pittsburgh, across the river and likewise dominated by the iron and steel industry, "capitalism armed to the teeth and carrying a chip on its shoulder. . . . lynch-law carefully codified by a trained legislature and carried out by uniformed desperadoes." The city, he suggested, was "an experiment in what might be called super-capitalism. It is a sociological experiment, akin (despite the oddity of the comparison) to the Utopias founded here and there from time to time by enterprising if unrealistic socialists. But instead of a poor, precarious, struggling, starved, doomed Utopia, it is a flourishing and, so far, absolutely triumphant Utopia. It is a Billion Dollar Capitalist Utopia."[45]

Often repression and paternalism mixed together. During the 1870s and 1880s, leading European iron and steel companies built industrial villages, including Krupp in Essen and Schneider in Le Creusot. Many American companies followed suit. Like the Lowell textile firms, steel mills in isolated locations needed to provide housing if they were to attract a workforce. When in the early 1890s the Pennsylvania Steel Company built its complex at Sparrows Point, an empty spit of land on the north side of Baltimore's harbor, it constructed a new town a half mile from the blast furnaces. Under an arrangement with the governor of Maryland, the company directly ran the community, without any local democratic structures. Rufus Wood, the company executive who designed the town, was the son of a foreman at the Boott cotton mill in Lowell. He modeled Sparrows Point on the Massachusetts city, though with mostly family accommodations rather than boarding-houses. Dwellings ranged in size and quality from an eighteen-room, three-story colonial for Wood himself down to small wooden houses without running water or indoor plumbing for black workers. As in Lowell, elaborate rules governed behavior not only on the job but in the housing, too.[46]

The most ambitious mill town scheme came in 1895, when the Apollo Iron and Steel Company decided to build a new mill a mile and a half from its existing plant in western Pennsylvania. It contracted with the firm headed by Frederick Law Olmsted, the foremost landscape architect and town planner in the country, to design a new town, Vandergrift, named after the Standard Oil partner who was the largest investor in the company. Cost concerns kept the Olmsted plan from being fully realized, but parts of the town featured curvilinear streets, wide boulevards, scattered small parks, and a village green, characteristic of the high-end suburbs that were beginning to surround older cities. But only the best-paid workers could afford those areas; most lived in a less attractive grid laid out on one side of town or in an unplanned hovel on another.[47]

Company officials saw housing as a way to retain workers. Some companies offered their employees rental housing at below market rates. Others sold them houses. Carnegie built housing for his workers just outside Homestead, offering low-interest loans that could be repaid through small deductions from their pay. Because many steel mills sat in essentially one-employer towns, home-owning employees, as their bosses knew, would be reluctant to jeopardize their jobs in any way, because without them they would be forced to move. Companies hoped that orderly, well-regulated communities—both Sparrows Point and Vandergrift banned the sale of alcohol—would produce orderly, disciplined workers.[48]

In the early twentieth century, when the world's largest company built the world's largest steel mill, it, too, built a company town. As the United States recovered from the depression of the 1890s, a wave of corporate mergers swept through the already highly concentrated steel industry. In 1901 Carnegie threatened to expand his finishing operations, in response to the backward integration of firms that had been purchasing his steel ingots. To avoid overcapacity and ruinous competition, J. P. Morgan, the country's leading financier, arranged a huge merger of steel concerns. For his interests, Carnegie received $226 million (the equivalent of several billion dollars today). The new entity, the United States Steel Corporation, controlled almost 60 percent of the output of the industry and was widely seen as the very embodiment of industrial capitalism.[49]

Four years after its formation, U.S. Steel bought nine thousand acres on the shore of Lake Michigan, just east of Chicago, where it erected a massive, sprawling integrated steel mill. To allow boat delivery of ore from its Minnesota mines, the company built a deep harbor next to the plant. It also laid out a new city, Gary, named after its chairman, Elbert H. Gary—American industrialists loved naming towns after themselves and each other—where it sold vacant lots and constructed rental housing. Vandergrift, by then part of the U.S. Steel empire, served as something of a model. But in building the new city, the company spurned the utopian pretensions of Lowell and Vandergrift, stating that it was not trying to create a model community, only building a necessary adjunct to its new facility.[50]

Scientific Management

Even with unions defeated and worker resistance tamed, steel companies still struggled to control labor in their mills and reduce labor costs, an imperative in periods of intense competition. With sprawling physical facilities and a large array of jobs, managers had difficulty even knowing what all their workers did, let alone how efficiently they were working. Skilled workers retained considerable autonomy, using knowledge accumulated through formal or informal apprenticeships to determine their methods of work, often effectively setting their own pace. Foreman pushed unskilled workers using threats and verbal abuse, with little planning or measurement of productivity.

Throughout American industry, factories had swollen in size and complexity without a proportionate increase in managerial personnel or sophistication. Well into the 1880s, many major firms still managed labor through the direct presence of top executives. Cyrus McCormick's brother and his four assistants long managed the giant McCormick Works in Chicago. Thomas Edison and three assistants personally supervised production at his factories in Harrison, New Jersey, and New York City.[51] But with the development of giant multiplant firms, such personal, informal control was no longer tenable.

"Systematic management," later more widely known as "scientific management," grew out of a quest for internal corporate controls and increased productivity, a sweeping effort at reorganizing production. Its development involved many different companies, engineers, and managers over an extended period, who instituted a series of incremental changes that together represented a substantial transformation in how manufacturing—and later office work—was carried out. But in the public mind, scientific management became largely associated with one man, Frederick Winslow Taylor, who emerged as its leading theoretician, ideologue, and publicist.

Taylor, the son of a prominent, liberal Philadelphia family, followed an unusual path in spurning college to become an apprentice machinist and patternmaker, before taking on a series of factory-management positions and then a career as an industrial consultant. (In 1876, he took six months off from his apprenticeship to work at the Centennial Exhibition.) Many of Taylor's key innovations took place during the 1880s at Midvale Steel Works, a Philadelphia producer of high-quality steel products, and then, during the last years of the nineteenth century, at the much larger Bethlehem Steel Company. Taylor had an intense interest in the mechanics of steel production and metalworking, particularly high-speed machine tools, making numerous technical advances. But his greater importance lay in applying a systematic, engineering mind-set to what commonly had been a seat-of-the pants, chaotic approach to managing manufacturing.

Taylor's contributions included improvements in cost accounting, inventory control, tool standardization, and shop floor layout. But his best-known innovations involved labor. Working among machinists, Taylor realized how commonly workers set a stint, a maximum output, designed to conserve their energy and spread out the work. Managers had no idea what the maximum output could be or what should constitute a full day's work. The first step to boosting productivity, Taylor came to believe, lay in the careful observation and measurement of workers as they did their jobs, using stopwatches and, later in the hands of his disciples, stop-action photography and motion pictures. Once managers understood the elements of any given task, they could determine the best way to carry it out and the time it should take to complete.

Critical to the Taylor method was the separation of the planning of work from its execution, breaking the hallmark of the skilled craftsperson, his or her ability to conceive of how to make various items and then to do the work themselves. All planning, Taylor believed, should be in the hands of management, in a specialized planning department (something previously all but unknown). Using knowledge of machinery and worker practices gathered through systematic observation, workers would be given detailed instructions about how to carry out each task (usually in the form of an instruction card). Pay would be calculated by a piecework system that rewarded with higher rates workers who met specified production norms and penalized those unable or unwilling to meet management-dictated standards.

For skilled workers in particular, like the machinists that steel companies employed to make finished products and maintain their equipment, Taylorism meant a loss of autonomy and an attack on craft pride, as well as an intensification of work, leading to fierce battles. But Taylor always claimed his system would benefit workers as well as company owners, because the increases in productivity that would result from scientific management were so great that workers could be given higher pay even as company profits rose. In an example Taylor repeatedly used in publicizing his system, a Bethlehem laborer he called Schmidt, who was loading pig iron into railroad cars, increased his daily tonnage to forty-seven tons, from a previous gang average of twelve and a half, by following precise instructions. For his increased output, Schmidt received a wage boost from $1.15 to $1.85 a day. His wages thus went up by roughly 60 percent, while output nearly quadrupled, a good deal for the company though also a gain, if much more modest, for the worker. At least in theory, scientific management, or Taylorism as it was sometimes called, made the struggle between workers and owners over wages no longer a zero-sum game. For this reason, in the eyes of many

Progressive Era reformers, scientific management held the promise of eliminating or at least ameliorating the class conflict that had come with industrialization and the giant factory, without fundamentally restructuring society.[52]

The Road to 1919

In practice, at least in the short run, scientific management failed to have much effect on the growing class tensions in the steel industry and the nation. Steelmaking remained a difficult and dangerous undertaking even with greater managerial presence and after mechanization eliminated some of the most arduous labor (along with many skilled, higher-paid positions).

After Carnegie's Homestead victory, the twelve-hour day became the norm in iron and steelmaking jobs. (Some positions, especially finishing work, had shorter hours.) Bessemer furnace workers and many others typically toiled for thirteen consecutive twelve-hour day or night shifts and then, after a day off, worked a "long turn" of twenty-four hours, which put them on the opposite shift for the next two weeks. The schedule wreaked havoc on their lives, making normal family life impossible and wearing men out at an early age.

During their shifts, workers had stretches of extremely demanding labor in almost unbearable heat. When open-hearth furnaces were tapped, the molten metal could be as hot as three thousand degrees Fahrenheit. Workers had to stand over giant ladles containing the liquid steel to throw in heavy bags of scrap metal and alloys to adjust the final chemical composition. In a sheet mill, John Fitch observed "men standing on floors so hot that a drop of water spilled would hiss like a drop on a stove." They wore special shoes with thick wooden soles to provide at least some protection.[53]

Long hours, giant mechanical devices, crisscrossing rail lines, and molten metal made steelwork extraordinarily dangerous. In just one

year, from July 1, 1906, to June 30, 1907, Allegheny County, Pennsylvania, which included Pittsburgh, Homestead, Braddock, and other
metalmaking towns, recorded 195 accident-related deaths in the iron
and steel industry. If death or maiming did not get a steelworker, occu-
pational disease very well might, like the fine dust that pervaded the air
in steel mills, ravaging workers' lungs, or the relentless noise that led to
widespread hearing loss.[54]

For years the companies showed little concern about the effect of
long hours and dangerous conditions on their employees, whom they
regarded—at least the unskilled workers—as easily replaceable. And
they were. Starting in the 1880s, a flood of Southern and Eastern European immigrants entered the steel mills (except in the Birmingham,
Alabama, area, the major southern iron and steel district, where African
Americans filled many of the unskilled jobs). In March 1907, in the
former Carnegie mills in Allegheny County, 11,694 of the 14,359 common laboring jobs were filled by Eastern Europeans. The immigrants
largely had been peasants or migrant laborers, coming to America without their families for what they expected to be a limited period, hoping
to make enough money to buy land, pay off a mortgage, or set up a shop
back home. Some eventually decided to stay, sending for wives and children, but many returned. As in Lowell, the rotating workforce provided
something of a safety valve for the owners, less likely to organize than
more permanent workers. Linguistic and cultural barriers between
the unskilled immigrants and the skilled workers, who generally were
native-born or from the British Isles, and among the immigrant groups,
also made organizing difficult, giving the steel companies free rein.[55]

Or at least for a while. In the early years of the twentieth century,
immigrant steelworkers began demonstrating their discontent in protests and strikes. Most were short and without union involvement. But
in 1909, five thousand immigrant and native-born workers conducted a
prolonged walkout at the Pressed Steel Car Company, a U.S. Steel subsidiary, in McKees Rocks, Pennsylvania. More than a dozen men died
as the company and local authorities repeatedly tried to bust the strike

through physical force. Ultimately, U.S. Steel was forced to give in to the strikers' demands, a sharp reversal after a string of company victories against organized labor and its declaration, just before the walkout, that going forward it would operate on a strictly nonunion basis in all of its plants.[56]

The episodic immigrant strikes led the steel companies to pay more attention to labor policies and to seek favor with their workers, especially because they occurred at the same time that the industry was coming under scrutiny from middle-class reformers. Their interest flowed from a broad concern with what came to be called "the labor question." Narrowly construed, the labor question meant how to maintain orderly relations between employers and employees and prevent the outbursts of labor warfare that had become common in the late nineteenth and early twentieth centuries. Between 1875 and 1910, state troops were called out nearly five hundred times to deal with labor unrest and at least several hundred persons died in strike-related violence. But for many labor activists, reformers, politicians, and even some business leaders, the question implied more. What place should workers have in American society? What say should they have in the workplace and in politics? And, most broadly, was democracy possible in an industrialized society with great economic inequality, and if so, what did it mean?[57]

By the start of the twentieth century, the factory-based, corporate-controlled Industrial Revolution had radically changed society. For millions of workers who entered the factory, leaving behind villages or farms in New England, Ireland, Italy, or Eastern Europe, wage work, industrial time discipline, and mechanized production were new and often troubling experiences. Just as industrial work was strange to them, the industrial worker was strange—and threatening—to many more prosperous Americans, especially employers. In 1889, three years before he crushed his workers at Homestead, Carnegie wrote, "We assemble thousands of operatives in the factory, and in the mine, and in the counting-house, of whom the employer can know little or nothing, and to whom the employer is little better than a myth. . . . Rigid castes are formed, and, as usual, mutual ignorance breeds mutual distrust."[58]

Early in the new century, a number of middle-class writers dressed up as workers and plunged into working-class life to report about a world utterly unfamiliar to better-off elements of society. Whiting Williams, a former personnel director at a Cleveland steel mill, spent nine months working undercover in steel mills, an iron mine, a coal mine, and an oil refinery to write *What's On the Workers Mind, By One Who Put on Overalls to Find Out*. Pioneer social workers and social scientists set out on a similar mission, in line with the Progressive Era belief in the reforming potential of exposure. Concerned about the brutalization of moral life brought about by industrialization and sympathetic to the plight of immigrant workers, these middle-class reformers nonetheless worried about the threat they posed unless assimilated to civil society and national culture.

The steel industry was a natural focus for their concern. The centrality of steel to the economy gave it special importance. The formation of U.S. Steel as the largest corporation ever created added to the sense that steel had to be a matter of public concern, not strictly a private endeavor. The juxtaposition of the toll steelmaking took on workers with the extraordinary rewards reaped by mill owners—the formation of U.S. Steel made Carnegie "the richest man in the world," Morgan told him—commanded the attention of not just unionists and political radicals, but a broad swath of the nation.[59]

In 1907 and 1908, several dozen investigators descended on the Pittsburgh area to conduct a massive study of work, workers, and civil life centered on the steel industry. Funded by the newly formed Russell Sage Foundation, the Pittsburgh Survey staff included some of the leading reform intellectuals of the day, like economist John L. Commons and Florence Kelly, a settlement house resident, suffragist, and consumer advocate, who had been the first to translate Engels's *The Condition of the Working Class in England* into English. The survey produced dozens of articles, six large books, and a photographic exhibition that documented life and labor in the Pittsburgh area, a model for the kind of foundation-funded social science that soon became prevalent. The picture painted by the survey was grim: families unable to live on an unskilled steelworkers' wages, poor housing, dangerous jobs, and a climate of repression.[60]

In the wake of the Pittsburgh Survey, the McKees Rocks strike, and a subsequent walkout at a Bethlehem Steel mill in South Bethlehem, the Senate launched an investigation of the steel industry and the Department of Justice filed an antitrust suit against U.S. Steel. The giant corporation, mindful of its precarious legal situation, sensitive to public opinion, and not facing the competitive pressures that characterized the industry before its creation, made some modest improvements in working conditions. It started giving more workers Sundays off, reducing the seven-day workweek to six days, but it clung tightly to the twelve-hour work shift, which it claimed a necessity (even though in other countries steel companies succeeded without it). Steel companies launched employee stock-purchase and pension plans and a safety campaign (responding not just to bad publicity but also to the growing crop of state laws requiring employers to provide accident insurance to their workers). Fundamentally, though, industry leaders stood pat, successfully repulsing all efforts, from workers and middle-class reformers alike, for basic change.[61]

Their most severe test came in 1919, when workers across the country mounted the most radical challenge to industrial capitalism in American history, part of a great, worldwide surge of reform and revolutionary sentiment. World War I transformed labor relations. The combination of a war-induced economic boom and an immigration cutoff created a labor shortage that left workers in a strong bargaining position, no longer fearful of losing their jobs since others could be easily found. With inflation pushing up prices, workers bounced from job to job, went on strike, and joined unions to better their lot. To keep labor strife from interrupting war production, the Woodrow Wilson administration, with strong input from the American Federation of Labor

(AFL), set up a series of administrative bodies and promulgated regulations designed to give workers new rights on the job. Companies were forced to end discrimination against union members and enter discussions with worker councils (though not unions per se). Under these circumstances, union membership increased by nearly 70 percent between 1917 and 1920, reaching just over five million. More than one out of every six nonagricultural workers carried a union card. Combined with the radical fervor set off by the Russian Revolution, a wave of near-millennial enthusiasm swept through working-class quarters. In 1918, the young leader of the clothing workers' union, Sidney Hillman, wrote to his infant daughter that "Messiah is arriving. He may be with us any minute.... Labor will rule and the World will be free."[62]

The steel industry was hit especially hard by changed labor market conditions and wartime federal progressivism. With immigration from Europe blocked by the fighting, the steel companies found themselves unable to tap their usual source for unskilled labor. In the spring of 1916, they began recruiting black workers from the rural South. But with the military draft soon pulling men out of their plants, a labor shortage remained, emboldening workers to launch a series of strikes. Meanwhile, under intense pressure from the federal government, the industry adopted the eight-hour day as its standard, though in practice that largely meant paying workers time-and-a-half for the last four hours of their twelve-hour shifts.[63]

With conditions favorable, unions decided to take another shot at organizing steel. This time the impetus came from two militant, Chicago-based unionists, John Fitzpatrick, the head of the Chicago Federation of Labor, and William Z. Foster, the future head of the American Communist Party, who had developed a new organizing model in their successful drive to unionize the meatpacking industry. Recognizing the impossibility of organizing large-scale industrial companies on a craft union basis, they convinced the AFL to set up the National Committee for Organizing Iron and Steel Workers, with which twenty-four unions affiliated. Organizing was centrally directed, with workers steered to the union appropriate for their job only after they had signed up with the National Committee.[64]

Launched in September 1918 in the Chicago region, the organizing drive, with the slogan "Eight Hours and the Union," got off to a fast start. Many immigrant steelworkers by then had decided to remain in the United States, giving them a greater stake in future job conditions. Unionists turned the democratic rhetoric of the war against industrial autocracy, giving the drive something of a patriotic air. Though short on money and organizers, it soon spread to Pittsburgh and other regions.

The end of the war made things much harder for unions. Companies began laying off workers and reverted to their hard-nosed antiunionism, defying government decrees that in theory remained in effect. In the Pittsburgh area, the National Committee had to wage a relentless battle simply to secure the right to assemble, as mill town officials, acting on behalf of the steel companies, forbid union meetings and even street rallies. It took mass arrests and national publicity to win modest cracks in the solid wall of antidemocratic practices. Still, in a measure of how much steelworkers resented company control over their lives and the new spirit ushered in by the war, more than a hundred thousand workers—the National Committee claimed a quarter of a million—signed up in the union drive.

The steelworker mobilization took place against the background of an extraordinary national wave of strikes, the largest in U.S. history proportional to the size of the workforce. Unions sought to maintain their wartime organizational gains and increase wages to keep up with inflation, while companies fought to roll back union advances and reestablish their dominance. Four million workers—one fifth of the workforce—took part in the strike wave, which included a general strike in Seattle, a police strike in Boston, a telephone operators strike in New England, an actors strike in New York, and, at the end of 1919, a strike by four hundred thousand coal miners. Everyone, it seemed, was walking off the job.[65]

The leaders of the steel-organizing drive hoped to avoid a strike, fully cognizant of the power of the companies. But Elbert Gary, the leader of U.S. Steel and effectively of the whole industry, rejected all requests for negotiations, even a private one from President Wilson. With workers increasingly restless and the companies firing activists, on September 22, 1919, the National Committee, feeling it had no alternative, launched the first national steel strike in American history. Within a week, some 250,000 workers—half the industry workforce—stopped work.

In a reversal of the usual past pattern, the strike was strongest among immigrant and unskilled workers, though many skilled workers supported it as well. In some regions, like Chicago, Buffalo, Youngstown, and Cleveland, the strike was nearly 100 percent effective, forcing the mills to shut down. But at the Bethlehem Steel plants, strike leader William Z. Foster estimated only about half the workers went out, while in the Pittsburgh area, the most important, the strike was 75–85 percent effective. In the South, it barely made a dent.

The companies fought back hard. Wherever they could, they kept plants in token operation, even if unprofitable, bringing in white scabs from Northern cities and black scabs from the South. State police, deputy sheriffs, private guards, and vigilantes operating on their behalf launched what Foster termed a "reign of terror." Pickets and organizers were arrested and driven out of town, mounted police attacked picketers and demonstrators and even a funeral procession, rallies were banned, strikers were shot. In Gary, the governor declared martial law and 1,500 regular Army troops occupied the city. A score of people were killed during the conflict, almost all strikers or their sympathizers, and hundreds were seriously injured.[66]

To a greater extent than in previous industrial battles, both sides recognized the importance of public opinion to the outcome. Company propaganda portrayed the worker action as not an industrial dispute but an attempted revolution, playing on the antiradicalism that came in reaction to the Russian Revolution. Foster's past record of radicalism was uncovered and widely publicized. The anti-Red, anti-strike campaign had a decidedly nativist tone, as it portrayed immigrant strikers as "un-American." The press generally supported the companies, while the Wilson administration, by then in sharp retreat from its wartime progressivism, failed to back the strikers, leaving them on their own in taking on the most powerful companies in the world.[67]

Slowly the employers began increasing production, as some workers, at first mostly skilled, began returning to their jobs and new workers were recruited and trained. Tens of thousands of strikers stuck it out into the winter. But on January 8, 1920, the National Committee acknowledged the futility of going on, ordering its members to return to work in what Foster himself called an "unconditional surrender."

The 1919 strike had been a test of the ability of organized labor to penetrate the great national manufacturing companies, backed and controlled by the most powerful financial interests. Its failure meant that for another generation the largest and most advanced factories in the United States would remain nonunion and their workers outcasts.

Yet even as it remained a fortress of industrial autocracy, the steel industry maintained its allure, even among those with little sympathy for the owners. The scale, power, and elemental processes of steelmaking commanded attention separate from the social arrangements that surrounded them. Neither the fierce discontent of labor, nor the well-documented dangers and difficulties of work in giant factories, nor the massive accumulation of power in the hands of the plutocracy that owned them dented the enthusiasm, cutting across the political spectrum, for the processes and products of the steel and other manufacturing industries, so proudly displayed at the world's fairs. Mary Heaton Vorse, who volunteered as a publicist for the strikers in 1919 and was a model for one of John Dos Passos's characters in his account of the clash in *The Big Money*, was far from alone when she wrote a year later, "I would rather see steel poured than hear a great orchestra." The steel mill had become the modern sublime.[68]

CHAPTER 4

"I WORSHIP FACTORIES"

Fordism, Labor, and
the Romance of the Giant Factory

In a 1926 entry in the *Encyclopedia Britannica*, Henry Ford (or the publicist who ghostwrote the article) defined "mass production" as "the modern method by which great quantities of a single standardized commodity are manufactured." If anyone knew about the manufacture of "great quantities of a single standardized commodity," it was Ford. His Model T, introduced in 1908, turned the automobile from a luxury plaything into a mass-consumer good. Prior to then, automobile companies typically manufactured at most a few thousand cars a year. By 1914, the Ford Motor Company was rolling out nearly a quarter of a million Model Ts annually. By the time the company stopped selling the iconic model in 1927, fifteen million had been produced.[1]

Henry Ford's worldwide fame stemmed as much from the methods his company used to make the Model T as from the car itself. To manufacture it, the Ford Motor Company built some of the largest factories that ever had been seen and introduced countless technical and organizational innovations, including the assembly line, which enormously increased the speed and efficiency of production. To control the tens of thousands of workers who populated its plants, the company devised new methods of labor management that extended beyond the factory walls into workers' homes and minds. Ford pioneered what amounted to a new political economy of inexpensive consumer products that transformed people's lives, high-volume factories to produce them, and high wages and strict controls to discipline the workforce. Before Ford himself popularized the term "mass production," commentators often spoke of "Fordism," "Ford methods," or the "Ford system," appropriate terms for the new production, distribution, and consumption regime, for it was Henry Ford and the Ford Motor Company that ushered in a new phase of industrialization and a factory scale that would be unsurpassed for nearly a century.[2]

Just as the "factory system" of early nineteenth-century England captured the interest and imagination of journalists, political activists, writers, and artists, so, too, did the "Ford system" of the twentieth century. Once again, it seemed like a new world was aborning. Part of what made Fordism so transfixing was the promise of a wholesale rise in the standard of living and amelioration of the class conflict that had been shaking the United States. In 1924 merchant and reformer Edward Filene wrote that in Fordism lay "a finer and fairer future than most of us have even dared to dream." Beyond the social implications of Fordism, many writers, painters, filmmakers, and photographers were entranced by the physical structures in which it unfolded. More than with earlier industrial production, artists and intellectuals explicitly linked Fordism to modernist trends in art and society. The great photographer Margaret Bourke-White, who through her work in *Fortune* and *Life* magazines did more than any other individual to popularize industrial imagery, captured the age when she bluntly declared "I worship factories."[3]

The Road to Mass Production

The Ford system was a culmination of past manufacturing practices and a radical break from them. Almost from the start, American facto-

Fuck u nick

ries had been engaged in the production of "great quantities of a single standardized commodity," be it the white sheeting made in Waltham or the rails that drove the expansion of the iron and steel industry. But automobiles were of an entirely different order of complexity. It was a long road to enable such complicated machinery to be produced on a mass scale.

Fordism built on two manufacturing innovations, interchangeable parts and continuous flow. Until the early nineteenth century, products with interacting metal parts, like guns or clocks, were individually made by skilled artisans, who spent a great deal of time fitting together parts, filing and adjusting them to make sure they worked together. No one finished product was exactly like the next.

The standardization of parts occurred first in the United States. Generally, introducing interchangeable parts initially *increased* the cost of production, since it required a huge investment in specialized machines, tools, jigs, and fixtures and a great deal of experimentation to achieve the tolerances that made it possible to assemble a product from a pile of parts without custom fitting. The key innovations took place before the Civil War in New England armories. The military greatly valued the ease of repair allowed by interchangeable parts and cared less about costs than private manufacturers. "Armory practice" slowly spread to the making of clocks, sewing machines, typewriters, agricultural equipment, bicycles, and other civilian products.[4]

American conditions promoted standardization and interchangeability. A mass market existed that justified heavy capital investment and that was hard to take full advantage of without uniformity. In 1855, 400,000 brass clocks were produced in the United States. During the Civil War, three million rifles were used.[5] A shortage of skilled workers and relatively high wages made it expensive and sometimes impossible to produce complex products in large quantities using traditional artisanal methods. With interchangeable parts, skilled workers were still needed to build specialized machinery and tooling, but less skilled workers could churn out parts and assemble them.[6]

None of this was easy to achieve. The Singer Manufacturing Company, one of the most celebrated manufacturers of its day, illustrated the challenge. Well before the Civil War, the company emerged as a leader in the sewing machine industry, selling a high-priced model made with traditional metalworking techniques. During the war, Singer began mechanizing, but it would take almost two decades before the company fully achieved interchangeable parts. In the interim, it expanded by hiring more and more workers to make parts using some specialized machinery and employing fleets of fitters, who filed and adjusted them. The factory Singer erected in Elizabethport, New Jersey, in 1873 was reportedly the largest in the United States making one product in a single building. Journalists wrote about it, tourists visited it, it appeared on postcards. Along with a second Singer plant in Scotland, it produced an extraordinary 75 percent of the world's sewing machines. Yet even when in 1880 the company was turning out a half million machines a year, they were still assembled, like almost all complex metal products at the time, by carrying all the needed parts to workstations where workers assembled one machine at a time, filing and finishing when less than true interchangeability had been achieved.[7]

Continuous flow operation ultimately led to a radically different approach to assembly. The idea of keeping material moving as workers conducted various operations first developed in industries handling liquid or semiliquid products, most notably oil refining. Grain milling, brewing, and canning came next. But the industry that apparently had the greatest influence on Ford was meatpacking, where the *disassembly* of animals was done by hanging newly killed carcasses on an overhead conveyor, moving them from worker to worker, each of whom made a particular cut or removed particular pieces, until the animal had been reduced to smaller chunks of meat that might then undergo further processing. Implicit in continuous flow processing was an intense division of labor; each worker performed just one or a few operations on something going by or momentarily standing still, rather than many operations on a stationary object.[8]

Ford began experimenting with continuous assembly in 1913, five years after introducing the Model T. Henry Ford had been born during the Civil War, to a farm family in Dearborn, Michigan, near Detroit. Beginning as a machine shop apprentice, he worked his way up through a variety of jobs before becoming the chief engineer in Detroit for the Edison Illuminating Company. He built his first car in 1896, proving his models' worth by racing them. He founded the Ford Motor Company in 1903 with investors who supplied the capital needed to take on the expensive business of making automobiles. In 1907 he wrested control of the firm from his partners. Aiming at rural America, Ford conceived of the Model T as a lightweight vehicle, sturdy enough to withstand the terrible roads that farmers depended on but simple enough for them to repair themselves and for him to produce at a price they could afford.[9]

Sold through a network of independent distributors, the Model T proved an instant hit. Sales zoomed from 5,986 units in 1908 to 260,720 in 1913, as the price of the touring model dropped from $850 to $550 ($13,629 in 2017 dollars).[10] Part of the reason Ford could make so many cars and sell them so cheaply was product standardization. "The way to make automobiles," Henry Ford said, "is to make one automobile just like another . . . . just like one pin is like another pin when it comes from the pin factory, or one match is like another match when it comes from the match factory." Ford, perhaps unconsciously, echoed Adam Smith's famous use of pin manufacturing in *The Wealth of Nations* to illustrate the savings that could come from the division of labor in producing a standardized product. From 1909 on, the Ford Motor Company only produced the Model T. The vehicle's different body styles all used the same chassis. For most of its history, it was available only in black.[11]

With just one model produced in high volume, Ford could invest heavily in equipment and experimentation to manufacture it as efficiently as possible. The tremendous profits the Model T generated freed him from depending on outside investors or Wall Street—which he despised—to expand his plants and add new machinery. Ford toolmak-

ers developed specialized fixtures and jigs to simplify and speed up operations. One machine simultaneously drilled forty-five holes into engine
blocks from four sides, replacing the numerous setups and operations
needed for the same result using traditional methods. The adoption of
single-purpose machinery also helped ensure that tolerances would be
met for interchangeability and easy assembly. The company boasted
that "You might travel round the world in a Model T and exchange
crankshafts with any other Model T you met enroute, and both engines
would work as perfectly after the exchange as before. . . . All Ford parts
of the same kind are perfectly interchangeable."

Specialized machines also were a strategy to deal with the severe
shortage, high wages, and union orientation of skilled workers in the
Detroit area as the automobile industry took off. Ford engineers called
their jigs and fixtures "farmers' tools," since they allowed new workers
to produce high-quality parts, lessening the need for skilled machinists
and their craft culture. (Preferring workers with no craft background
had a long history among American manufacturers; arms maker Samuel Colt once said "the more ignorant a man was, the more brains he
had for my purpose.") The Ford company also made extensive use of
stamped parts, a practice adopted from the bicycle industry, cheaper
and easier than casting and machining.[12]

For most of the nineteenth century, standard machine shop practice had been to group machines together by type—all lathes in one
area, drill presses in another, and so on—which required a significant
expenditure of manpower to move pieces from one area to another
as the production process proceeded. By the early twentieth century,
the most advanced manufacturers, including the Olds Motor Works,
which made the Oldsmobile, and Ford began what Ford called "the
planned orderly progression of the commodity through the shop." Placing machine tools, carbonizing furnaces, and other equipment in the
sequence in which they were used reduced the time spent on transporting unfinished parts and made immediately obvious where hold-
ups were occurring. Here was a spatial embodiment of the logical flow

Marx saw in the mid-nineteenth century when he wrote that in a "real machinery system" "[e]ach detail machine supplies raw material to the machine next in order."

At Ford, progressive placement of machinery went hand in hand with an ever-greater division of labor. Each workstation was manned by a worker who did only one or a few tasks, usually simplified by the creation of equipment designed to do just those operations, over and over again. The gains in productivity were enormous. In 1905, with three hundred workers, Ford produced twenty-five cars a day; three years later, with some five hundred workers, it rolled out one hundred.[13]

Next came installing mechanical devices to move parts from one workstation to another, rather than doing so by hand, applying continuous flow processing to complex manufacturing. In 1913, Ford began experimenting with a conveyor system in its foundry and with slide rails and tables for assembling magnetos and transmissions, having workers stand still while parts for processing or assembling moved past them. Before the new system was installed, it took a single worker about twenty minutes to assemble a magneto at a stationary workbench. After Ford introduced what would become called an assembly line, splitting up the process into twenty-nine separate steps, it took fourteen workers a cumulative time of five minutes to make a magneto, a fourfold increase in productivity.[14]

Inspired by the enormous savings, Ford engineers turned to the assembly of chassis and finished cars. Originally, Ford assembled its cars following the standard practice for manufacturing complex machinery: "we simply started to put a car together at a spot on the floor," Ford recalled, "and workmen brought to it the parts as they were needed in exactly the same way that one builds a house." Other early automakers also used the "craft method" of assembling vehicles on stationary sawhorses or wooden stands.

With the Model T, Ford moved from having a team of workers assemble an entire automobile to breaking down the assembly process into many discrete steps. At stationary stands, arrayed in a large circle,

Figure 4.1 The magneto assembly line at Ford's Highland Park factory in Detroit in 1913.

cars were put together piece by piece, with parts carried to the stands as they were needed. But rather than working on one car until it was completed, workers walked around the circle, at each stand doing just one particular operation—attaching the frame to the axles or fitting in the engine or installing the steering wheel. After the last operation (fitting in the floorboards), the completed car was removed for testing and shipment and the first parts for a new vehicle were laid out at the station. In mid-1913, the Model T assembly area had a hundred stations, with five hundred assemblers cycling around them and another hundred workers bringing them parts.[15]

From there it was just one small step, but a world-historic revolution, to keeping the workers stationary and moving the vehicles as they were being assembled. In August 1913, Ford engineers tried pulling chassis frames through a corridor of preplaced parts, with assemblers walking along with the vehicles installing them. Then they switched to positioning stationary workers along the path of the vehicles, having them attach parts to the chassis being slowly pulled past by a chain drive below. By April 1914, the assembly line had reduced the labor time needed for final assembly of a car from twelve and a half hours to ninety-three minutes.

The success of the final assembly line led to a burst of innovation, as Ford engineers introduced gravity slides, rollways, conveyor belts, chain-driven assembly lines, and other material-handling systems to various subassembly operations, everything from putting together motors to upholstering seats. Many of the subassembly lines fed directly into the final line, delivering engines, wheels, radiators, other components, and, ultimately, finished bodies to the appropriate spots for their installation on the moving chassis. Just as at the Derby silk mill and the Waltham cotton mill, a new system of production came together in a remarkably short period of time. In less than two years after the first experiments with the assembly line, Ford had installed the system for all phases of Model T production. The factory had become one huge, integrated machine.[16]

Ford Labor Problems and the Five Dollar Day

Some of the productivity gain of the assembly line came from the greater efficiency of material handling. Some came from the increased division of labor. But much of it came from the sheer intensification of work, the elimination of the ability of workers to wander around looking for a part or tool, to slow down while a foreman wasn't watching, or to store up finished parts to allow resting later on. For assembly-line workers, work was relentless and repetitive, a single task or just a few done over and over again, every time a new part or subassembly or chassis appeared before them.[17]

In the late nineteenth and early twentieth century, management experts considered "soldiering" (workers deliberately working at less than a maximum possible pace) the paramount obstacle to efficiency and profits. To counter it, they devised all sorts of schemes, from elaborate systems of piecework pay to Frederick Winslow Taylor's "scientific management." The assembly line provided an alternate solution to the same problem, having machinery set the pace of work rather than foremen or incentives. Well before Ford adopted the assembly line, packing house managers saw the possibilities in mechanically pacing production; in 1903, a Swift supervisor said, "if you need to turn out a little more, speed up the conveyers a little and the men speed up to keep pace."[18]

Assembly-line work proved physiologically and psychologically draining in ways other types of labor were not. More than ever before, workers were extensions of machinery, at the mercy of its demands and its pace. One worker complained, "The weight of a tack in the hands of an upholsterer is insignificant, but if you have to drive eight tacks in every Ford cushion that goes by your station within a certain time, and know that if you fail to do it you are going to tie up the entire platform, and you continue to do this for four years, you are going to break under the strain." Another said, "If I keep putting on Nut No. 86 for about 86 more days, I will be Nut No. 86 in the Pontiac bughouse." Ford workers complained that assembly-line work left them in a nervous condition they dubbed "Forditis." Speed, dexterity, and endurance, not knowledge and skill, were the attributes needed for assembly-line work. Men aged quickly on the line, no longer considered desirable workers well before middle age.[19]

The swelling sales of Model Ts left the Ford Motor Company with a voracious appetite for labor, especially "operators," unskilled workers who by 1913 constituted a majority of the workforce. From about 450 employees in 1908, the company leaped to roughly 14,000 in 1913. The Highland Park factory, where Model Ts were made, averaged 12,888 workers in 1914, a size that surpassed even the largest nineteenth-century plants.

Highland Park was not unique. Big and very big factories were becoming more common in the United States. In 1914 there were 648 manufacturing establishments with over one thousand workers. By 1919, there were 1,021 (54 of which made automobiles or automobile parts or bodies), which together employed 26.4 percent of the manufacturing workforce. Rising demand led firms to expand existing facilities, as many companies preferred to keep manufacturing centralized near their administrative headquarters, expediting supervision and coordination. General Electric had 15,000 workers at its Schenectady, New York, complex and 11,000 at a plant in Lynn, Massachusetts. Pullman and International Harvester each employed 15,000 workers at their Chicago plants. Goodyear Tire and Rubber had 15,500 employees in Akron, Ohio.

With its best-selling car and assembly-line operations, Ford soon leaped to a whole new scale. In 1916, Highland Park averaged 32,702 workers; in 1924, 42,000.[20] Photographs of the inside of the plant show workers standing literally elbow to elbow, a density of human labor unlike anything seen in textile or steel mills or other types of manufacturing. They were crammed together not just because of their sheer numbers but by design. Ford engineers wanted workers and machines placed as close to one another as possible, to minimize the time and effort needed to transport parts and subassemblies.[21]

When Ford introduced the assembly line, extraordinarily high turnover added to the company's difficulty in meeting its ever-growing need for workers. Turnover was a general problem for American industry in the late nineteenth and early twentieth centuries. Skilled workers were loyal to their craft, not their employer, often changing jobs to learn new skills or try a different environment. Unskilled workers left their jobs to seek higher pay, to take a vacation (in an era before employers provided any), when they had a dispute with a foreman, or for myriad other reasons. Staying put had no particular benefit.[22]

Ford methods pushed the turnover rate through the roof. Many workers hated Ford's extremely routinized, repetitive work and the stressful pace of production, quitting often after only short tenures. Most simply walked away, never formally resigning. In 1913, the year the assembly line was introduced, Ford had an astounding turnover rate of 370 percent. To maintain a workforce of a bit less than 14,000, that year the company had to hire more than 52,000 workers. Absenteeism added to the difficulties; on any given day, 10 percent of Ford workers did not show up.

Ford had other labor problems, too. Increasingly, the labor pool in Detroit was made up of immigrant workers, especially in the unskilled ranks. In 1914, foreign-born workers made up 71 percent of the Ford workforce, from twenty-two different national groups. A babel of languages meant that workers often could not communicate with foremen or one another. One supervisor recalled that "every foreman had to learn in English, German, Polish and Italian" to say "hurry up." Ethnic tensions sometimes exploded into fistfights. In January 1914, the company fired over eight hundred Greek and Russian workers for staying home to celebrate what by their Orthodox Christian calendar was Christmas but for the company was just another production day.

Detroit automakers, including Ford, also worried about unions. The introduction of the assembly line coincided with a national surge of labor militancy. In Detroit, both the radical Industrial Workers of the World and the new Carriage, Wagon, and Automobile Workers' Union, affiliated with the more moderate American Federation of Labor, launched organizing drives in the auto industry, leading a few short strikes. Their gains were modest, but their specter haunted employers.[23]

Ford responded to its labor problems with a program of higher pay and shorter hours, "The Five Dollar Day." Already, the company had begun instituting policies to retain employees and increase their productivity. In 1913, it introduced a multitiered wage plan that boosted pay as workers' skills grew and, with longevity, a spur to self-improvement and steady employment. In early January 1914, the company went farther, shortening the workday from nine hours to eight (six days a week), which reduced the strain on workers while allowing Highland Park to go from two shifts to three. And more dramatically, it announced that it would effectively double the wages of unskilled workers, from somewhat below $2.50 to $5.00 a day. The wage boost set a precedent for mass production, especially automobile manufacturing, to be a high-wage system. Supporters hailed high wages for allowing workers to buy the kinds of goods they made, creating the mass purchasing power necessary to keep mass production going.

But the Five Dollar Day was more ambitious and more complicated than just a wage boost. Technically, it was not a pay increase at all but a possibility for workers to get what was dubbed a profit-sharing payment that would bring their daily income up to five dollars. Qualification was not automatic; women were not eligible (at least initially), male workers generally had to be over twenty-one, and, most importantly, they had to abide by a set of standards and regulations the company set, aimed not only at behavior in the factory but away from it, too. Workers had to be legally married to their partners, "properly" support their families, maintain good "home conditions," demonstrate thrift and sobriety, and be efficient at their jobs. Ford established a "Sociological Department" to investigate if workers were eligible for the profit sharing and to guide them in behavioral change if they were not.

Fifty investigators, often accompanied by translators, made home visits to Ford workers to assess their qualifications for the plan. After an initial round of investigations, 40 percent of the workers eligible by age and sex were deemed deficient in some respect to receive the payments. Failure to rectify their behavior within a given period led to dismissal, but improvements could win retroactive profit-sharing.

Ford was particularly concerned with "Americanizing" immigrant workers. Sociological Department agents encouraged them to adopt American habits and teach their children American ways. Workers who did not speak English were heavily pressured to attend an English school the company established, which taught "industry and efficiency" and American customs and culture along with language. Some 16,000 workers graduated in 1915 and 1916 alone, reducing the non-English-speaking component of the workforce from 35 percent in 1914 to 12 percent in 1917.[24]

There were precedents for many aspects of the Ford labor policies. The Lowell-style mills had their own elaborate regulations for behavior on and off the job. Like Ford, the mill owners had the challenge of establishing behavioral norms and worker self-discipline necessary for the collective, integrated nature of factory work. And like Ford, they had moral concerns that extended beyond the factory walls. In the late nineteenth and early twentieth century, a new wave of behavior-shaping programs began as many companies, especially manufacturers with large plants, initiated "welfare work" to increase worker productivity and reduce turnover. Companies built cafeterias, libraries, and "rest rooms"; offered recreational activities, health services, and pensions; established savings and insurance plans; and occasionally introduced the type of social work Ford imposed.

But the comprehensiveness of the Ford program, its intrusiveness, and its link to a doubling of wages put it at the forefront of employer efforts to shape the behavior and mindset of employees to make them fit into a factory regimen. S. S. Marquis, who became head of the Sociological Department in late 1915 (renaming it the Educational Department in response to widespread worker criticism of the home investigations), wrote: "as we adapt the machinery in the shop to turning out the kind of automobile we have in mind, so we have constructed our educational system with a view to producing the human product in mind."[25]

Ford executives would have agreed with Italian communist leader Antonio Gramsci when he wrote, "In America rationalization has determined the need to elaborate a new type of man suited to a new type of work and production process." Henry Ford's rural Protestant moralism, with its stress on thrift, sexual rectitude, and spurning of alcohol and tobacco, prescribed a way of life that Ford executives— and Gramsci—saw as necessary for the physical and psychological demands of mass production. As the Italian communist, sounding like an auto executive, noted, "The employee who goes to work after a night of 'excess' is no good for his work." "The enquiries conducted by the industrialists into the workers' private lives," Gramsci cautioned, "and the inspection services created by some firms to control the 'morality' of their workers are necessities of the new methods of work. People who laugh at these initiatives . . . and see in them only a hypocritical manifestation of 'puritanism' thereby deny themselves any possibility of understanding the importance, significance and objective import of the American phenomenon, which is *also* the biggest collective effort to date to create . . . a new type of worker and a new type of man."[26]

Ironically, by the time Gramsci wrote his essay "Americanism and Fordism" (in prison after his 1926 arrest by the fascist Italian government), Henry Ford already had abandoned his effort to create "a new type of man." As part of a cost-cutting drive during the 1920–21 recession, Ford shrank the responsibilities of the original Sociological Department until it effectively disappeared. He also abandoned his profit-sharing scheme, switching to a basic wage rate of six dollars a day (an income boost less than inflation), with bonuses based on skill and longevity. Deeming paternalism and welfare work too expensive and a threat to the control of the factory by production officials, Ford instead turned to an elaborate spy system and autocratic management to control labor. The "Service Department," into which he folded the remnants of the Sociological Department, was headed by a Harry Bennett, a former boxer with extensive ties to the police and organized crime, who used spies and brute force to maintain discipline, hiring many ex-convicts to do the job.[27] But if Ford himself abandoned the link between mass production and the creation of a "new man," the idea itself would live on for decades, including in some very different places.

Alfred Kahn and the Modern Factory

To make the Model T, Ford created not only a new production system but also new types of factory structures, which became templates for generations of giant factories around the world. Their technical and visual legacy remains strong today.

Ford's first factory, on Mack Avenue in Detroit, had been a small, one-story, wood-framed building. His second, completed in 1904 on Piquette Avenue, was considerably larger, a handsome, three-story brick building. But in design it differed little from an early nineteenth-century textile factory: long and narrow, with large windows and wooden columns, beams, and floors.[28]

Even before Model T production began, Ford anticipated that his company would soon outgrow Piquette Avenue, purchasing land in nearby Highland Park for a new plant. To design the factory he hired Detroit architect Albert Kahn, who would become the foremost factory designer of the twentieth century. Kahn stumbled into industrial architecture early in his career, somewhat by chance. Eclectic in his commissions and styles, Kahn, a German Jewish immigrant, met Henry B. Joy, the head of the pioneer automaker Packard Motor Company, who helped him get a number of nonindustrial commissions before asking him to design a new factory complex for his firm.[29]

The first nine buildings Kahn designed for Packard were conventional. But the tenth was a radical departure, made not of wood and brick but of reinforced concrete. In designing it, Kahn worked closely with his brother Julius, who had developed a system for reinforcing concrete with a particular type of metal bar.

Reinforced concrete, first used in Europe during the 1870s and in the United States not long after, was strong, resistant to vibration, inexpensive, and fireproof. It allowed for large, uninterrupted spaces and a greater window area than older construction methods. A concrete shoe factory, built in Massachusetts in 1903–04, brought the material to the attention of industrial architects. Kahn's 1905 reinforced concrete Packard Plant Number 10, with its large window area and orderly layout, attracted much attention, as did a plant he built the following year in Buffalo for the George N. Pierce Company, which incorporated overhead cranes and rail platforms for loading, unloading, and moving materials.[30] So when Ford hired him, Kahn already had begun building a reputation as an innovative factory designer.

The Highland Park complex extended Kahn's earlier work. The exterior walls of the main four-story factory building were mostly glass, allowing in so much light that observers dubbed it the "Crystal Palace," a reference to the London exhibition hall built over a half century earlier. Kahn convinced Ford to allow him to use metal window sashes, at the time so unusual that they had to be ordered from England, which gave the building a particularly clean, modern look. Inside, the large open spaces facilitated the experiments that led to the assembly line.

But in some ways, the initial Highland Park buildings still harkened back to traditional factory design. The long, narrow main building, with stairs, elevators, and toilets in four external towers, had the proportions and layout of a Lowell mill, even if much larger. The adjacent one-story machine shop, with its sawtooth roof, resembled an English weaving shed. Even after the assembly line had been installed in the factory, some material, including car bodies, was moved by horse-drawn cart.[31]

Kahn's 1914 addition to Highland Park, the "New Shop," represented a more radical break from the past. Almost immediately after Highland Park opened, Ford began adding more Kahn-designed buildings to the tightly clustered complex, including an administration building and a large power plant. It soon needed new assembly space as well. The company decision to begin making parts that it previously had bought from outside suppliers, along with the growing volume of production and a growing workforce, left the main factory crowded almost as soon as it was completed. Furthermore, the assembly line and the rapid pace of production made material handling an ever-greater priority, as large quantities of raw materials, parts, and subassemblies needed to be delivered to particular points along various assembly lines at a pace that avoided pileups of inventory or shortages that stopped production.

Kahn's solution in the New Shop was to build two parallel six-story factory buildings, connected by an 842-foot-long, glass-roofed shed. Along the bottom ran railroad tracks, so that trainloads of supplies could be brought directly into the plant. Along the top ran two overhead cranes that could lift loads of up to five tons to some two hundred platforms jutting out from all levels of the adjacent buildings. From the platforms it was only a short distance to any place within the new buildings, allowing workers to use hand trucks to quickly deliver supplies to the many workstations within. Strikingly modern, the craneway, with concrete and glass buildings making up its walls, the staggered pattern of the jutting platforms, and its glass roof, was a new kind of space, resembling more the great nineteenth-century shopping arcades, like the Galleria Vittorio Emanuele II in Milan, stripped of ornamentation, than a traditional factory.

Inside the New Shop, the foundry and machine shop were positioned on the top floor rather than on the bottom level, the usual practice, possible because of the strength of the reinforced concrete construction. Production could then flow downward, as parts and subassemblies were lowered from floor to floor by gravity slides and conveyor belts, until reaching the final assembly line on the ground level. Air circulation was accomplished through ducts inside hollow concrete columns, an approach reminiscent of that used in English factories by Lombe, Arkwright, and Strutt over a century earlier.[32]

The Highland Park factory almost immediately became the object of enormous worldwide attention for its design, its assembly line, its experiment in high-pay paternalism, and the Model Ts that came out of it. Ford sought the attention, using the building complex as an advertisement for his firm. (Manufacturers had been doing variations of this for decades, designing handsome factories adorned with large signs, putting engravings of their plants on their stationery, allowing postcards of them to be issued, and sometimes welcoming journalists.[33]) The freestanding administration building was handsomely designed and carefully landscaped. The nearby power plant had plate glass windows, allowing passersby to look in at the giant generators. Henry Ford insisted that the plant have five chimneys, so giant letters spelling out Ford could be positioned between them, though fewer chimneys would have sufficed. In 1912, the company began conducting public tours of the plant. By the summer of 1915, three to four hundred people a day

Figure 4.2 An aerial view of Ford's Highland Park factory in 1923.

were visiting. To further publicize the factory, Ford issued a booklet detailing its operations, with pictures from its own, in-house Photographic Department (which also produced weekly short films to distribute to Ford dealers and local theaters).[34]

Among the most important visitors to Highland Park was Giovanni Agnelli, the chairman of the Italian automaker FIAT, who came away determined to adapt Ford methods to the European auto industry, which still largely made cars through handcrafting. To accommodate the Ford system, he commissioned a new factory in the Lingotto district of Turin, which opened in 1923. The plant—one of the great landmarks of modernist architecture—was Highland Park turned on its head. Like the New Shop, it had two long, linked, parallel buildings for assembly operations, each five stories high and over a quarter mile long. In the huge courtyard between the buildings, two spiral ramps connected all of the floors to the roof. In an opposite procedure from Highland Park, raw materials were delivered on the ground floor and production proceeded upward until finished cars were driven onto a test track on the roof, with banked curves that allowed high speeds. Then the cars were driven down a ramp for delivery. (In a ricochet, when

Kahn designed an eight-story service center for Packard on the West Side of Manhattan, he included two interior ramps that allowed access to a rooftop test track.)[35]

Highland Park positioned Kahn as the leading architect for the automobile industry. He was soon designing factories for the Hudson Motor Company, the Dodge brothers, Fisher Body, Buick, and Studebaker, as the industry rapidly adopted both the assembly line and reinforced concrete construction. Ultimately his firm designed a wide range of industrial buildings, not only in North America but in South America, Europe, Asia, and Africa as well. Kahn also designed office buildings for the auto industry and other industrial firms, including the massive General Motors Building in midtown Detroit (the largest office building in the world when it opened in 1922), and the adjacent, opulent headquarters for Fisher Body. And he designed homes for auto executives, including lakefront mansions in Gross Pointe for Henry Joy and Henry Ford's son, Edsel. He even designed the Henry Ford Hospital. The extraordinary productivity of his firm, which by the late 1920s had four hundred employees, and the rapidity with which it could complete designs, rested on a high degree of division of labor, with various departments performing specialized functions, an application to professional, white-collar work of some of the principles Ford perfected for manufacturing. To track work, Kahn's firm used forms similar to those used by Ford at Highland Park.[36]

River Rouge

Even as Kahn's practice grew, Henry Ford remained his most important client. Together they designed what became the next flagship of industrial giantism, Ford's River Rouge plant. Almost as soon as the New Shop was completed, Ford began planning a much larger complex in nearby Dearborn, buying massive tracts of land. Some was used for Ford endeavors besides the car company, including a separate firm that produced Fordson tractors. But most of it was devoted to making the Model T. Ford decided to advance to the extreme his effort at vertical integration, seeking to make not only parts but also basic materials like steel, glass, and rubber for his cars, eliminating the possibility of suppliers raising prices or not fulfilling orders when inventories were tight. The Dearborn property, along the Rouge River, allowed the direct delivery of bulk goods, including iron ore, coal, and sand, from Great Lakes ships and had plenty of water for industrial processes. Also, the sparsely populated Dearborn suburb gave Ford greater control over his environment than Detroit, with its heterogeneous population and episodic labor activism.[37]

Ford began constructing a blast furnace at River Rouge in 1917. It was followed by a series of other processing plants, including coke ovens, open-hearth furnaces, a rolling mill, a glass factory, a rubber and tire plant, a leather plant, a paper mill, a box factory, and a textile mill. Ford put great effort into integrating the various plants and reusing byproducts. Impurities from the blast furnaces, for example, were sent to an on-site factory to be made into cement. Ford also began buying coal and iron mines and vast tracts of forest land in the Upper Peninsula of Michigan, where he built sawmills, kilns, and factories to make wooden parts for the Model T. Sawdust and scrap lumber were used to make the charcoal briquettes, sold under the Kingsford brand, which to this day fuel barbecues and family happiness across America. His grandest effort at backward integration was a vast rubber plantation in the Amazon Basin that proved a costly failure.[38]

Complete Model Ts were never produced at River Rouge, which initially served as a feeder plant for Highland Park. Engines, tires, windows, and other components were taken from the Rouge to Highland Park for final assembly. But with the high volume of Model T production, even the feeder operations were vast. The River Rouge foundry, where engine blocks were cast from molten iron conveyed from adjacent blast furnaces, was the largest in the world, employing ten thousand men.[39]

When final assembly operations did begin at the Rouge, it was, ironically, to make boats, not cars. During World War I, Henry Ford contracted with the Navy to build 112 submarine chasers using assembly-line methods. The Navy paid for a new plant to produce them, the "B Building," designed by Kahn. Freestanding, it was the largest factory ever built, 300 feet wide and 1,700 feet—a third-of-a-mile—long, a huge shed with walls composed almost entirely of windows. As tall as a three-story building but open inside to accommodate boat production, it was designed to allow the later addition of intermediate floors. When the last of the Eagle Boats left the building in September 1919 (none were completed in time to be used in combat), floors were added and the building was used to assemble Model T bodies, which previously had been purchased from outside contractors.

The B Building represented the beginning of a shift in factory design principles for Ford and Kahn, moving away from the ingenuous architectural machine that they had just developed at the New Shop. Kahn helped lead not one but two revolutions in industrial architecture. Rather than multistory buildings, at the Rouge Kahn and Ford erected very large single-story factories to avoid the cost of hoisting materials and to allow bigger uninterrupted spaces, since columns to support upper floors were no longer needed. The expansive, open areas gave engineers flexibility in machine placement, aided by the company decision to stop using overhead shafts and belts to power machinery, instead deploying individual electric motors. Single-story plants also avoided the need to punch holes between floors when assembly lines were repositioned. In 1923, Ford switched its standard design for branch plants from multistory to single-story as well.

With the move to single-story factories, Kahn abandoned reinforced concrete, no longer needing its vibration dampening qualities. Instead he used steel frames, which allowed structures to be put up more quickly and expanded more easily. Kahn's new buildings had, if anything, even more glass on the walls than his earlier structures, and he generally used roof monitors—raised structures with glass facing in varied directions—rather than sawtooth roofs, which provided more diffuse natural light.

The loft-style, concrete buildings Kahn helped popularize continued to be built for manufacturing and storage. Resistant to water damage and strongly constructed, they can be found in large numbers in older American industrial districts, sometimes still used for manufacturing, sometimes abandoned, sometimes converted to warehouses or offices, and occasionally turned into trendy apartments. But Kahn himself almost never returned to the style.

Instead, Kahn embraced sleek surfaces of glass and metal in buildings both functional and beautiful. Over the course of two decades, he created a bounty of industrial buildings of great modernist design— clean, light, spare, seemingly endless. Many of Kahn's Rouge buildings were expressions of almost pure form—tall cylindrical chimneys, long glass walls, shapely monitor roofs—unsullied by ornamentation. The Engineering Laboratory, completed in 1925, where Henry Ford had his office, had a particularly striking interior, with a long central space flanked by smaller galleries, with two levels of monitor windows on both sides flooding it with light. Some of Kahn's later designs, like his Chrysler Half-Ton Truck Plant, are widely recognized as among the greatest industrial buildings ever erected, modernist masterpieces.

Yet neither Kahn nor Ford thought of themselves as modernists. In a 1931 speech, Kahn gave a nuanced but largely negative appraisal of modernist architecture. Kahn criticized the extreme functionalism and lack of ornamentation of architects like Walter Gropius and Le Corbusier (arguably traits that characterized his own factory designs). "What we call modernism today is largely affectation, a seeking for the radical, the extreme." In his nonindustrial projects, Kahn drew on a variety of historical styles, designing often handsome but rarely pathbreaking buildings. Henry Ford was even more explicitly antimodernist at the very moment he was creating a new industrial modernity. Concurrent with the creation of the Rouge, he continued to add to his collection of old machines, furniture, and buildings, which he eventually installed in Greenfield Village, near the Rouge plant, a recreation of an earlier, small-town America. Even as his cars and factories promoted urbanization and cosmopolitanism, Ford remained deeply nostalgic about the parochial, rural world he grew up in and chose to leave.

Buildings continued to be added at the Rouge all through the 1920s and 1930s. The Press Shop, completed in the late 1930s, became the largest single factory building in the world, with a floor area of 1,450,000 square feet. Ford spaced the Rouge buildings far apart to allow for later expansion, having plenty of room on the 1,096-acre site. An elaborate system of rail lines, roads, 142 miles of conveyors, monorails, and an elevated "High Line" with an automatic transport system moved raw materials, parts, and subassemblies within and between buildings. Employee parking lots ringed the vast, isolated complex, but many workers arrived at special streetcar and bus terminals. Fences, railroad tracks, and guarded gates restricted access to the plant, which came to resemble a fortress, in contrast to Highland Park, which was situated in a busy urban neighborhood, with public sidewalks alongside the factory buildings.[40]

Ironically, while the Rouge was being built out to produce everything needed to make a Model T, the car itself was becoming obsolete. By the mid-1920s, other car companies, including General Motors and Chrysler, had introduced more technically advanced and varied models than Ford, which still only sold the Model T (though it offered luxury cars under the Lincoln nameplate). By 1927, as sales diminished, it became evident that something had to be done. Abruptly, Ford stopped making the Model T, even before finalizing the design of its replacement, the Model A. For six months, Ford factories sat idle, while the company replaced 15,000 machine tools and rebuilt 25,000 more. New molds, jigs, dies, fixtures, gauges, and assembly sequences had to be created. Meanwhile, the layoff of 60,000 Detroit-area Ford workers created a social crisis, as relief agencies, free clinics, and child-placement agencies struggled to meet the huge demand for their services.

The underbelly of the Ford system had been exposed. Extreme standardization had allowed other companies to win over consumers on the basis of style and change, what General Motors president Alfred P. Sloan, Jr., called "the 'laws' of Paris dressmakers . . . in the automobile industry." Single-purpose, specialized machinery, which made it inexpensive to produce particular parts, made it expensive to switch over to new products (a problem that went all the way back to the high-speed but inflexible machinery used in the early Lowell mills). The changeover from the Model T to the Model A cost the Ford Motor Company $250 million ($3.5 billion in 2017 currency) and first place in sales to General Motors. Vertical integration had its downside, too, evident when the economy and auto sales tanked just a few years after the introduction of the Model A; Ford had a harder time cutting costs than the other major automakers, which bought most of their parts from outside suppliers. Over the course of the decade starting in 1927, Ford had a cumulative net loss, while General Motors made nearly $2 billion in after-tax profits.

The introduction of the Model A completed the transfer of the center of the Ford empire from Highland Park to River Rouge. The final assembly line for the new car was set up in the B Building, which was so large that it also could house at various times an assembly line for Fordson tractors, a trade school, fire department, and hospital. The geographical move was accompanied by a purge of pioneer Ford engineers and executives, most of those remaining from the team that had created the Model T, the assembly line, and the Ford system. With Harry Bennett and Charles Sorenson, a long-time, very tough Ford production manager, effectively running the Rouge, an autocratic, chaotic, and brutal culture came to characterize the plant. Workers decried harsh discipline for petty offenses, arbitrary, ever-changing rules, and tyrannical foremen. One Rouge worker complained that "The bosses are thick as treacle and they're always on your neck, because the man above is on their neck and Sorenson's on the neck of the whole lot—he's the man that pours the boiling oil down that old Henry makes. . . . A man checks 'is brains and 'is freedom at the door when he goes to work at Ford's."

The Rouge—"that self-sufficing industrial cosmos, a masterpiece of ingenuity and efficiency," Edmund Wilson called it—embodied an extreme strategy of industrial concentration. Ford set up dozens of branch plants in the United States to assemble kits of parts shipped from Highland Park and later Dearborn, but manufacturing remained highly centralized at the major complexes. During the 1920s and 1930s, the company built a series of "village industry" factories in rural southeastern Michigan. Powered by small hydroelectric dams, the plants produced small parts for use at Highland Park and the Rouge—starter switches, drill bits, ignition coils, and the like. Henry Ford conceived of the plants as providing work for farmers during the slack winter season. Again, as at Greenfield Village, he seemed to be embracing an idealized vision of a decentralized Jeffersonian society, even as his life's work undermined it. But with a combined workforce at their height of only some four thousand workers, the village factories were not much more than an ideological gesture in the shadow of the giant Ford plants.

Other automakers also built very large plants. The complexity of manufacturing an automobile, with its hundreds of different parts; the cost of transporting bulky components like frames, axles, motors, and bodies; and the heavy investment needed to build and equip an automobile plant made concentration of production a widely shared strategy. The Dodge Main plant in Hamtramck (an independent enclave within Detroit) began as a parts supplier for Ford, but the Dodge Brothers later expanded it to produce their own car. Albert Kahn designed the first buildings; Smith, Hinchman, & Grylls, another Detroit architectural firm, many additional buildings, most of them multistory structures made of reinforced concrete. Under the Dodges and later Chrysler, which bought the company after its founders' deaths, the factory became a fully integrated manufacturing and assembly plant, larger in floor space than Highland Park, its nearest equivalent. It had some 30,000 workers in the late 1930s and even more during World War II, remaining in operation until 1980. General Motors became famous for its divisional structure and decentralization, but in Flint, Michigan, it, too, had a huge production complex, several really. In the late 1920s, the gigantic Buick plant (yet another Kahn design) had 22,000 workers; a cluster of Chevrolet factories employed 18,000 workers; Fisher Body, by then a GM subsidiary, had 7,500 workers; and still more workers could be found in the factories of AC Spark Plug, another GM subsidiary.

But nothing touched the Rouge in sheer scale. Historian Lindy Biggs characterized it as "more like an industrial city than a factory." In 1925 it had 52,800 workers, still trailing Highland Park, where the workforce had swelled to 55,300. With the Model A, though, the Rouge moved ahead. It peaked at 102,811 workers in 1929, a level of employment entirely unprecedented at a single factory complex. To this day, at least in terms of the size of its workforce, it remains unmatched in the United States. It was, simply, the largest and most complicated factory ever built, an extraordinary testament to ingenuity, engineering, and human labor.[41]

Celebrating Ford

Ford methods attracted widespread interest among industrial professionals as soon as they were introduced. Henry Ford welcomed reporters, especially from the technical press, into his factories, openly sharing details about his latest innovations, a departure from the usual wariness among manufacturers about releasing information about their techniques. Trade journals like *American Machinist*, *Iron Age*, and *Engineering Magazine* ran extensive articles about the methods developed to produce the Model T. Other American automobile companies and consumer goods manufacturers quickly adopted the assembly line.[42]

The general public was likewise fascinated by the Ford system, especially the assembly line. Henry Ford realized that public interest in the methods of making Ford cars could help sell them. In addition to providing tours of the Highland Park plant, he took the assembly line on the road. At the 1915 Panama-Pacific International Exposition in

San Francisco, just two years after the assembly line had been introduced, a Ford exhibit included a working production line that turned
out twenty Model Ts a day. When in 1928 Ford unveiled the Model
A at Madison Square Garden, the company put up displays of every
facet of the production process, from dioramas of Ford iron and coal
mines to workstations for making glass and upholstery. At the 1933–34
Chicago Century of Progress Exposition, part of the Ford Exposition
Building, designed by Albert Kahn and later moved near the entrance
to the Rouge plant, showed "the complete production of the car in all
its parts." In 1938, nearly a million people visited the display. And they
flocked to the Rouge itself, too. In the late 1930s, Ford offered a two-
hour tour of the complex starting every half hour. Other manufacturing firms, including Chrysler and General Motors, also opened their
plants and set up exhibits for a public endlessly fascinated with how
things were made, especially with the complex, wondrous choreogra-
phy of the assembly line. The Kahn-designed General Motors Exhibit
at the Chicago Exposition featured a model production line, which
allowed visitors on an overlooking balcony to watch workers assembling vehicles.

The public romance with the giant factory and the assembly line
proved long-lasting. In 1971, 243,000 people visited the Rouge, a record
number. A few years later, the U.S. Department of Commerce published a list of plants in the United States that offered tours. It ran to
149 pages, with everything from distilleries to steel mills, including a
dozen auto plants.[43]

Intellectuals and political activists were caught up in the allure of
Fordism, too. Perhaps surprisingly, given Ford's later reputation as a
union-hating, conservative autocrat, some prominent leftists at first
praised the Ford system. In early 1916, after visiting the Highland
Park plant, Kate Richards O'Hare, a well-known socialist leader, published two articles in *The National Rip-Saw*, a mass circulation socialist monthly, praising Henry Ford. O'Hare saw the Five Dollar Day,
the Sociological Department, and the Ford English School as advanc-

ing the lot of workers (along with Ford's decision to take the power to fire away from foremen). Using a jarringly racist simile, she wrote that as a result of Ford's policies "men freeze to a job in the Ford plant like a negro to a fat possum." "If every Capitalist in the United States were to suddenly become converted to Ford's ideas . . . it would not solve the social problems, eliminate the class struggle or inaugurate the co-operative commonwealth, BUT it would advance the cause of social justice, demonstrate the soundness of the socialist theories and bring the mighty pressure of education to hasten the final and complete emancipation of the working class."[44]

Later that same year, John Reed, soon to be the most important chronicler of the Russian Revolution and a founder of the American Communist Party, wrote a similarly glowing if more sophisticated portrait of Ford in the left-wing journal *The Masses*. Ford's strategy of low prices and high wages, especially the profit-sharing built into the Five Dollar Day, for Reed represented a huge step forward from normal industrial practices. Reed detailed the difference high wages made in the lives of Ford workers. Beyond that, after interviewing Ford, he came to believe that the auto giant was moving toward some sort of new form of corporate control that would give workers a say; the Five Dollar Day was "turning into something dangerously like a real experiment in democracy, and from it may spring a real menace to capitalism." This was why, Reed believed, "capitalists hate Henry Ford," an echo of Ford's own perception of himself, in the Populist idiom he grew up around, as a producer of value having to fight off the parasitic financiers of Wall Street.[45]

Left-wing praise for Henry Ford diminished over time, in part in response to changes in his company's practices and his rabid anti-Semitism during the 1920s; Edmund Wilson, writing fifteen years after Reed, dubbed him the "despot of Dearborn." But Fordism struck a strong chord with a group that during the New Deal would ally with elements of the left, businessmen and their supporters who saw mass consumption as critical to maintaining prosperity and profits. Edward

Filene, who made his money in department stores, was perhaps the most outspoken member of those who have been dubbed "proto-Keynesians" for seeing the need for mass purchasing power to maintain economic growth. Unlike in the past, Filene wrote in 1924, businesses needed to produce "prosperous customers as well as saleable goods." Fordism, with its promise of high wages and cheaper products, was a way to create a virtuous circle of mass purchasing power, mass consumption, mass production, and economic growth. Unlike O'Hare and Reed, Filene acknowledged the monotony of Fordist labor, but saw shorter hours as partially ameliorating the problem. And, in any case, "every man is not an artist, every man is not a creative craftsman." "Poverty brings a monotony a thousand times more deadly to body and mind than the monotony of factory routine," he added in a comment reminiscent of W. Cooke Taylor's remark about child labor eighty years earlier.[46]

Novelists, too, saw in Fordism a startling development, a step into a new type of world. John Dos Passos profiled Ford in *The Big Money* (1936), which concluded his great three-volume portrait of the country, *U.S.A.*, writing not only about the Model T and the exhausting labor used to produce it but also the automaker's many contradictions, his pacifism, war profiteering, and anti-Semitism, his revolutionary inventions and antiquarianism. (Alfred Kazin shrewdly observed that *U.S.A.*, with its complex structure composed of different types of narrative building blocks, was itself a "tool," "another American invention— an American *thing* peculiar to the opportunity and stress of American life.")[47] Louis-Ferdinand Céline, who visited a Detroit Ford factory in 1926, included a scene of working on the company assembly line in *Journey to the End of the Night* (1932). Upton Sinclair wrote a not very good novel about Ford, *The Flivver King: A Story of Ford-America* (1937). And most famously, Aldous Huxley's *Brave New World* (1932) depicts a dystopia of Fordism, a portrait of life A.F.—the years "Anno Ford," measured from 1908, when the Model T was introduced—with Henry Ford the deity.[48]

Dos Passos, Sinclair, Céline, and Huxley all wrote about Ford and

Fordism during the 1930s, well after the initial burst of journalistic and industrial excitement over mass production. Their work was colored by the Great Depression and the Ford Motor Company's violent anti-union actions, which radically changed the public image of Ford and the Fordist project. By contrast, the key visual depictions of Fordism began earlier, during the 1920s. More than in the written word, it was in the visual arts that Fordism and the giant factory were celebrated.

Giant Factories and the Visual Arts

Factories had been portrayed from their earliest days in drawings, lithographs, and paintings. But only in the twentieth century did the factory become an important subject for artists. It is difficult to think of a truly great eighteenth- or nineteenth-century artistic representation of a factory, but there are plenty of great twentieth-century factory paintings, photographs, and films. For many artists during the 1920s and 1930s, the factory represented modern life—secular, urban, mechanical, overwhelming—a break from the rural landscape or intimate domestic interior. And it provided a vehicle for modernist modes of artistic representation, moving toward abstraction. While in the nineteenth century, novelists and other writers played a major role in shaping public perceptions of the factory and the factory system, in the twentieth century, visual artists came to the fore.

Photography, in particular, took the lead in influencing public perceptions of the giant factory. Itself a product of the Industrial Revolution that created the factory system, photography allowed the easy reproduction and dissemination of imagery, while painting remained an inherently elite form, largely created for private viewing by collectors or museum goers. It was fitting that photography and film, so well suited to the creation of unlimited identical products, proved the most important media for the representation of mass production.

Early in the twentieth century, a number of American photographers, including Paul Strand, Alfred Stieglitz, and Alvin Langdon Coburn, began taking pictures of machinery, machine parts, and industrial landscapes. By the 1920s, photographers and artists elsewhere—purists in France, futurists in Italy, Bauhaus affiliates and *Neue Sachlichkeit* photographers in Germany, constructivists in the Soviet Union—also had turned to industry for visual ideas, symbols, and a machine aesthetic.[49] But photographing actual factories, especially their interiors, presented formidable technical problems in an era of large, heavy cameras, a limited choice of lenses, slow film, and primitive lighting devices. The photographer who first overcame many of the challenges and did more than any other to disseminate images of giant industry was Margaret Bourke-White.

Bourke-White's father, an engineer and inventor, worked for a printing press manufacturer. He often took Margaret, while a child living in New Jersey, to the plants where presses were being made or installed. She later wrote of the first time he took her to a foundry, "I can hardly describe my joy. To me at that age, a foundry represented the beginning and end of all beauty." Her lifelong fascination with industry was linked to her intense feelings for her father, who died when she was only eighteen. "I worshipped my father," she wrote. "Whenever I go on a job, I always see machinery through my father's eyes. And so I worship factories."

Bourke-White moved to Cleveland in the mid-1920s to try to make a go of it as an architectural photographer, documenting upscale homes and gardens. But she found herself drawn to the Flats, the smoky, dirty, noisy district in the heart of the city that housed heavy industry. "Fresh from college with my camera over my shoulder, the Flats were photographic paradise."

Soon Bourke-White was selling exterior shots of industry to a local bank for its house publication. But getting inside factories was another story; Cleveland industrialists, like most factory owners, had no interest in allowing outsiders inside. Her break came when the head of Otis Steel gave her access to his mill. With a confidence beyond her years, she pronounced to him "that there is a power and vitality in industry that makes it a magnificent subject for photography, that it reflects the age in which we live." She had come to believe that "Industry . . . had evolved an unconscious beauty—often a hidden beauty that was waiting to be discovered."

After five months of experimenting with camera positions, lighting, film, and darkroom technique, Bourke-White managed to capture the drama of molten steel being poured. Otis Steel bought her prints, and other industrial commissions began coming her way. For the stage set of Eugene O'Neill's play *Dynamo*, she photographed the generators at the Niagara Falls Power Company. Years later, when she reprinted the image, she wrote in the caption, "Dynamos were more beautiful to me than pearls," quite a statement for a woman devoted to stylish looks and expensive clothes.[50]

In 1929, Henry Luce, the publisher of *Time*, hired Bourke-White for his new business publication, *Fortune*. A lavish, heavily illustrated magazine, with some of the top writers and designers in the country, *Fortune* provided sophisticated documentation, celebration, and analysis of American business. Its photographers, including Bourke-White, had access to the largest and most advanced industrial complexes in the country. In 1930, she photographed the Rouge. Four years later, she took pictures at Amoskeag Mills, where years earlier Lewis Hine had photographed child workers.

Bourke-White's audience expanded exponentially when Luce shifted her to his new "photo-magazine," *Life*. The cover of the first issue, dated November 23, 1936, was a Bourke-White photograph of the spillway of the world's largest earth-filled dam, the Fort Peck Dam in eastern Montana, a masterpiece of formal, nearly abstract composition and human-dwarfing scale. Within months, *Life* was selling a million copies a week, with Bourke-White one of its stars.

In her early industrial photographs, Bourke-White displayed little interest in workers. Often they are totally absent. When present, they seem negligible compared to the huge structures and machines that dominate her pictures. This effacing of workers from industrial imagery was a common characteristic of photographs and paintings during the 1920s and early 1930s (in Europe as well as the United States), a sharp contrast to the earlier work of Hine. Though Hine sometimes showed machines dwarfing humans, emphasizing their large scale and abstract shapes, the bulk of his work centered on the human experience of labor, on the faces, bodies, and expressions of the workers who inhabited the industrial realm. For Bourke-White, at this stage of her career, it was not the worker who held her interest, nor the products being made, but the abstract forms of industry. "Beauty of Industry," she wrote in 1930, "lies in its truth and simplicity."[51]

Charles Sheeler, who beat Bourke-White to the Rouge, shared her credo. "I speak in the tongue of my times," he said in 1938, "the mechanical, the industrial. Anything that works efficiently is beautiful." "Our Factories," he declared, "are our substitutes for religious expression." A precisionist painter from Philadelphia, whose early work included the magnificent, abstracted urban landscapes *Church Street El* (1920) and *Skyscrapers* (1922), Sheeler took up photography as a way to support himself while painting. His commercial work included photographs for a Philadelphia advertising agency, N. W. Ayer & Son, which the Ford Motor Company engaged to promote the introduction of the Model A. Vaughn Flannery, the Ayer art director, working with Ford, decided to sell the new car by portraying the giant machines and factories used to manufacture it. Flannery sent Sheeler off to the Rouge, where he spent six weeks producing an extraordinary portfolio of images. Most of the photographs depict steelmaking and stamping processes, with their giant equipment and elemental drama. There are no photographs of assembly operations. Many of the images appear nearly abstract, with chimneys, conveyors, pipes, and cranes cutting across the picture plane, often at dramatic angles. Workers are entirely absent in many photographs and barely visible, at the edges of the frame, in others. As in some of Bourke-White's photographs, when humans are present they serve to make evident the massive scale of the equipment and buildings near

Figure 4.3
Charles Sheeler's striking photograph of the Ford River Rouge factory, *Criss-Crossed Conveyors— Ford Plant, 1927.*

them (not dissimilar to the relationship between man and machine in illustrations of the Corliss engine at the Centenary Exhibition).

"The Flannery Ford campaign," wrote architectural historian Richard Guy Wilson, "was the first to portray a beauty and heroism in the manufacturing process in order to spur sales. The Rouge ads started a fad, as many advertisers found that industrial views could be used in popular, mass-circulation magazines as well as in trade journals." Flannery shrewdly realized that the giant factory, with its Promethean grandeur, represented a modernity with which consumers would want to associate themselves.[52]

While Ford made use of Sheeler's Rouge photographs for advertising, some were presented as art objects. Sheeler himself used them in a photomontage exhibited at the Museum of Modern Art in 1932. He also produced a series of paintings, drawings, watercolors, and prints of the Rouge. The best-known paintings, *American Landscape* and *Classic Landscape*, were not studies of individual factory buildings but vistas of the complex. Both realistic and abstract in their concentration on form, line, and light, the near absence of people in Sheeler's depictions of an industrial plant which had tens of thousands of workers gives an eerie air to the paintings. The critic Leo Marx wrote of *American Landscape* that Sheeler "eliminated all evidence of the frenzied movement and clamor we associate with the industrial scene. . . . This 'American Landscape' is the industrial landscape pastoralized."

In depicting few people on the Rouge site, Sheeler was being literal. Other observers noted that, counterintuitively, very few people could be seen outside the factory buildings in many parts of the highly mechanized complex. But Sheeler also was making choices about what to depict. After World War II, he did a series of paintings of the by then-shuttered Amoskeag Mills. Hine's Amoskeag photographs portrayed young workers. Bourke-White's captured the symmetry and repetitive patterns of the machinery. Sheeler's Amoskeag paintings were again landscapes, with no person in sight.[53]

Art historian Terry Smith criticized Bourke-White and Sheeler for "banishing productive labor, excluding the human, implying an autonomy to the mechanical, then seeking a beauty of repetition, simplicity, regularity of rhythm, clarity of surface. This is the gaze of management at leisure, marveling at the new beauties which its organizational inventiveness can create." Smith has a point. After all, Bourke-White's first clients were business leaders who wanted beautiful images of the buildings and facilities they controlled, before she moved on to a broader audience of business readers at *Fortune*. Edsel Ford bought *Classic Landscape*. Abby Aldrich Rockefeller, wife of John D. Rockefeller, Jr., bought *American Landscape*.[54]

But to leave it there is to miss the greatness of this art. Bourke-White's subject was not the control of industry by capital; it was the grandeur of the structures of industry and the processes of production. Her photographs celebrate the power and creativity of humanity as manifested in industrial forms and the transformation of intractable materials. In her early work, the creations of workers effaced the workers themselves or at least diminished them. But over time, her interest in workers and the impact of industry on them grew. For *Fortune*, she photographed not only factories but skilled artisans, laborers, and industrial workers. At the Rouge, she had groups of workers informally pose for her. Her cover story for the first issue of *Life* documented not only the Fort Peck Dam but also the boomtown that grew up for the workers building it. One of her most striking images is of workers relaxing at a local bar. Her 1938 *Life* photographs of a Plymouth factory documented men at work.[55]

In his Rouge photographs, Sheeler was even more concerned with form and geometry than Bourke-White, creating stunning formal compositions (some of which *did* include workers). He, too, had a central concern with power, as *Fortune* recognized when it commissioned him to create six paintings on the theme for its December 1940 issue. But if Sheeler's industrial photographs have a cool, triumphal feel, his industrial paintings, with their near absence of humanity, have a melancholy air, reminiscent of Edward Hopper in their light, treatment of shadow, and emotional tenor. These are far deeper and more ambiguous images than simple celebrations of possession.[56]

During the 1920s and 1930s, other painters besides Sheeler found a rich subject in large-scale industry, many lumped together under the label of precisionism, including Elsie Driggs (who did a painting of the Rouge in 1928), Charles Demuth, and Louis Lozowick. Lozowick, a self-conscious leftist who had extensive contact with the European and Soviet avant-garde, defended the portrayal of industrial machinery "more as a prognostication than as a fact" of the time when "rationalization and economy" would be "allies of the working class in the building of socialism." Other painters, like Stuart Davies and Gerald Murphy, adopted what has been dubbed a "machine aesthetic," though they never made industrial structures themselves their subject. But the artist who best captured the world of heavy industry, and the Rouge in particular, was not a precisionist but rather a Mexican muralist, Diego Rivera.[57]

Diego Rivera and *Detroit Industry*

Automaking turned Detroit into a boomtown. As workers poured in to take factory jobs, the population more than tripled, from 466,000 in 1910 to 1,720,000 in 1930, and the city sprawled. The newly enriched industrial captains built their mansions in lakeside suburbs and took it upon themselves to endow the city with the civic and cultural institutions that mark centers of power. Among them was the Detroit Institute of Arts, owned by the city but overseen by a small board, which was headed by Edsel Ford and included Albert Kahn and Charles T. Fisher of Fisher Body.[58] In 1930, the ambitious museum director, William Valentiner, commissioned Diego Rivera to paint two murals in the courtyard of its new building. The artist, already well known in international art circles, at the time was working on his first murals in the United States. Valentiner convinced Edsel Ford, whom he tutored in art history, to finance the project.

By the time Rivera and his wife, Frida Kahlo, arrived in Detroit in April 1932, it was a very different place than when Sheeler had taken his photographs five years earlier. The Depression had hit the city hard, with mass unemployment in the auto industry and severe deprivation in the working-class neighborhoods. Radical movements had swelled, demanding jobs, relief, and unionization. On March 7, 1932, Ford guards and Dearborn police opened fire on a march of unemployed workers and their supporters, killing four and wounding many others. A funeral procession for the slain attracted sixty thousand marchers.

Though a self-identified Marxist and sometimes communist, Rivera (and Kahlo, too) seemed oblivious to the ferocious class conflict. Instead, he was entranced by Henry Ford and the industrial empire he had built. "My childhood passion for mechanical toys," he later wrote, "had been transformed to a delight in machinery for its own meaning for man—his self-fulfillment and liberation from drudgery and poverty." Rivera admired the photographs of industrial equipment that Kahlo's father, a prominent Mexican photographer, had taken. The

Figure 4.4 Left to right: Albert Kahn, Frida Kahlo, and Diego Rivera at the Detroit Institute of Arts on December 10, 1932.

artist toured a variety of Detroit-area factories, but like for so many others it was the Rouge that captured his imagination and became the centerpiece of his work. Rivera grew so enthusiastic that Valentiner and Edsel Ford agreed to enlarge the commission to cover all four walls of the museum courtyard (at double the original fee), with twenty-seven panels providing space for a huge pictorial program, which, in accordance with Edsel's wish, included not only the Rouge but also scenes from other locally important industries.[59]

Rivera completed the murals in mid-March 1933, the very low point of the Great Depression. While he and his assistants had worked on them from heavy scaffolding, groups of visitors had watched, much like the tourists at River Rouge, whom Rivera incorporated into one of his panels. Even before they were unveiled, the murals were subject to attacks of all kinds. But they proved immensely popular—thousands came the first week to see them—and they have remained one of Detroit's premier attractions ever since.[60]

Detroit Industry is one of the triumphs of twentieth-century art, the most fully realized visual representation we have of the factory system. The two largest panels depict with remarkable visual compression the complex process of automobile manufacturing at the Rouge. The north wall panel shows the production of transmission housings and

Figure 4.5 A detail from the north wall of *Detroit Industry*, a series of frescoes completed by Diego Rivera in 1933.

V8 engines (just recently introduced by Ford), from the blast furnace through casting, drilling, and assembly. The south wall portrays the stamping and finishing of steel car bodies and the final assembly line. Visually dense, with conveyors, pipes, cranes, and balconies serpenting through the panels, Rivera's Rouge, unlike Bourke-White's or Sheeler's, teems with people: workers toiling, supervisors and tourists watching, and Henry and Edsel Ford, Valentiner, Rivera himself, and—thrown in for good measure—Dick Tracy all standing by.[61]

As remarkable as the Rouge panels are, they are only part of a larger array, epic in its conceptual and visual sweep. Other panels depict the miracle of modern medicine, the constructive and destructive sides of the aviation and chemical industries, huge figures representing each of the races, fruits and vegetables illustrating the bounty of the earth, and even the earth itself, with its stratifications and fossils and a fetus within it. While most of the Rouge workers have the faces and bodies of European Americans or African Americans, other figures, including two remarkable giant portraits of nude women representing the bounty of agriculture (in the upper corners of the east wall) are indigenous Mexicans in face and body, a fusion of two countries and two cultures in Rivera's vision of modernity.

Human labor and machines co-dominate the Rivera mural. The toll that Fordism took on workers is evident in a predella panel of their tired bodies trudging across an overpass on their way home. But in its totality, the mural celebrates the strength of man and machine, the power seized from nature by mankind and harnessed in the giant factory.

Only in one tiny detail does an explicit critique of Ford appear, a hat worn by one worker that reads "We Want," no doubt a reference to the union movement then gaining power in Detroit and ferociously resisted by the company. Rivera, though, could not contain his disdain for capital (though not for the Fords, father and son, whose company he seemed to genuinely enjoy). As soon as he finished *Detroit Industry*, he headed to New York to create a mural in the newly completed Rockefeller Center. His refusal to remove portraits of Lenin and of John D. Rockefeller, Jr., with a drink in hand and women nearby led the Rockefellers to destroy the work.

Rivera also had been commissioned to create a mural entitled *Forge and Foundry* for the Kahn-designed General Motors exhibit at the upcoming Century of Progress International Exposition in Chicago. The architect, who initially had not been enthusiastic about commissioning the Rivera murals at the Institute of Art, had come to strongly defend them. But after the Rockefeller Center controversy, General Motors ordered him to fire Rivera. Kahn promised the artist to "do my best to get permission for you to proceed," but the auto company did not relent. Rivera told the press, "This is a blow to me. I wanted to paint men and machinery." Returning to Mexico, he hardly ever did again. Fordism and the giant factory lost their greatest chronicler.[62]

Ironically, and tellingly, today the most widely seen image of the Rouge in high culture is probably neither the Rivera murals nor Sheeler's work but a painting by Frida Kahlo. When she came with Rivera to

Detroit, Kahlo was almost completely unknown as an artist, but while in the city she produced a number of works that eventually came to overshadow Rivera's mural in the global art world, just as her overall reputation came to overshadow his. In her best-known work of the period, the extraordinary painting *Henry Ford Hospital,* the Rouge appears as visual and topical background to the central image of a bleeding Kahlo lying in bed after the miscarriage she had in Detroit (probably induced as an abortion). Among other things, her painting is a premonition of the shift of cultural interest in North America and Europe away from industry toward intensely personal, inward concerns.[63]

The Tramp in the Factory

In terms of sheer popularity, the premier visual representation of Fordism and the giant factory was not a painting or photograph at all, but Charlie Chaplin's film *Modern Times*, released in 1936. Mass production had long fascinated the filmmaker, by then one of the country's best-known celebrities. In 1923 he had visited Detroit, touring the Highland Park powerhouse and assembly line with Henry and Edsel Ford as his guides. Years later, trying to come up with a way to cinematically deal with the misery caused by the Great Depression and more broadly with the machine age, the Ford factory provided inspiration. In the last major silent film to be made in Hollywood, Chaplin utilized what already was an archaic technology to critique mass production, mass consumption, and the capitalist crisis. (The film has a sound track, but the only voices heard come from mechanical devices until, near the very end, we finally hear Chaplin's voice, singing a nonsense song with no intelligible words.)

From the very first frame—a picture of a clock face—Chaplin presents the demands of industrial discipline. In a long early sequence, his character, the Tramp (his long-standing film persona, though in this film identified as "A Factory Worker"), works on an assembly line tightening bolts for a never-seen product. Funny and horrifying, the workers struggle to keep up with the line while the Tramp mischievously tries to subvert the system. The company president, from his office (where he is doing a jigsaw puzzle), can see everything in the factory, including the bathroom, through a television system (in real life then still in an experimental stage), which he uses to issue commands to speed up the line. The dehumanization of the worker in the service of productivity reaches its climax when the Tramp is used as a guinea pig for a machine designed to feed workers while they continue to work. It malfunctions, forcing bolts into the Tramp's mouth and assaulting him with food and a mechanical mouth wiper. Soon, the endless repetitive motion of the assembly line has the Tramp uncontrollably twitching and eventually going mad, a comedic representation of the "Forditis" workers suffered when Ford introduced the assembly line.

As the film proceeds, it broadens out to encompass the ills of the whole society—mass unemployment, inequality, hunger, labor unrest, and heartless government authorities. The Tramp returns for a second stint in the factory, this time as a mechanic's helper, to find himself literally dragged into the bowels of the machinery. Chaplin is not oblivious to the rewards of Fordism; at one point the Tramp, out of a job again because of a strike, and his companion, the beautiful Gamin played by Paulette Goddard, fantasizes life in a well-furnished worker's bungalow, with modern appliances and a cow that furnishes milk on demand. But in the end, there is no satisfactory place for the Tramp and the Gamin in *Modern Times*, in the world of the giant factory. The film concludes with the couple walking down a rural road toward sunset and an unknown future, with a touch of hope provided by the final title, "Buck up—never say die. We'll get along."

Chaplin's film is a critique of Depression-era capitalism, but it is also a critique of the fundamental characteristics of the mass-production factory. For Chaplin, the only solution to the soul-deadening drudgery and monotony of the giant factory is literally to walk away. In this regard, *Modern Times* is different and far more radical than the work of other left-wing chroniclers of the giant factory, including Rivera, who saw it as advancing humanity, even if, as Louis Lozowick had written, it might only be in the future that "rationalization and economy" would be "allies of the working class in the building of socialism." Left-wing labor leader Louis Goldblatt told Chaplin his film was "Luddite." Machines, Goldblatt asserted, were necessary for improving living standards of the working class.

At least publicly, though, the left largely applauded *Modern Times.* Chaplin had become friendly with Boris Shumyatsky, the head of the film industry in the Soviet Union, during his visit to the United States, and Shumyatsky's public praise for the film made it hard for those in the communist orbit to do otherwise. (A *Daily Worker* review did say that in *Modern Times* "machinery turns out to be a gadget for comic use, like a trick cigar.") Much of the mainstream press hailed the film as a triumphant comeback for Chaplin, who had not made a movie for five years.

As Edward Newhouse noted in *Partisan Review,* few critics, even as they praised him, acknowledged Chaplin's radical message. *Modern Times* became a favorite of the cineastes and leftists for decades. It was shown in cinemas in the Soviet Union and, after the Cuban Revolution, when mobile projection crews brought motion pictures to remote villages where they had never been seen, the first film they showed was *Modern Times.* But communist leaders, like capitalists, had no desire to walk away from factory modernity, the way the Tramp did in Chaplin's masterpiece. To the contrary, at the very moment the film premiered, the Soviet Union was well into a crash industrialization program, building giant factories that used Ford methods, even as in the United States workers were finally finding a way to tame them. [64]

Unionizing Mass Production

"Jesus Christ, it's like the end of the world." So mouthed a tirebuilder at the huge Firestone tire factory in Akron, at 2 a.m. on January 29,

1936, when the workers began one of the first major sit-down strikes in American history. It was a chilling moment, as Ruth McKenney reconstructed it in her book *Industrial Valley*, when a tirebuilder pulled a handle to shut down the production line:

> With this signal, in perfect synchronization, with the rhythm they had learned in a great mass-production industry, the tirebuilders stepped back from their machines.
>
> Instantly, the noise stopped. The whole room lay in perfect silence.... A moment ago there had been the weaving hands, the revolving wheels, the clanking belt, the moving hooks, the flashing tire tools. Now there was absolute stillness.

When the silence broke, the men began cheering. "We done it! We stopped the belt!" Then they sang "John Brown's Body." Out the windows they chorused "He is trampling out the vintage where the grapes of wrath are stored."[65]

It was like the end of the world, or at least the beginning of the end of the world of industrial autocracy that had been part and parcel of factory giantism. The great labor upheaval in the United States during the late 1930s and 1940s transformed the giant factory, the lives of industrial workers, their families and communities, and the nation itself. With unionization, an industrial system that had once brought so much misery now brought unprecedented working-class upward mobility, security, and well-being. The unionized giant factory helped create what many Americans look back at as a golden era of shared prosperity, when children did better than their parents and expected their children to do better than themselves.[66]

Workers had tried to unionize large-scale industry before the 1930s, but repeatedly they had been repulsed, unable to overcome the physical fortresses and financial resources of the giant manufacturing concerns. But by the mid-1930s conditions had changed. The Great Depression robbed big business and its allies of political legitimacy and popular support. Financially pressed, companies eliminated many of the welfare programs they had introduced in the early twentieth century. Wage cuts, speedup, and layoffs further angered workers. Various left-wing groups, though small, provided ideas and leaders to disaffected workers, by this time less divided by ethnicity and language as a result of the restrictions on immigration that came during and after World War I. And crucially, the New Deal and its state-level equivalents provided symbolic and practical support for workers trying to unionize. In 1935, a group of veteran unionists, seeking to capitalize on the new circumstances, founded the Committee for Industrial Organization (CIO), dedicated to organizing the mass-production industries across the board, bringing skilled and unskilled workers into the same organizations.[67]

The largest industrial facilities, like the U.S. Steel plant in Gary and the main plants of the Big Three automakers—General Motors, Ford, and Chrysler—initially remained impervious to significant union gains. Instead, industrial workers generally first made organizational advances in smaller or peripheral plants. In the automobile industry, unions progressed among skilled tool- and die-makers; in parts plants, like Electric Auto-Lite in Toledo, Ohio, struck in 1934; and at smaller firms outside of the industry's Michigan heartland, like White Motors in Cleveland and Studebaker in South Bend, Indiana. In the electrical-equipment industry, early labor success largely came at smaller companies, like Philco Radio in Philadelphia and Magnavox-Capehart in Fort Wayne, Indiana. At the number-two company, Westinghouse, unionists established a toehold at the East Springfield, Massachusetts, plant, but at the company's giant East Pittsburgh facility, scene of bitter battles in earlier years, management maintained firm control. General Electric, the industry giant, had a more liberal labor policy, allowing small unions to start up at its giant complexes in Schenectady, New York, and Lynn, Massachusetts, but they had little real power.

By 1936, with an economic recovery under way and the CIO providing support, industrial unions began making progress even in some factory goliaths. In Akron, where the nation's tire-making capacity was highly concentrated in a few large factories, a prolonged strike at Goodyear followed the Firestone sit-down. In the auto industry, the CIO-affiliated United Automobile Workers (UAW) began building a base in the General Motors empire.[68]

The UAW picked General Motors—which operated 110 factories and had more employees than any other manufacturing enterprise in the world—as its primary target in its effort to break into the Big Three. The contest between the infant union and what by some measures was the largest corporation anywhere seemed absurdly lopsided. But UAW organizers understood that a high degree of centralization and the tight integration of the company production processes left it vulnerable to a militant minority. In particular, only two sets of dies for making the bodies for the newest GM model existed, one in Cleveland and the other in Flint. Stopping those factories would shut down most of the company's domestic car-making.

Franklin Roosevelt's reelection in November 1936, in a campaign marked by sharp class rhetoric and massive labor support for the president, gave a boost to organizing efforts. UAW leaders hoped to launch a national strike against GM in early 1937, but outbreaks of worker militancy forced their hand sooner. In mid-November, workers at the GM plant in Atlanta began a sit-down strike. A month later, so did GM workers in Kansas City. Then, on December 28, workers in the GM plant in Cleveland sat down, too.

In Flint, the heart of the GM production system, after several years of effort the union still had signed up only a small minority of the forty thousand workers. But when on December 30 a union activist saw body dies being loaded to ship out, apparently to factories in areas with less union strength, the workers sat down in the small Fisher Body Plant No. 2 and the seven-thousand-worker Fisher Body No. 1, blocking the removal of the equipment. In the days that followed, workers at more GM plants in Indiana, Ohio, Michigan, and Wisconsin followed suit. With the production of car bodies and other key components halted, within a week the whole GM national operation began grinding to a halt, with roughly half the workforce idled. The efficiencies and strategic advantages of the giant factory had come back to haunt the company, as a minority of workers, by seizing key choke points, leveraged power far beyond what one might expect from their modest numbers (which the sit-down tactic help disguise).

During the forty-four days strikers stayed inside the Flint plants, the giant factory turned from a site of managerial control to an arena of worker self-expression. The strikers organized themselves into committees in charge of overall leadership, security (including making sure no machinery was damaged), sanitation, and food. Makeshift sleeping quarters were built in car bodies and on factory floors, using car cushion stuffing to provide a touch of comfort. Cards, games, radio, Ping-Pong, and classes on labor history and parliamentary procedure helped ease the boredom and fear. So did dancers, theater troupes, and other sympathetic outsiders who entered the plants to provide entertainment.

The GM strike captured national attention, closely reported by newspapers, radio, and newsreels. The tense confrontation included an effort by company guards and Flint police to evict the occupiers of Fisher No. 2, repulsed by workers heaving heavy door hinges out second-story windows and training high-pressure water hoses on the police (who during their retreat opened fire on union backers); the mobilization of strikers' wives and other family members to physically defend the occupied plants and provide the sit-downers with food and supplies; the seizure of an additional Flint plant, the gigantic Chevy No. 4 factory, which made every engine used in a Chevrolet; the mobilization of the Michigan National Guard, which surrounded the occupied factories; and, ultimately, negotiations involving GM officials, CIO President John L. Lewis, Michigan governor Frank Murphy, and federal officials, all the way up to President Roosevelt. The agreement that ended the strike, in itself, constituted but a modest union gain, a written company pledge that for six months it would recognize the UAW as the representative of its members in the struck plants. But as huge crowds cheered the haggard, bearded, smiling men who marched out of the occupied Flint plants, everyone knew that the world had changed; workers had shown that they could bring one of the most powerful corporations in the world to its knees by shutting down the giant factories in which they labored.[69]

The UAW victory set off a wave of strikes and union organization everywhere from giant factories to local retail stores. Nearly five million workers took part in walkouts during 1937, including four hundred thousand sit-downers. For its part, General Motors gave its workers a 5 percent pay hike and agreed with the UAW to a shop-steward system and the use of seniority in layoffs. Meanwhile, the auto union won agreements with smaller car companies, with parts makers, and, after a month-long sit-down in Dodge Main and six other factories, with Chrysler. In the electrical-equipment industry, the United Electrical Workers signed a contract with RCA covering the nearly ten thousand workers (three-quarters female) at its Camden, New Jersey, plant while General Electric agreed to a national contract that covered most of its largest plants, including its sprawling complex in Schenectady.[70]

The most remarkable breakthrough came in the steel industry, what Lewis called "the Hindenburg line of [American] industry." Less than a week after the end of the General Motors strike, Lewis signed an agreement with Myron Taylor, the chairman of U.S. Steel, which granted workers a wage increase, the forty-hour week, time-and-a-half for overtime, and a grievance procedure. The CIO had created the Steel Workers Organizing Committee (SWOC) to try to unionize the industry, but the going had been slow. Nonetheless, Taylor apparently decided that, given the union victory over GM and the pro-labor sentiment in Washington and in the statehouses of key steelmaking states, unionization was inevitable. Rather than allowing a prolonged battle that would mobilize the rank and file and perhaps interrupt production, Taylor cut a deal with Lewis, with no involvement of local activists or even SWOC leaders.[71]

As impressive as it was, the CIO offensive failed to sweep the field, as a number of key operators of very large industrial facilities successfully resisted unionization. The worst setback came in steel, as the so-called

"Little Steel" companies, giants except in comparison with U.S. Steel, refused to recognize SWOC. In response, their workers walked out in late May 1937, but the strike ended in defeat; as in the past, the companies mobilized local governments, police, and the press against the strikers. Eighteen workers died during the battle, including ten shot by police during a peaceful protest in front of Republic Steel's South Chicago mill. Just days earlier, when the UAW sent organizers to pass out leaflets outside the Rouge, they were set upon by Ford thugs and beaten mercilessly. Westinghouse, Goodyear, International Harvester, and, most importantly, Ford all dug in their heels and refused to sign contracts with the CIO, weakened as it was by the Little Steel defeat and a downward plunge of the economy that began in mid-1937. The victory of industrial unionism was not yet assured.[72]

But World War II allowed the American labor movement to complete the unionization of large-scale industry. Even before the United States entered the conflict, a defense buildup revived the economy, tightening labor markets and bolstering worker confidence. Also, the 1935 National Labor Relations Act, which gave workers the right to join unions without reprisal and established a mechanism for their legal recognition, finally began forcing employers to change their ways. By late 1941, through a combination of legal challenges, worker mobilization, strikes, and federally supervised recognition elections, SWOC succeeded in organizing Little Steel. Westinghouse, International Harvester, Goodyear, and other holdouts fell to the CIO as well.[73]

The largest and symbolically most important victory came at Ford. In the fall of 1940, the UAW relaunched its stalled effort to organize the company. By the end of the year, the union had won substantial backing at the Rouge and a Lincoln plant in Detroit, filing for recognition elections. On April 1, 1941, a strike broke out at the Rouge after the company fired members of a union grievance committee in the rolling mill. As the number of strikers swelled, union leaders called a full-scale walkout at all Ford plants. To keep scabs out of the Rouge, with its immense perimeter, the strikers supplemented traditional picketing with a motorized encirclement of the plant and even aerial surveillance. In a reversal of the past pattern, Ford "servicemen" working for Harry Bennett found themselves being beat up by unionists. After ten days, the company agreed to end the strike by reinstating the fired workers and holding union recognition elections. At the Rouge, seventy-four thousand workers cast ballots in one of the largest such elections ever held, with 70 percent supporting the UAW. The union won decisive victories at Highland Park, the Lincoln plant, and other Ford factories as well. Then, in a startling and somewhat inexplicable move, the company agreed to one of the most generous contracts that any CIO union had achieved, including a provision that required all new employees to join the union, a checkoff of union dues (which the company took out of workers' pay and gave to the union), disbanding Bennett's Service Department, strengthened seniority and grievance systems, the rehiring, with back pay, of workers fired for union activity, and even allowing smoking in designated areas at the Highland Park and Lincoln plants, repudiating Henry Ford's imposition of abstinence on his employees.[74]

The swelling of the union movement continued during the war itself. To check inflation, the federal government kept wage rates at prewar levels, but gave unions a boost by granting them "maintenance of membership," requiring all workers at unionized plants to join unless they took advantage of a brief opt-out window. Virtually every new hire at unionized firms automatically became a union member, a flood of duespayers as defense payrolls soared. Other new members came through organizing campaigns, which unions, aligning themselves with the war effort, often portrayed as patriotic endeavors. Union membership, which jumped from 3.6 million at the start of the Great Depression to 10.5 million in 1941, reached 14.8 million in 1945, with roughly one out of three nonagricultural workers carrying a union card. With only a few notable exceptions, the giant factory had been placed under the roof of the house of labor. Fordism had revolutionized the American economy and society; the uprising of industrial workers gave mass production a new, more democratic meaning.[75]

CHAPTER 5

"COMMUNISM IS SOVIET POWER PLUS THE ELECTRIFICATION OF THE WHOLE COUNTRY"

Crash Industrialization in the Soviet Union

IN DECEMBER 1929, PHILIP ADLER, A REPORTER FOR THE *Detroit News*, visited Stalingrad, on the Volga River in southwestern Russia (until 1925 called Tsaritsyn), where the government of the Soviet Union was erecting a huge new tractor factory on a muddy, treeless field that had been used for growing melons. The factory held special interest for Motor City readers because American companies and workers—many from Detroit—were heavily involved in its planning and operation. Albert Kahn served as the overall architect, the Frank D. Chase Company laid out the foundry, and R. Smith, Incorporated, designed the forge shop. McClintic-Marshall Products Company fabricated the beams and trusses. Most of the production equipment was made in the United States, and the Soviets hired several hundred Americans to work at the plant, in many cases as foremen or supervisors.

Before going to the factory site, a half hour out of town, Adler visited the city center, where in the market he found "the familiar figures of the tinker, the cobbler and the dealers in second hand clothing and furniture who employ the most primitive methods of manufacture and salesmanship. The ox team, the camel and the biblical ass rival the horse as mediums of transportation." From a minaret among the church cupolas

Figure 5.1 Margaret Bourke-White's iconic 1931 photograph, *Stalingrad Tractor Factory.*

came the cry "'Allah Ho Akbar!'—Allah is powerful!" But when Adler got to the construction site, the watchword everywhere was "'Amerikansky Temp' or 'American tempo'" and the slogan plastered about was "To catch up with and surpass America." The following summer, with the plant beginning to turn out its first tractors, Margaret Bourke-White arrived after an arduous journey, taking what would become one of her most iconic photographs, of three workers on a newly finished tractor coming off an assembly line.[1]

The *Tractorstroi* ("tractor factory") in Stalingrad was part of a feverish drive by the Soviet Union to rapidly industrialize, boosting its standard of living and increasing its defensive capacity on the road to creating a socialist society. Most Bolshevik leaders believed that a socialist or communist society could be achieved in Russia—a poor and economically backward country—only after significant industrial development. Seizing political power was not enough. "There can be no question of . . . communism," Vladimir Lenin declared in 1920, "unless Russia is put on a different and a higher technical basis than that has existed up to now. Communism is Soviet power plus the electrification of the whole country, since industry cannot be developed without electrification." And it was a particular type of industrialization Lenin and his comrades had in mind, "large-scale machine production."[2]

It took time before the Soviet Republic could launch a major industrialization effort, but by the late 1920s a detailed plan had been adopted. In 1929, on the twelfth anniversary of the October Revolution, Joseph Stalin wrote, "We are advancing full steam ahead along the path of industrialization—to socialism, leaving behind the age-old 'Russian' backwardness. We are becoming a country of metal, a country of automobiles, a country of tractors."

Industrial behemoths were key to the Soviet effort to leap from "The ox team, the camel and the biblical ass" to "a country of metal, a country of automobiles, a country of tractors." The First Five-Year Plan, begun in 1928, centered on a series of very large scale factory and infrastructure projects, including three huge tractor factories, a big automobile plant in Nizhny Novgorod, immense steel complexes at Magnitogorsk and Kuznetsk, the Dnieporstroi hydroelectric dam, the Turksib railway connecting Kazakhstan with western Siberia, and the Volga-Don canal. Lacking the technical expertise and industrial resources for creating and equipping projects of such size and sophistication, the Soviets turned heavily to the West, especially the United States, for engineers, construction and production experts, and machinery, adopting the techniques of scientific management and mass production and in some cases creating virtual replicas of facilities in the United States. As Stephan Kotkin wrote in his landmark history of Magnitogorsk, for the communists, "The dizzying upheaval that was Soviet industrialization was reduced to the proposition: build as many factories as possible, as quickly as possible, all exclusively under state control." In the Soviet Union, just as in the United States, the giant factory came to be equated with progress, civilization, and modernity.[3]

But the Soviet Union was a very different place than the United States. Would the factory itself be different there? Would it have a different social significance? In 1927, Egmont Arens, an editor of the left-wing journal *New Masses*, reviewing a play, *The Belt*, which demonized assembly line production, remarked, "*The Belt* is something that has got to be faced even by advocates of a workers' state. Right now Russia is installing modern industrial plants of her own. Are the horrible things that *The Belt* does to minds and bodies of workers inevitable? Or is there a difference between high pressure production in Socialist Russia and Henry Ford's Detroit?"[4]

The factory had developed largely as a means for industrialists and investors to make money for themselves. Though it was sometimes freighted with moral imperatives and claims of social good, its physical design, internal organization, technology, and labor relations were determined primarily by the desire to maximize profits.[5] What did it mean to have a factory in a society where profits, in the usual sense, did not exist, where all large-scale productive entities belonged to a government that, at least in theory, served as the agent of the people, especially the working class? Could and should the capitalist factory, as a technical, social, and cultural system, simply be moved into a socialist society? Were methods like scientific management and the assembly line, designed to increase efficiency and labor productivity in order to boost profits, appropriate for a society in which the needs of workers and the well-being of the entire population were declared paramount?

The Soviet Union differed from the United States not only in its ideology but also in its level of economic development. Before the 1917 revolution, Russia had been an overwhelmingly agricultural society. What industry it did have was severely disrupted by the revolution and the civil war that followed. Could large, technically advanced industrial facilities successfully operate in such an environment, short-cutting the long process of economic development that had occurred in Western Europe and the United States? Could a heroic effort to leap directly to large-scale industrialization stimulate broad economic growth, or would chaos ensue from the lack of needed material inputs, logistics, and worker and managerial skills?

Questions about the role of the giant factory in economic development and social structure remain alive today, both in the few remaining countries that call themselves communist—most importantly China and Vietnam—and in the capitalist world. With much of the world's population still living in poverty, the issue of how to raise living standards remains a central economic, political, and moral concern. What role should the giant factory play in the effort to achieve broad material and social well-being? What price should industrial workers pay for social abundance?

Some of the answers to these knotty questions began to emerge during the 1930s, from the muddy fields on the outskirts of Stalingrad and from other sites like it across the Soviet Union. The experience with the American-style giant factory proved crucial not only in shaping the history of the Soviet Union but also in defining a path for development for much of the world in the decades after World War II. Stalinist industrial giantism, for better and for worse, became one of the main paths for trying to achieve prosperity and modernity, a Promethean utopianism that mixed huge social ambitions with enormous human suffering.

"Marxism Plus Americanism"

In the twentieth century, American production techniques and managerial methods—what came to be called "Americanism"—commanded considerable interest in Europe. Some of it was technical, in high-speed machining and the high-strength metals it required, the standardization of products, the use of various kinds of conveyance devices, and the mass-production system that these developments made possible. But interest was at least as great in the ideology associated with advanced manufacturing, the promise that with productivity gains the income of workers could go up even as profits rose, thereby dissipating class conflict and social unrest.[6]

As avatars of scientific management and mass production, Frederick Winslow Taylor and Henry Ford became well-known and well-regarded figures in Europe. By the early twentieth century, Taylor's writings had been translated into French, German, and Russian. In the early 1920s, Ford displaced Taylor as the icon of Americanism, as worker criticism of Taylorist management grew and the wonders of the assembly line and the Model T became better known abroad. In Germany, Ford's autobiography, *My Life and Work*, translated in 1923, sold more than two hundred thousand copies.

Though Americanism as a technical and ideological system had considerable influence all across Europe, perhaps surprisingly its greatest impact occurred in the Soviet Union. The groundwork was laid before the revolution. What industry Russia had tended to be highly concentrated, with quite a few large factories, some foreign-owned and operated with the help of foreign experts who were aware of the latest trends in management thinking, including those associated with Americanism. In addition, at least a few Russian socialists, most importantly Lenin, knew about scientific management and thought about its implications.

In his first comments about scientific management, while in exile in 1913, Lenin echoed critiques common among American and European unionists and leftists, seeing its "purpose . . . to squeeze out of the worker" more labor in the same amount of time. "Advances in the sphere of technology and science in capitalist society are but advances in the extortion of sweat." Three years later, he plunged deeper into scientific management in preparation for writing *Imperialism: The Highest Stage of Capitalism*, reading a German translation of Taylor's book *Shop Management*, a book on the application of the Taylor system, and an article by Frank Gilbreth on how motion studies could increase national wealth. In the end, he never discussed management techniques in *Imperialism*, but his notes from the time indicate a view of scientific management in keeping with the general tenor of the book, in which capitalist advances, whatever their motives, were portrayed as laying the basis for a socialist transformation, in line with Marx's portrayal of capitalism as an antechamber to a socialist economy.[7]

The 1917 revolution radically changed the context for Russian thinking about scientific management. Instead of critiquing existing social arrangements and defending workers, Russian communists and their allies now found themselves facing the almost overwhelming challenge of restoring the economy of a country depleted and disrupted by war and revolution to the point of famine, even as they fought a civil war and tried to consolidate their power. For Lenin, scientific management became a necessary tool to increase productivity and overcome economic backwardness, a prelude to establishing a socialist society:

> The Russian is a bad worker compared with workers of the advanced countries. Nor could it be otherwise under the tsarist regime and in view of the tenacity of the remnants of serfdom. The task that the Soviet government must set the people in all its scope is—learn to work. The Taylor system, the last word of capitalism in this respect, like all capitalist progress, is a combination of the subtle brutality of bourgeois exploitation and a number of its greatest scientific achievements in the field of analyzing mechanical motions during work, the elimination of superfluous and awkward motions, the working out of correct methods of work, the introduction of the best system of accounting and control, etc. The Soviet Republic must at all costs adopt all that is valuable in the achievements of science and technology in this field. . . . We must organize in Russia the study and teaching of the Taylor system and systematically try it out and adapt it to our purposes.

Lenin even suggested bringing in American engineers to implement the Taylor system.[8]

Lenin's backing helped legitimize scientific management as a practice and ideology in the new Soviet Republic. Exigency accelerated its adoption. One of its earliest adoptions came in railroad shops and armaments factories during the civil war, when keeping train engines operating and producing arms were literally matters of life and death for the revolution. As Commissar of War, Leon Trotsky embraced Taylorism as a "merciless" form of labor exploitation but also "a wise expenditure of human strength participating in production," the "side of Taylorism the socialist manager ought to make his own." Desperate to increase production, the Soviet government adopted piecework as a general practice and set up a Central Labor Institute to promote means of increasing labor productivity, including time and motion studies and other forms of scientific management.[9]

The embrace of Taylorism did not go unchallenged. As in the West, many workers and trade unionists opposed the imposition of more stringent work norms through piecework and so-called scientific methods, especially if workers themselves did not play a role in establishing and administrating them. And there were more sweeping ideological objections, too, centered on the relationship between building a new kind of society and using capitalist methods.

On the one side were trade unionists, "Left Communists," and, later, members of the "Workers Opposition" within the Communist Party, who believed that a socialist society required different social structures of production than had developed under capitalism, with greater worker participation and authority on the shop floor, in managing enterprises, and in determining methods of production. These critics of scientific management wanted to devise ways to increase productivity without further exploiting workers, opposing the extreme division of labor that transformed "the living person into an unreasoning and stupid instrument." To simply adopt methods workers had long criticized under capitalism would negate the meaning of the revolution.

On the other side were those who viewed capitalist production methods as simply techniques that could be used to any end, including the creation of wealth that would be the property of the whole society under a socialist regime. Alexei Gastev, a one-time worker-poet who became the secretary of the All-Russia Metal Workers' Union, the head of the Central Labor Institute, and the leading Soviet proponent of scientific management, wrote in 1919, "Whether we live in the age of super-imperialism or of world socialism, the structure of the new industry will, in essence, be one and the same." Like other Soviet supporters of scientific management, Gastev saw in Russian culture, especially among peasants and former peasants who had entered industry, an inability to work hard at a steady pace, instead alternating spurts of intense labor with periods of little if any work (the same complaint early English and American factory owners had about their workers). American methods and an American sense of speed would provide a cure. Trotsky gave intellectual and political weight to the case for adopting capitalist methods, advocating the use of the most advanced production techniques, regardless of their origin. Labor compulsion, necessary during the transition to socialism, he contended, had different significance when used in the service of a workers' state than for a capitalist enterprise (an argument that made little headway with many Soviet trade unionists).[10]

The dispute over scientific management was largely resolved at the Second All-Union Conference on Scientific Management, held in March 1924. The participation by top communist leaders in the extensive public debate that preceded it was a measure of the importance of the question of the use of capitalist management methods in the Soviet Union. By and large, the conference came out in support of Gastev and the wide application of scientific management, reflecting the demographic and economic circumstances of the period. The prerevolutionary and revolutionary-era skilled working class, the natural center for opposition to Taylorism, had been all but decimated by war, revolution, and civil war, with many of its survivors co-opted into leadership positions in the government and the party. The main challenge in trying to raise Soviet productivity was not squeezing more labor out of experienced, skilled workers but getting useful labor out of new workers with little or no industrial experience, for which scientific management, with its stress on the simplification of tasks and detailed instructions to workers, seemed well suited.[11]

It is not clear how much actual impact the endorsement of scientific management had on Soviet industry, at least in the short run. The Soviet Union lacked the experts, equipment, and experience to implement the methods advocated by Taylor and his disciples. Gastev's institute, the center of scientific management, did not have even basic equipment, conducting simplistic experiments of little practical significance. Much of its work consisted of exhorting workers: "Sharp eye, keen ear, alertness, exact reports!" Gastev urged. "Mighty stroke! Calculated pressure, measured rest!" Many Soviet managers adopted piecework pay, but unless accompanied by detailed studies and reorganization that did nothing to increase efficiency, instead simply inducing workers to work harder using existing methods. Some scientific management techniques did become common, like the use of Gantt charts for production planning, as over time Soviet management journals and training institutes spread the Taylorist gospel. But the immediate importance of the endorsement of scientific management lay not in the field but in opening the door to a broader embrace of Western methods and technologies, which would soon lead to a crash program to re-create the American-style giant factory.[12]

An early experiment came in the textile industry, in cooperation with an American labor union. In 1921, Sidney Hillman, the president of the Amalgamated Clothing Workers (ACW), after meeting with top Bolshevik leaders and Soviet trade unionists, signed an agreement to set up the Russian-American Industrial Corporation (RAIC), a joint enterprise with the Russian Clothing Workers syndicate, which ended up controlling twenty-five garment and textile factories that employed fifteen thousand workers. The deal came just as the Soviet Union was abandoning "War Communism," the direct state control and partial militarization of the economy during the civil war, turning to a partial restoration of private ownership and market relations under the "New Economic Policy (NEP)."

The ACW proved a perfect partner for what in effect was a state-sponsored cooperative enterprise, meant to deploy the most advanced American equipment and management techniques in the restoration of the Russian garment industry. Many members and leaders of the heavily Jewish ACW, including Hillman, had emigrated from the Russian Empire, infected with the same radicalism that culminated in the revolution. Under Hillman's leadership, though, the ACW had become increasingly practical in its policies, seeing in scientific management a way to improve productivity in a fragmented, often technologically primitive industry, creating the basis for upgrading worker living standards. The trade-off the ACW insisted on was union involvement in setting production norms and piecework rates and a system of neutral arbitration to resolve grievances. But the union affinity with scientific management was not strictly pragmatic; as Hillman's biographer Steve Fraser wrote, "the ACW elite was firmly implanted in those socialist traditions that affixed the tempo and timing of socialism to the inexorable rhythms of industrial and social developments under capitalism."

Through RAIC, the ACW brought to the Soviet garment industry not only Western capital but more importantly advanced equipment and expertise, including leading proponents of scientific management, factory managers the union had dealings with in the United States, and skilled workers familiar with joint union-management efforts at Taylorization. In short order, RAIC could boast of factories that matched the most advanced plants in the United States in their equipment, productivity, and progressive labor relations.[13]

Flirting with Ford

The NEP, which RAIC was part of, reanimated the Soviet economy. But it failed to fully restore Soviet industry to prerevolutionary levels of production, let alone fulfill the promise of the revolution to improve life for tens of millions of workers and peasants. In October 1925, Soviet industry still produced only 71 percent of pre–World War I Russian output. Fairly small investments under NEP were able to boost industrial output because there was considerable unused capacity. But by the mid-1920s, with utilization much higher, fewer possibilities remained for quick gains and possible reversals loomed; little capital investment for a decade meant that much of the industrial machinery in the country had reached or exceeded its expected service life. Further advances would require heavy investment in plant renovation, construction, and equipment.[14]

For most Soviet planners and political leaders, that meant staking the future of the revolution on large-scale industrial and infrastructure projects, though they disagreed sharply about the means and pace of investment. The Marxist tradition had long associated progress and modernity with the concentration of capital and mechanization. The prerevolutionary Russian industrial experience also influenced the Soviet sense of scale. In 1914, over half of Russian factory workers were employed in plants with more than five hundred workers, compared to less than a third in the United States. On the eve of the revolution, Petrograd had a cluster of very big government-controlled armament factories, some with well over ten thousand workers, as well as a few giant private plants, including the Putilov metalworking complex, with around thirty thousand workers (where a strike helped kick off the revolt against the Tsar).[15]

Many Soviets credited the success of the United States, which they saw as an exemplar, to its adoption of standardized products and large industrial complexes. As in Western Europe, Henry Ford was well known in the Soviet Union, seen as a living embodiment of the most advanced social, technical, and economic developments. By 1925, the Russian translation of *My Life and Work* had gone through four printings. But even more important in spreading Ford's fame was his tractor, the Fordson.

Before World War I, there were only about six hundred tractors spread across the vast domains of Russia. Believing the upgrading of agricultural productivity central to the revolution, starting in 1923 the Soviet Union began importing tractors in growing numbers, largely Fordsons. By 1926, 24,600 orders for Ford tractors had been placed. The Soviet Union also imported some Model Ts. A pipeline from River Rouge to the Russian steppes and cities had been opened.

In 1926, the Soviet government asked Ford to send a team to see how it could improve the maintenance of tractors, a large percentage of which, at any given time, were inoperable because of poor servicing, a lack of quality replacement parts, and inefficient labor. Also, the Soviets wanted to explore the possibility of Ford setting up a tractor factory in Russia. Already, they were trying, not very successfully, to produce knock-offs of the Fordson on their own. After spending four months touring the Soviet Union, a Ford delegation recommended against building a factory, fearful of political interference in operations and possible future expropriation. Undeterred, Soviet officials still hoped for Ford-style factories to make much-needed agricultural equipment and motor vehicles.[16]

By then, Ford methods did not evoke great controversy in the Soviet Union. The debate over Taylorism already had led to the endorsement of the use of capitalist methods. Also, Fordism less directly challenged the small but influential cadre of skilled metalworkers than scientific management, since, even with assembly lines, craftsmen would be needed to make tools and dies and maintain machinery. After touring the Soviet Union in 1926, William Z. Foster reported "revolutionary workers are . . . . taking as their model the American industries. In Russian factories and mills . . . . It is all America, and especially Ford, whose plants are generally considered as the very symbol of advanced industrial technique." "Fordizatsia"—Fordisation—became a favorite Soviet neolism.[17]

Still, there was some opposition to Fordism by left-wing critics who thought the adoption of methods designed to extract more labor from workers went against the fundamental socialist project of diminishing the exploitation and alienation of the working class. One of the sharpest ripostes to them came from Trotsky, a leading advocate of the adoption of Ford methods, just as he had been a leading advocate of the adoption of scientific management. In a 1926 article, he bluntly declared, "The Soviet system shod with American technology will be socialism.... American technology . . . will transform our order, liberating it from the heritage of backwardness, primitiveness, and barbarism."

Trotsky thought the assembly line, or conveyor method as he called it, would supplant piecework as a capitalist means of regulating labor, replacing an individualized mode with a collective one. Socialists needed to adopt the conveyor, too, he argued, but under their control it would be different, since the pace and hours of work would be set by a workers' regime. Nonetheless, he acknowledged that by its very nature the assembly line degraded human labor. In perhaps the most powerful argument ever made in defense of the Fordist factory, at least from a point of view other than that of those who profited from it, Trotsky answered a question he had been asked, "What about the monotony of labor, depersonalized and despiritualized by the conveyor?" "The fundamental, main, and most important task," he replied, "is to abolish poverty. It is necessary that human labor shall produce the maximum possible quantity of goods. . . . A high productivity of labor cannot be achieved without mechanization and automation, the finished expression of which is the conveyor." Just like Edward Filene, Trotsky claimed "The monotony of labor is compensated for by its reduced duration and its increased easiness. There will always be branches of industry in society that demand personal creativity, and those who find their calling in production will make their way to them." Then came a final flourish: "A voyage in a boat propelled by oars demands great personal creativity. A voyage in a steamboat is more 'monotonous' but more comfortable and more certain. Moreover, you can't cross the ocean in a rowboat. And we have to cross an ocean of human need."[18]

Embracing the Giant Factory

Just how to cross that ocean of need became the subject of an intense debate among Soviet leaders in the mid-1920s. The Bolshevik assumption always had been that the survival of their revolution would depend on the spread of socialism to advanced countries in Western Europe, which would then help Russia develop. But by a half-dozen years after World War I, it was clear that in the near future there would be no triumphant revolutions elsewhere. For economic development, the Soviet Union would have to depend on its own very limited resources.

Some Soviet leaders, including Nikolai Bukharin, argued that under the circumstances the best road forward lay in modest, balanced growth, driven by upgrading the agricultural sector. Increased peasant income would expand the market for consumer goods, which could be met through investments in light industry. Heavy industry would have to grow slowly.

Others wanted heavy industry to take the lead, with a faster pace of industrialization and economic growth. In part, they were driven by fears that the Western powers would again use their military forces to try to overthrow the Soviet regime, as they had during the civil war, necessitating the rapid development of an industrial base that could support a powerful army. They also feared placing the fate of the economy in the hands of a peasantry that wavered in its allegiance to the Soviet regime, withholding grain and other goods when prices were low or when there were too few consumer goods available to spend their money on. Instead, advocates of rapid industrialization, including Trotsky, sought to extract more wealth from the peasantry, if need be through levies, selling grain and raw materials abroad to finance industrialization.

A Communist Party congress in late 1927 balanced the two positions. But during the next two years, as a detailed Five-Year Plan for the economy was worked out, policy shifted toward the "super-industrializers" and then went far past even their most ambitious goals. The final plan called for a pace of industrialization unprecedented in human history, in half a decade doubling the fixed capital of the country and increasing iron production fourfold.

The swing coincided with Joseph Stalin's victory over his rivals in the battle for the leadership of the Communist Party that followed Lenin's death in January 1924. Having outmaneuvered his most formidable opponent, Trotsky, Stalin appropriated his program of rapid industrialization and vastly accelerated it. Stalin feared that boosting the wealth of the peasantry would increase its political power. To free the party and state once and forever from being held hostage, he sought to diminish the economic resources of the peasantry and ultimately transform it by collectivizing farm production. Wealth squeezed out of agriculture would finance the growth of heavy industry and, with it, an enlarged working class.

The call for very rapid industrialization was central to what historians have dubbed Stalin's "revolution from above." Its success was predicated on a revival of the heroic spirit and mass mobilization of the revolution and the civil war. Stalin's "vision of modernity" embodied in the First Five-Year Plan, wrote historian Orlando Figes, "gave a fresh energy to the utopian hopes of the Bolsheviks. It mobilized a whole new generation of enthusiasts," including young workers and party activists, for whom the industrialization drive was to be their October. By sheer willpower, the Soviet Union would seize modernity and catch up to and pass its capitalist rivals.[19]

The giant factory played a pivotal role in the effort. One Soviet planner said that preparing a list of needed new factories was "the soul of five-year plans." While some funds were invested in renovating and expanding existing plants, building new ones provided the opportunity for installing the most advanced technology available. Some experts proposed adopting European methods and machine designs, their smaller scale and lesser demand for precision standardized parts as more appropriate to the existing state of Soviet industry than American mass production. But Soviet leaders decided to adopt the American

model, seeing investment in a few, very large factories, where great economies of scale could be achieved through rationalization, specialization,
and mechanization, as a better use of precious investment funds than
spreading them thinly to build more, smaller, less technically advanced
plants. When one critic of that approach questioned the availability of
trained labor to operate American machinery, asking "Maybe you want
to breed a new race of people," Vassily Ivanov, the first manager of the
Stalingrad tractor plant, replied "Yes! That is our program!"

The First Five-Year Plan incorporated a few big projects already begun
or planned, like the Dnieporstroi dam and hydroelectric project, and
proposed massive new ones, like the Magnitogorsk iron and steel complex and the several tractor and automobile factories. These landmark
projects would create the transportation and power infrastructure,
iron and steel industry, and tractor and vehicle output to transform
the whole country and lay the basis for new defense production.

The epic scale of the First Five-Year-Plan projects reflected the utopianism associated with it and the need to stir popular imagination for the
massive mobilization and sacrifice it required. Giantism was as much
an ideological matter as a technical one. The very scale of the planned
industrial complexes made achieving modernity, measured against the
most advanced nations, and doing so quickly, seem a palpable possibility, something worth suffering to achieve. Tempo was deemed an
existential issue. "We are 50–100 years behind the advanced countries,"
Stalin declared in 1931. "We must make up this distance in ten years.
Either we do this, or they will crush us."[20]

Turning to the West

The Soviet Union lacked the engineering cadre, experience, and capital goods manufacturing capacity to build the Five-Year Plan projects
on its own. Necessity forced it to turn to personnel and machinery
from the capitalist world. Some foreign experts already were working

in the Soviet Union, but their role greatly expanded once the First Five-Year Plan got under way. Not only did the Soviet Union have too few engineers, industrial architects, and other specialists experienced with large-scale projects; equally important, the Bolsheviks distrusted the experts they had, most of whom had begun their careers working for private firms, had not supported the revolution, and were seen as lacking knowledge of the newest industrial developments and the boldness and initiative found abroad, especially in the United States. The largest group of foreign experts the Soviets recruited came from Germany, with Britain and Switzerland also providing significant numbers of engineers and technicians. But in terms of their role, American companies and consultants were most important, taking on outsized roles in the leading Five-Year Plan projects. Though unsympathetic to the Russian Revolution, American businesses did not hesitate to take advantage of the commercial possibilities it presented.[21]

The influx began with work on Dnieporstroi, the huge dam and hydroelectric project in the Ukraine, the largest in Europe when it opened in 1932. In 1926, a Soviet delegation visiting the United States signed a contract with Hugh L. Cooper, who had supervised the construction of the dam and power station in Muscle Shoals, Tennessee, to play a similar role for Dnieporstroi. For one or two months a year, Cooper worked on site, while a small group of engineers from his firm stayed year-round. The Soviets purchased much of the heavy equipment for the project in the United States. The Newport News Shipbuilding and Drydock Company built nine turbines for the dam, the largest ever manufactured, and sent engineers to supervise their installation. General Electric built some of the generators, part of its very extensive involvement in Soviet electrification and industrialization during the late 1920s and 1930s.[22]

American involvement in the Stalingrad tractor plant was even more extensive. The tractor held almost mythical importance in the Soviet Union; Russian-American writer Maurice Hindus, who traveled frequently in his native land, declared the tractor the "arbiter of the peasant's destiny," "not a mechanical monster, but a heroic conqueror." Tractors almost never were sold to individual peasants but rather used as inducements and support for collective cultivation. The tractor station, which housed equipment for use on nearby collective farms, became a key Soviet institution, not only supplying mechanical power but also collecting grain for the state and serving as a symbol of modernity and Bolshevik power. Rebellious villages did not get access to tractors.[23]

Already spending heavily to import tractors, the Soviet government made their domestic production an investment priority. Having been spurned in its request to Henry Ford to set up a Russian tractor plant, the government turned to the next best thing, Ford's favorite architect, Albert Kahn. Soviet leaders knew of Kahn because of his work at River Rouge. But in planning what would become the Stalingrad plant, they did due diligence, in November 1928 sending a delegation of engineers to the United States to study tractor production and visit equipment manufacturers and engineering and architectural firms, including Kahn's. In early May 1929, Amtorg, a trading company controlled by the Soviet government, signed a contract with the Detroit architect to design a factory capable of producing forty thousand tractors a year (a target later raised to fifty thousand). Kahn also agreed to lay out the site, supervise the construction, help procure building materials and equipment from U.S. companies, and supply key personnel for the start-up of the plant.[24]

Upon signing the Amtorg contract, Kahn presented the problems the Soviet Union faced as technical, with it having many of the same challenges and opportunities as the United States. As would be true in most of his statements about the U.S.S.R., he never mentioned communism and avoided politics. Perhaps to forestall criticism from anticommunist businesses, he portrayed the Soviet Union as a large potential market for U.S. equipment manufacturers.[25]

In choosing the Kahn firm, Soviet leaders threw in their lot with a company capable of operating at the rapid pace at which they hoped to carry out industrialization. Within two months of signing the contract, two Kahn engineers arrived in the Soviet Union with preliminary drawings for the main buildings. John K. Calder had worked on building Gary, Indiana, and been the chief construction engineer at River Rouge, a role he essentially reprised at the Stalingrad *Tractorstroi*, working alongside Vassily Ivanov. Leon A. Swajian, another Rouge veteran, assisted him. Other Kahn representatives and engineering recruits soon joined them.

But if leading Bolsheviks and the Kahn firm were largely in tune about pace—if anything the Russians wanted to go faster—as Calder quickly discovered conditions on the ground were anything but conducive to rapid progress. Modern equipment for transportation and construction was all but absent—camels were used to move materials—while many Soviet construction officials objected to the fast-track methods Calder introduced. Ivanov later wrote that he had to confront "the sluggish inertia of Russian building methods" in what became a political as well as technical battle over the all-important issue of "tempo." A popular play by Nikolai Pogodi, entitled *Tempo*, would portray the struggle, with a character based on Calder overcoming many obstacles, including bureaucracy and lack of discipline, to push the project forward.

Remarkably, the basic construction at Tractorstroi, which became the largest factory in the Soviet Union, with an assembly building a quarter mile long and large adjacent foundry and forge buildings, was completed in just six months, though it took another half year for all the equipment to arrive and be installed. Meanwhile, factory officials set up a recruiting office in Detroit and hired some three hundred and fifty American engineers, mechanics, and skilled workers to help start up the plant, including fifty from the Rouge, a process made easier by the beginning of the Great Depression. At the same time, young Soviet engineers were sent to collaborate with the Kahn firm on design work and to various American factories to gain experience with the kind of machinery that would be used in the plant. Ivanov himself traveled to meet with equipment suppliers in the United States, where "The straight

Figure 5.2
The Stalingrad Tractor Factory Is Open; the celebratory cover of a 1930 issue of a Soviet magazine.

roads, the abundance of machines, the whole technical equipment ... convinced me of the correctness of the course we had chosen."[26]

On June 17, 1930, just fourteen months after Amtorg signed its contract with Kahn, tens of thousands of spectators gathered in Stalingrad to watch the first tractor, decorated with red ribbons and placards, come off the assembly line. By then, a start-up workforce of 7,200 had been assembled, 35 percent female. Stalin sent his congratulations to the workers, declaring, "The fifty thousand tractors which you are to give the country every year are fifty thousand shells blowing up the old bourgeois world and paving the way to the new socialist order in the countryside." He ended, less bombastically, by giving "Thanks to our teachers in technique, the American specialists and technicians who have rendered help in the building of the Plant."[27]

While work on the Tractorstroi was proceeding, Amtorg went on a buying spree in the United States, signing technical assistance and equipment purchase agreements with some four dozen companies. The most important agreement was with Ford. When Kahn signed his contract, Henry Ford seemed to regret not being involved in the great experiment of Soviet industrialization. Publicly, he offered Kahn help and asked him to tell the Soviets "anything we have is theirs—our designs, our work methods, our steel specifications. The more industry we create no matter where it may be in the world, the more all the people of the world will benefit." Privately, he asked Kahn to signal to the Soviets that he was now willing to make a deal.

Nine months earlier, the Soviet government had set up a commission to build up its vehicle industry, which at the time consisted of only two small factories producing fewer than a thousand trucks a year. In the spring of 1929, the decision had been made to build a giant vehicle plant near Nizhny Novgorod, 250 miles east of Moscow. By then, the Soviets had approached both Ford and General Motors about assistance, but without much progress. Impatient, Stalin personally intervened behind the scenes, demanding that Amtorg speed up negotiations. Ford's new interest was thus a godsend, and by the end of May Amtorg signed an agreement with his firm.

The pact did not revive the idea of Ford setting up a plant in the Soviet Union. Instead, it called for massive assistance to the Soviets in building up an automobile industry under their own aegis. In a nine-year contract, Ford agreed to help design, equip, and run a plant at Nizhny Novgorod capable of manufacturing seventy thousand trucks and thirty thousand cars a year, as well as a smaller assembly plant in Moscow. Ford granted the Soviets the right to use all of its patents and inventions and produce and sell Ford vehicles in the country. It pledged to provide detailed information about the equipment and methods used at River Rouge and to train Soviet workers and engineers at its Detroit-area plants. The agreement also called for the Soviet Union, during the period its own plants were being started up, to buy seventy-two thousand Ford cars, trucks, and equivalent parts. (The vehicles were sent as knocked-down kits to be assembled at Soviet plants.) Although Ford later claimed that it lost money on the agreement, it served both sides well, giving the U.S.S.R. a huge boost in setting up a modern car and truck industry while providing Ford with work during the depth of the Depression and allowing it to sell off the tools and dies for the Model A as it switched to its new V8 model.[28]

To design the Moscow assembly plant and a temporary assembly plant in Nizhny Novgorod, the Soviets again turned to Kahn. But for the main Nizhny Novgorod factory, which was to be the largest automobile plant in Europe—conceived of as a scaled-down version of the Rouge, a fully integrated, mass production facility—and for a nearby city to accommodate thirty-five thousand workers and their families, the Soviets signed a contract with the Cleveland-based Austin Company, one of the leading industrial builders in the United States, which had recently erected a huge Pontiac factory for General Motors. If Kahn's firm was noted for its design innovations, Austin was best known for its one-stop approach, planning, building, and equipping complete industrial facilities using standardized designs and highly rationalized techniques. Though experienced with big projects, the Soviet commission was larger than anything it had ever undertaken.[29]

Like the Kahn engineers in Stalingrad, the first fifteen Austin engineers to arrive at Nizhny Novgorod—there would be forty at the peak—faced challenges quite unlike anything they had known. Living conditions were difficult and good food scarce. Chronic shortages of materials and labor delayed construction (though at the height of the effort forty thousand workers—40 percent female—were on the job). Water, heat, and power facilities and systems for transporting and storing equipment and supplies had to be built from scratch. The Soviets lacked the managerial experience or tools for a project of this scope. Expensive imported equipment was lost, misplaced, left outside to deteriorate, and stolen, while primitive machinery and brute force were used in its stead. Layer upon layer of bureaucracy, competition among organizations involved in the project, and constant personnel changes made decisions torturous and their implementation difficult. Cost-cutting forced last-minute design changes and the redoing of carefully worked out plans. And then there were the natural conditions, months and months of extreme cold, springtime floods, and massive fields of mud.[30]

Austin largely retained control over the design and engineering of the factory complex, but the Soviets ultimately took over planning the adjacent city. The urban center would be one of the first new cities built in the Soviet Union and as such became an opportunity to envision what a socialist city should look like. A design competition led to a plan that included extensive communal facilities and, in some sections, no traditional living units.

The first phase of the city had thirty four-story residential buildings. Most were divided into individual apartments housing several families each (already the urban norm in the face of a massive national housing shortage), but some buildings were designed for an experiment in social reorganization. Clusters of five of these buildings, connected by enclosed elevated walkways, were to be living and social units for a thousand persons each. Each unit had its own clubhouse with social, educational, and recreational facilities and a large communal dining room, where it was anticipated that most meals would be consumed. Showers were clustered communally and there were library, reading, chess, and telephone rooms and special spaces for the study of political matters, military science, and science experimentation (to encourage innovation and technical expertise, allowing the country to free itself of dependence on foreigners). Kindergartens and nurseries allowed parents to leave their children as long as they chose, including, essentially, full time. Living spaces were small, meant largely for sleeping, with no individual cooking facilities. The top floors of the "community units" had larger rooms designed for "communes" of three or four young people who would live, work, and study together.

The utopianism of the auto city quickly floundered in an ocean of need and the desire of construction workers and later automobile workers for individual apartments. Even before the first residential buildings were completed, they were flooded with squatters, workers who had been living in tents, dugouts, and other improvised structures through a long winter. Cots and little individual stoves appeared everywhere. Planners expected that communal living would become more popular, allowing them to convert buildings divided into traditional apartments to the community unit model, but in the end the conversions went the other way, as workers sought more private, individualized spaces. Also, cost-cutting meant that after the first buildings were completed, designs for communal facilities were reduced, and eventually the whole master plan for the city was abandoned. Still, even in its truncated form, the new workers' city represented a particularly elaborate realization of a broader effort to provide extensive social, cultural, and recreational programs and benefits through the workplace, with factories all over the Soviet Union taking responsibility for housing and feeding their workers and their families, educating them, and uplifting their cultural level. The Soviet welfare state centered on the large factory.[31]

In spite of all the obstacles, the huge auto complex at Nizhny Novgorod, soon to be renamed Gorky, was essentially completed in November 1931, just eighteen months after the first American engineers arrived (though construction of the accompanying city lagged behind). Specialists from the United States and the application of American methods accounted for some of the success. But much of the credit had to go to Soviet government and party officials, who, in spite of their inexperience, bureaucratic ways, and frequent ineptitude, proved able to mobilize heroic efforts by Soviet workers. They could do so because they could capitalize on a reservoir of deep commitment by at least some workers, particularly young ones, to crash development—industrialization as a form of revolution. Engaged in what they understood as a world-historic project and defense of the revolution, Soviet workers made extraordinary sacrifices, living in miserable circumstances, volunteering to work unpaid Saturdays, joining "shock brigades," accepting dangerous worksite conditions, and putting up with the bumbling and arrogance of officials in charge of the big Five-Year Plan projects. For at least a brief moment, many Soviet workers saw the factories they were building as theirs, as the means to a brighter future, to a different kind of society, and were willing to do whatever was necessary to complete them.[32]

The Kahn Brothers in Moscow

The Stalingrad tractor plant and the Gorky automobile factory were among the best-known Soviet projects in the West, receiving extensive coverage in the American press. The *New York Times*, *Detroit Times*, *Detroit Free Press*, *Time*, trade journals, and other publications regularly ran stories about them.[33] But there were many other large Soviet projects with Americans involved, too. Du Pont helped set up fertilizer factories, Seiberling Rubber Company assisted in constructing a large tire factory, C. F. Seabrook built roads in Moscow, other companies advised on coal mines, and the list went on and on.[34]

Albert Kahn took on an expanded role after work on the Tractorstroi started. In early 1930, his firm signed a two-year contract with Amtorg that made it the consulting architect for all industrial construction in the Soviet Union. Under the agreement, twenty-five Soviet engineers worked with the firm in its Detroit offices. But more importantly, it established a Kahn firm outpost in Moscow within a newly created, centralized Soviet design and construction agency. Albert's younger brother Moritz led a team of twenty-five American architects and engineers in the new Russian office, not only designing buildings but also teaching Soviet architects, engineers, and specialists the methods of the Kahn firm.

The contract with the Soviet Union provided a boon for Kahn, enabling his firm to survive through the trough of the Great Depression, when virtually no new construction took place in the United States. But more than just expediency, the Soviet-Kahn partnership grew organically from a shared vision of progress through physical construction and rationalized methods. Moritz relished the opportunity to apply the "standardized mass production" system of the automobile industry to construction—a notoriously chaotic industry making custom products—which would be possible in the U.S.S.R. because there would be one centralized design agency and one customer, the Soviet government, allowing the development of designs for particular types of factories that could be used over and over. Moritz pointed out that government ownership would eliminate the costs associated with advertising, sales promotion, and middlemen and allow the rationalization of transportation and warehousing, all of which appealed to his technocratic sensibility. Albert was more patronizing; he told the *Detroit Times*, "My attitude toward Russia is that of a doctor toward his patient."[35]

The joint Moscow design center proved challenging but ultimately successful. There were few qualified Soviet architects, engineers, or draftsmen available when it began and a lack of basic supplies, from pencils to drafting boards, with only one blueprint machine in all of Moscow. Nonetheless, in two years the Kahn team supervised the design and construction of over five hundred factories across the Soviet Union, using the Fordist methods the firm had perfected in Detroit. Equally important, some four thousand Soviet architects, engineers, and draftsmen were trained by the Kahn experts, including in formal classes taught in the evenings. They, in turn, took the approach to design and construction developed by Kahn, in collaboration with Ford and other U.S. manufacturing firms, and spread it throughout the country. Kahn's methods, according to Sonia Melnikova-Raich, who chronicled his Soviet collaboration, "became standard in the Soviet building industry for many decades."[36]

Kahn also did more Soviet design work in his Detroit office, including two new tractor plants to meet the insatiable demand for mechanized agricultural equipment. A plant in the Ukraine, on the outskirts of Kharkov, was virtually a copy of the Stalingrad plant, designed to produce the same model tractor and varying only in the greater use of reinforced concrete, as the Soviets diminished their expensive steel imports from the United States. Leon Swajian, after finishing up as number two at the Stalingrad plant, served as general superintendent for the construction (receiving the Order of Lenin for his role). The other plant was the biggest yet. Located in Chelyabinsk, some 1,100 miles due east of Moscow, just east of the Urals near the border between Europe and Asia, it was designed to produce tractors with metal crawlers rather than wheels. The buildings in the complex, looking like a chunk of Detroit industry planted in the Russian wilds, had a combined floor area of 1,780,000 square feet, laid out on a tract of 2,471 acres (twice as large as the Rouge). Though the Soviets began building the plant without American advisors on site, when things got bogged down, American engineers, including Calder and Swajian, were called in to help.[37]

Starting Up

If building the *gigant* Soviet factories had been an enormous challenge, getting them to actually produce goods proved even more difficult. Their start-up became a moment of truth for the idea that the Soviet Union could leapfrog into modernity by adopting the most advanced capitalist methods on a giant scale, building a socialist society without going through an extended process of industrialization like the United States and the Western European powers had experienced.

The Stalingrad *Tractorstroi* was the first test. Stalin's June 1930 message congratulating the tractor-factory workers on beginning production of fifty thousand tractors a year proved wildly premature. During the first month and a half, the factory produced only five tractors. During its first six months, only just over a thousand. During all of 1931, 18,410.

Not all the equipment had arrived and been installed when the plant opened. But the bigger problem was the utter unfamiliarity of the vast bulk of the workers and Russian supervisors with basic industrial processes, let alone advanced mass production. When Margaret Bourke-White visited the factory during its first summer of operation, she reported, "the Russians have no more idea how to use the conveyor than a group of school children." In the plant, "the production line usually stands perfectly still. Half-way down the factory is a partly completed tractor. One Russian is screwing in a tiny bolt and twenty other Russians are standing around him watching, talking it over, smoking cigarettes, arguing."[38]

The American workers, engineers, and supervisors hired to help start up production and teach the workforce necessary skills had their hands full. Henry Ford's dictum, that mass production could occur only if parts were so standardized that no custom fitting was required, immediately proved a trial. The skilled Russian workers the plant did have largely had been trained in craft ways. Plant manager Vassily Ivanov raced around the factory in a rage when he saw foremen using files to fit together parts (probably because some parts were not truly interchangeable, a problem at Highland Park as late as 1918). As usual in the Stalinist universe, the metaphor of war was used to describe the situation: "We were fighting our first battle," Ivanov later said, "against handicraft 'Asiatic' methods," making the traditional Marxist equation of Asia with backwardness and Europe with modernity.

Unskilled workers posed, if anything, a greater problem. Many had just arrived from small peasant villages, never having seen a telephone, let alone a precision machine tool. Frank Honey, an American toolmaker, described the first worker sent to him to train as a spring maker as "a typical peasant . . . dressed as he was in some strange, countrified sort of clothes." Such workers did not have any notion of basic factory procedures. Bearings in expensive new machines were quickly damaged because they did not know to keep oil free of dirt. Discipline was often lax, with a great deal of standing around doing nothing. It required a slow, painstaking process to teach the new workforce, which swelled to fifteen thousand, how to operate the sophisticated machinery, especially as the American instructors had to work through translators.

Furthermore, the Soviet Union lacked the well-developed supply chains on which Fordism rested. High-speed machine tools required steel of precise specifications, but when the tractor factory could get the raw materials and supplies it needed at all (which was often not the case), the composition and quality varied from batch to batch, making for spoiled parts, damaged tools, and long delays.

Fordism also required complex coordination, which the plant management had no experience in achieving. Workers and managers spent endless time in consultations and meetings, but nonetheless things did not arrive where and when they were expected. When Sergo Orjonikidje, the Commissar for Heavy Industry, in charge of implementing the Five-Year industrialization plan, visited the factory as political pressure mounted to get production going, he reported, "What I see here is not tempo but fuss."

With Stalin personally monitoring daily production figures—a measure of how important the plant was seen to the future of the country—personnel changes came quickly. Ivanov was replaced by a more technically knowledgeable communist official to work alongside a new top engineering specialist. The Soviet Automobile Trust sent yet another American engineer to the plant, an expert on assembly-line production, to try to straighten out the mess. To help establish order, the plant cut back from three daily shifts to just one.

Slowly, production began to improve, though product quality remained a problem. Much of the advance came from the growing experience of the workforce and skills gained though a massive training and education effort. The peasant newcomer whom Honey schooled eventually became a skilled worker and later foreman of the spring department. (Rapid promotions for such workers, though, created more problems, as their replacements needed to be trained.) During the first six months of 1933, the plant turned out 15,837 tractors, a significant improvement, but, after three years of operation, still well below the projected annual production of "fifty thousand shells blowing up the old bourgeois world."[39]

At the Nizhny Novgorod auto plant, managers tried to avoid the start-up problems encountered at Stalingrad. They sent hundreds of workers to Detroit to learn production techniques at Ford, while recruiting hundreds of Americans to come help get the plant going. (The presence of a female Soviet metallurgist studying heat treatment at Ford merited a headline in the *New York Times,* part of an unending fascination among American reporters and engineers with Soviet women holding blue-collar jobs that in the United States were strictly male.) Production was begun gradually, first just assembling car and truck part kits sent from Detroit before beginning to make all the needed parts on site. Still, the plant took longer than expected to get up to speed.[40]

Again, shortages of supplies and managerial ineptitude were part of the problem, but a shortage of labor, especially skilled labor, would have made a rapid start-up impossible under the best conditions. Larger than the Stalingrad *Tractorstroi,* what was soon named GAZ (*Gorkovsky Avtomobilny Zavod* ["Gorky Automobile Factory"]) had thirty-two thousand workers. Few had any industrial experience or much work experience of any kind. When the plant opened, 60 percent of the workers were under age twenty-three and only 20 percent over age thirty. Nearly a quarter of the manual workers were female. It was almost like being in an early British or American textile mill, in a world of the young.

New workers and their foreign teachers confronted difficult conditions. Living quarters were primitive, if somewhat better for the Americans, and meat, fish, fresh fruit, and vegetables nearly impossible to find. When Victor and Walter Reuther, auto union activists from Detroit, arrived at the plant in late 1933 to work as tool- and die-makers, most of the complex had no heat. They were forced to perform and teach precision metalworking in temperatures far below freezing, periodically going into the heat-treatment room to warm their hands.

As at Stalingrad, political pressure quickly mounted to get production going. Even before the plant opened, ineptitude became criminalized; nine officials were tried for "willful neglect and suppression" of suggestions made by American workers and technical specialists. After a show trial in Moscow before several thousand spectators, light sentences—at most the loss of two months' pay—were handed out, in a warning to other managers. Three months after production began, Orjonikidje came to inspect, accompanied by Lazar Kaganovich, like him a member of the Politburo, the top communist ruling body. The pair blamed local communists and unionists for mismanagement and slandering engineering and technical personnel, resulting in the firing of some plant and regional party officials.

But slowly production improved, a measure of the eagerness of the young workforce to learn new skills and what amounted to a whole new way of life and their resilience in the face of hardship. By the time the Reuther brothers headed back to the United States after eighteen months at GAZ, most of the other foreign workers already had departed, the skill level of the native workforce had enormously improved, more food and consumer goods were available, and cars and trucks were steadily coming off the line. *New York Times* Moscow reporter Walter Duranty, a big booster of Stalinist industrialization, in declaring his confidence that GAZ would quickly get up to speed, chided that "Foreign critics sometimes fail to realize two things about Russia today—the astonishing capacity for bursts of energy to get the seeming impossible accomplished and the fact that Russians learn fast." When two Austin engineers returned to the plant site in 1939, they were "dumbfounded" to see that a city of 120,000 people had grown up around the core residential area they had constructed, with six- to eight-story apartment buildings, paved streets, "quite a few flowers," and people who "looked better."[41]

As a cadre of skilled workers developed, other start-ups became easier. When the Kharkov tractor plant began operations in the fall of 1931, it benefited from a large group of experienced workers who were transferred from its twin in Stalingrad. Also, rather than immediately having to manufacture the 715 custom parts that went into its tractors, the plant could begin assembling vehicles using some parts shipped over from the Stalingrad factory.[42]

By contrast, the construction and initial operation of the Magnitogorsk Metallurgical Complex made the Stalingrad tractor factory and the Gorky automobile plant look like easy sailing.[43] Before the revolution, Russia had only a small iron and steel industry. The First Five-Year Plan called for a huge leap in metal production. Key to the effort was to be a massive integrated steel plant forty miles east of the Urals, next to two hills which contained so much iron ore that they affected the behavior of compasses, giving them the name Magnetic Mountain (*Magnitnaia gora*) and the city that was to arise with the plant the name Magnitogorsk. By some accounts, Stalin personally called for the creation of the complex after learning about the U.S. Steel plant in Gary, Indiana. Like Gary, the plant was to include every phase of the production of steel products, including blast furnaces, open-hearth converters, rolling mills and other finishing plants, coke-making furnaces, and equipment to make chemicals out of coke by-products. Unlike Gary, the complex included its own iron mine.

Magnitogorsk—"The Mighty Giant of the Five Year Plan," as one Soviet periodical dubbed it—was but one component of an even larger scheme, a *Kombinat*, an assemblage of functionally and geographically related facilities, which stretched all the way to Kuznetsk in Central Siberia, the source of most of the coal initially used in the steel complex, and which included the Chelyabinsk tractor factory, 120 miles northwest of Magnitogorsk. Even some of the less-heralded *Kombinat* factories were huge, like the railroad car plant in Nizhny Tagil, north of Chelyabinsk. A prominent part of the Second Five-Year Plan, which began in 1933, the sprawling factory complex employed forty thousand workers and had its own blast furnaces and open-hearth department.[44]

Foreign experts helped design Magnitogorsk, but unlike in Stalingrad and Nizhny Novgorod no one firm coordinated the whole effort, creating myriad problems. In 1927 the Soviets retained the Freyn Engineering Company of Chicago as a general advisor in developing its metallurgy industry, and it did some initial planning for Magnitogorsk. Then the Soviets hired the Cleveland firm of Arthur G. McKee & Company to do the overall design, but amid much rancor the company proved unable to churn out plans at the rate the Soviets desired. So its role was cut back and other U.S. and German firms were brought in to design particular components of the complex, with various Soviet agencies playing a part, too. As a result, in the words of American John Scott, who spent five years working at Magnitogorsk, its elements were "often very badly coordinated." The whole project was late in getting going and took far longer to complete than originally projected.

Even if the planning had been better managed, the scope of work and the challenges of the site would have made the "super-American tempo" the Soviets claimed was being maintained impossible to achieve. When work at Magnitogorsk began, there was nothing in place, no buildings, no paved roads, no railroad, no electricity, insufficient water, no coal or trees to provide heat or energy, no nearby sources of food, no cities within striking distance. Out of the dust of the steppe, Soviet officials and foreign experts had to conjure up a vast industrial enterprise, and do so in the cruel weather east of the Urals, where summers were short and winters exceedingly long and cold. In January and February, the low temperature *averaged* below zero degrees Fahrenheit. Some winter mornings it was thirty-five degrees below zero. John Scott, while working as a welder on blast furnace construction, once came upon a riveter who had frozen to death on the scaffolding.[45]

Much like the first English textile factory owners, Magnitogorsk managers had to recruit a workforce to build and operate the complex, which by 1938 had twenty-seven thousand employees, and come up with ways to house it, feed it, and take care of all its needs in an isolated spot where there never had been a large assemblage of people. Some workers came voluntarily, swept up in enthusiasm for the effort to leap forward to modernity and socialism or simply looking for an escape from their village or an unpleasant situation. Others were assigned by their employers to go to Magnitogorsk, like it or not. But such workers were not enough, especially since they flowed out of Magnitogorsk almost as quickly as they flowed in, put off by the extremely primitive living conditions and difficult work. So, again, like the early English mill owners, the Soviets turned to unfree labor, on a huge scale.

The Soviets used forced labor at many big projects, including the Chelyabinsk tractor factory, the Dnieprostroi Dam, and, most famously, the White Sea–Baltic Canal, constructed almost entirely by prisoners. At Magnitogorsk, by Scott's account, in the mid-1930s some fifty thousand workers were under the control of the security police, the GPU (after 1934, the NKVD), most doing unskilled construction work but some employed in the steel plant itself. Even more than the early English textile mills, Magnitogorsk refuted simple correlations between industrialization, modernity, and freedom.

Forced laborers in Magnitogorsk fell into several categories. Common criminals made up the largest group, over twenty thousand workers, most serving relatively short sentences, living in settlements (including one for minors) surrounded by barbed wire, going to work under guard. A second group consisted of peasants dispossessed during the collectivization drive, so-called kulaks, deported to the steel city. In October 1931, there were over fourteen thousand former kulak workers and twice that number of their family members living in "special labor settlements," initially enclosed by barbed wire, too. Even by Magnitogorsk standards, conditions for the forced migrants were appalling, with 775 children dying in one three-month period. (By 1936, most restrictions on these workers were eased.) Finally, there were veteran engineers and technical experts, trained under the old regime, who had been convicted of crimes but nonetheless worked as specialists and supervisors, in some cases, especially in the early days, holding very responsible positions, generally indistinguishable from other managerial personnel except for their legal status.[46]

The use of prison labor constituted just one part of the intertwining of the national security apparatus with the crash industrialization. In Magnitogorsk, as construction and production delays and difficulties stretched on and on, the NKVD became ever more involved with the steel complex, a shadow force with more power than the factory administration and the local government and, at some points, even than the local Communist Party. Problems stemming from poor planning, incompetent management, untrained workers, supply and transportation shortages, and the wear on machines and workers from politically driven crash efforts were increasingly attributed to failure to follow the Communist Party line, to deliberate wrecking and sabotage, and eventually to conspiracies involving foreign powers and internal oppositionists, like the "Trotskyite-Zinovievite Center" and the "Polish Military Organization," which were alleged to be operating in Magnitogorsk. Starting in 1936, all industrial accidents became subjects of criminal investigations. "Often they tried the wrong people," Scott commented, "but in Russia this is relatively unimportant. The main thing was that the technicians and workers alike began to appreciate and correctly evaluate human life."

But if technicians and workers developed a greater appreciation of human life, the police and judiciary became ever more cavalier in their treatment of workers and managers, as arrests, interrogations involving "physical measures," fabricated evidence, detentions, and executions became common. Top factory managers, state officials, and party functionaries toppled into the abyss as real and perceived failures were attributed to treachery and counterrevolution, until finally even the leaders of the Magnitogorsk NKVD, who led the terror, themselves fell to it. Though no exact count is available, according to Scott, in 1937 the purge led to "thousands" of arrests in Magnitogorsk. And it was similar elsewhere; at the Gorky auto plant, during the first six months of 1938, 407 specialists were arrested, including almost all the Soviet engineers who had spent time in Detroit and some of the few Americans who still remained at the factory.[47]

Watching on the ground, Scott saw the fury of charges, counter-charges, and arrests impede production, but in his view only temporarily and to a limited extent. Overall, as managers and workers slowly mastered their jobs, supply and transportation problems were ironed out, and new components of the complex came on line, Magnitogorsk's output of iron ore, pig iron, steel ingots, and rolled steel all moved upward, as did productivity.[48] Some of the *gigants* built during the 1930s never reached their projected output, but, overall, the First Five-Year Plan (which was accelerated to be finished in four years) and the Second Five-Year Plan that followed led to an enormous leap in Soviet industrial output. Estimates vary, but between 1928 and 1940 total industrial output increased at least three-and-a-half-fold and by some accounts as much as sixfold. The greatest gains were in heavy industry. Iron and steel production more than quadrupled. Machine production increased elevenfold between 1928 and 1937, and military production twenty-five-fold. By the latter year, motor vehicle production approached two hundred thousand vehicles. Electrical power increased sevenfold. Transportation and construction also swelled. By contrast, output of consumer goods—a low priority in the First Five-Year Plan—rose only slightly. Stalin was premature in 1929 when he said, "We are becoming a country of metal, a country of automobiles, a country of tractors," but a decade later there was much truth to his claim.[49]

Making Socialist Citizens

The giant Soviet factories were conceived of not only as a way to industrialize and protect the country but also as instruments of culturalization, which would create men and women capable of operating these behemoths and building socialism. Communist leaders often described this cultural project as fighting backwardness—the illiteracy, ignorance of modern medicine and hygiene, and unfamiliarity with science and technology that characterized the bulk of the population of the prerevolutionary Russian Empire. Many Bolsheviks, especially Lenin, defined culture in traditional European terms, as literacy, knowledge of science, appreciation of the arts. Civilization meant novels, chess, Beethoven, indoor plumbing, electricity. But some communists, and to some extent the party and state as a whole, at least through the early 1930s, believed that a distinctly *communist* culture and civilization should be created out of the revolution. The factory was an instrument to realize socialist modernity.[50]

The simple act of coming to a factory could launch the process of cultural change. This was especially the case for men and women from peasant villages, and even more so for migrants from nomadic regions of the country. Many newcomers had never seen a locomotive, indoor plumbing, electric lights, even a staircase. Walking into a factory for the first time could be terrifying, just as it had been in earlier years in England and the United States. A. M. Sirotina, a young woman who came to the Stalingrad tractor factory from a village near the Caspian, remembered, "There was an awful roaring and hammering of machines and there were motor-cars whizzing to and fro over the shop. I dodged to one side in fright and took refuge behind a stand."[51]

That a young woman was on the shop floor of the Tractorstroi reflected the profound change in gender roles and family relations that accompanied the gearing up of heavy industry. After the revolution, the Communist Party and the Soviet government promoted women's equality and new familial arrangements, but the changes were especially dramatic in the budding industrial centers, where there was no old order that had to be overthrown. At the start of the First Five-Year Plan, 29 percent of industrial workers were female; by 1937, 42 percent. Women held many types of positions for which they would never even be considered in the United States or Western Europe, such as crane and mill operator. Still, old ways died hard, as some men refused to allow their wives to work, abused them, and abandoned their families without alimony or child support.[52]

Learning utterly unfamiliar jobs took time. To hasten the process, the Soviets launched a massive educational effort. In addition to informal shop-floor training by skilled workers, supervisors, and foreign experts, formal classes were held after work to teach skills for specific jobs. Victor Reuther recalled that the Gorky auto plant "was like one huge trade school." Rollo Ward, the American foreman of the gear-cutting department at the Stalingrad tractor plant, noted that while in the United States factory owners tried to keep workers from fully understanding the machinery they operated, in the Soviet Union workers were encouraged to learn everything about the equipment, beyond just what was needed to perform their own particular tasks.[53]

The educational push was not narrowly vocational. In the new industrial cities, crash efforts were made to build enough kindergartens and elementary schools for the flood of incoming and newborn children. Adult literacy courses were heavily attended, in Magnitogorsk enrolling ten thousand students. To raise the political level of activists, there were schools teaching Marxist theory and Soviet economic and social structure. For workers who had mastered basic skills, there were advanced schools in engineering, metallurgy, and the like. Women made up 40 percent of the students at the Magnitogorsk Mining and Metallurgical Institute.

Many of these programs had some day students receiving stipends and many more students coming after work. John Scott, who attended night school most of the time he lived in Magnitogorsk, reported that virtually everyone in the city between ages sixteen and twenty-six was studying in some sort of formal program, which took up almost all of their spare time. "Every night from six until twelve the street cars and buses of Magnitogorsk were crowded with adult students hurrying to and from schools with books and notebooks under their arms, discussing Leibnitz, Hegel, or Lenin, doing problems on their knees, and acting like high-school children during examination week in a New York subway." For worker-students, the tremendous dedication needed to get to class, stay awake, and then do homework after a hard day's work opened a path of upward mobility. For Soviet leaders, the education push, especially in technical fields, liberated the country from dependence on foreign and old-regime expertise.[54]

Factories tried in other ways, too, to transform worker culture. "Red

Figure 5.3 The workers' cafeteria at the Gorky auto plant.

corners" were common. Somewhat like the reading rooms in nine-teenth- and early twentieth-century English and American union halls, these designated spaces had books, pictures of Lenin and other commu-nist leaders, political posters, and room for meetings. Many enterprises sponsored literary groups, with worker-writers producing wall newspa-pers and broadsheets that were posted at worksites. At the Gorky auto plant, management sponsored a competition among departments for ideas to elevate the cultural level. One department brought in artifi-cial palm trees from Moscow, which were placed alongside the assem-bly line. The department in which the Reuther brothers worked made dies to punch out metal spoons, considered a cultural advance over the wooden, peasant-style spoons workers used in the plant cafeteria and at home. In Stalingrad, the factory manager, inspired by what he saw in the United States, had trees and grass planted around the tractor plant to keep down dust that might damage machinery and to create a more attractive view for workers as they arrived and left.[55]

The cities that arose alongside the giant factories were at least as important in promoting new habits and values as the plants themselves.

Generally, in the U.S.S.R., local Soviets—the government—owned housing and other urban facilities. But in the industrial boomtowns, factories often filled that role, taking charge of almost all aspects of their employees' lives. Much as in the early English textile industry, many factories owned stores and farms to supply them, with workers spending a substantial proportion of their wages at factory canteens and shops (with special shops with better goods and lower prices for foreign workers and later on for party officials, top managers, "shock workers," and other favorites).

In Magnitogorsk, the steel company had four thousand employees in its department of "Everyday-Life Administration," in charge of housing and an array of social and cultural programs. The factory controlled 82 percent of the living space in the city and sponsored many of its cultural institutions, which included a large theater, two theater troupes, eighteen movie houses, four libraries, a circus, and twelve workers' clubs, among them one for ironworkers and steelworkers, the Palace of the Metallurgists, which featured a large auditorium, marble hallways, chandeliers, and an elegant reading room. The largest movie house in town, the Magnit, showed foreign as well as domestic films, including Chaplin's *Modern Times*, which the local press hailed as "a rarity in bourgeois cinema—a great film," perhaps missing the irony of its radical critique of Fordist production. Physical culture was not neglected, with two stadiums, many gymnasiums and skating rinks, and an aeronautical club that offered flying and parachuting lessons, popular activities in the Soviet Union. What the city did not have was a single church.[56]

By necessity and design, life in the *gigant* factory towns was more communal than in Western industrial centers. Especially at isolated sites like Magnitogorsk, but even in Stalingrad, workers initially lived in barracks, without private kitchens or toilets (or often indoor toilets of any kind), sleeping together in large, poorly heated rooms. Town construction lagged behind factory construction; in Magnitogorsk, in 1938, when the population had grown to nearly a quarter of a million, half of the people still lived in barracks or other temporary housing. Planning the steel city turned into something of a fiasco, as a team of modernist German architects, headed by Ernst May, and various Soviet officials went back and forth over designs, while on the ground building began haphazardly, with no plan at all. The first permanent housing in Magnitogorsk, as in the town adjacent to the Gorky auto plant, had utopian-communal features: small living spaces in large buildings with shared toilets and baths and meals to be either eaten at public cafeterias or prepared in a single kitchen serving a whole structure. But a tilt toward more traditional family structures, coming from below and above, led to the adoption of communal apartments as the new norm, with several families, rather than a whole building, sharing kitchens and toilets.[57]

Within a few years, the most radical cultural ideas associated with the First Five-Year Plan were abandoned. Nonetheless, the giant factory transformed the workforce. The story of G. Ramizov, a die forger at the Stalingrad tractor plant, captured the national arc. From a poor peasant family, he arrived with just the clothes he wore, one change of underwear, and a basket holding all his worldly possessions. His earnings soon allowed him to buy his very first toothbrush, a towel, his first suit and tie, and a winter coat. As time went on, and he switched from construction to production work, he was able to obtain furniture, books, a clock (a symbol in the U.S.S.R., just as it had been in England and the United States, of modernity and industrial discipline), a stove, dishes, and pictures to decorate his living quarters (including portraits of Lenin and Stalin). Conventional, ordinary, unimpressive, unless one came from the poverty, illiteracy, and cultural isolation that was the lot of the vast bulk of people in the Russian Empire before the revolution and the crash industrialization it sponsored.[58]

Celebrating the *Gigant*

Soviet writers, artists, and government officials relentlessly celebrated big factories during the decade prior to World War II. Joining outsized infrastructure and huge collective farms in a cult of giantism, they were at the core of national self-understanding and state propaganda. Artists and propagandists commonly equated the struggle to build socialism with the drive to industrialize, making the factory a central site for the fight against backwardness and the plunge into a new type of future.

In literature, the machine often appeared as a metaphor for society. It also appeared more literally. The title character in Lydia Chukovskaya's novella *Sofia Petrovna*, written at the end of the 1930s, comments that in Soviet stories and novels "there was such a lot about battles and tractors and factory shops and hardly anything about love."[59]

But where the factory truly came to the forefront was in the many documentary projects of the era. The Soviets were in the vanguard of what was a heyday of documentary art and literature in Europe and the United States, helping to inspire and shape the broader movement. While elsewhere documentary art and writing often focused on social ills, including those stemming from the Great Depression, in the Soviet Union documentary work had a celebratory tenor, highlighting the great progress being achieved across the vast nation.[60]

The most innovative work combined photography and journalism in elaborate, visually striking accounts of the advance of Soviet society through large-scale industry, infrastructure, and collectivization. The magazine *USSR in Construction* featured many of the most outstanding visual artists in the Soviet Union. Produced by the State Publishing House of the Russian Soviet Republic, with an editorial board that included Maxim Gorky, the large-format journal came out each month from 1930 through 1941 in four editions: Russian, English, German, and French, with a Spanish edition added in its last years. It specialized in the photo-essay, an innovative format often attributed to *Life* magazine but actually developed earlier in the Soviet Union. The title page of the fifth issue captured the editorial agenda in its subhead: "More Iron! More Steel! More Machinery!" In issue after issue, photo-essays appeared on dams, canals, hydroelectric plants, railroads, auto factories, tractor factories, tractors arriving at collective farms, paper mills, woodworking plants, garment factories, a match factory, shipyards, workers' housing, technical institutes, and workers' clubs. Whole issues were devoted to the Chelyabinsk tractor factory, Magnitogorsk, the Nizhny Tagil railroad car factory, and GAZ ("The Soviet Detroit").

USSR in Construction bore no resemblance to the engineering and trade journals in the United States that documented industry, like *Scientific American* (in its early years) or *Iron Age*, with their technical language, tightly packed text, and diagrammatic illustrations. The Russian magazine was gorgeous, with innovative design, selective use of color, and arresting layouts of photographs taken by leading Soviet photographers, including Max Alpert, Arkady Shaikhet, Georgy Zelma, Boris Ignatovich, Semyon Fridlyand, Yevgeny Khaldei, and, perhaps most notably, Alexander Rodchenko. Gorky himself wrote some of the text for the early issues. But the true auteurs of the magazine were the designers, who included Nikolai Troshin, Rodchenko and his wife Varvara Stepanova, and the married team of El Lissitzky and Sophie Küppers. Layouts were complex, varied, and innovative, juxtaposing picture and text in ever-changing ways, making use of unusual typography and montage. Sometimes the layouts were conceived before the photos were taken, with the photographers instructed on the kinds of images that would be needed for the assemblage. Over time, the designs became ever more elaborate, as the magazine began using horizontal and even vertical foldout leaves, inserts, maps, pop-ups, irregularly cropped photographs, and transparent overlays. One issue, devoted to a new airplane model, had an aluminum cover.[61]

Some of the same photographers and designers were involved in books that used a similar format to exhaustively document the industrialization effort. *USSR stroit Sotzsialism* ("USSR Builds Socialism"), a 1933 volume, was organized by industry—electric, coal, metallurgy, and so forth—with exquisite photography, montages, and other graphic devices. Like *USSR in Construction*, it was in part designed as propaganda promoting Soviet achievements abroad, with the main text in Russian but captions in German, French, and English as well. But the most important audience for the celebration of industrialization and the giant factory was at home. In its early years, the press run for the German and English editions of *USSR in Construction* peaked just over ten thousand (with fewer copies in French), but the Russian edition had issues with print runs exceeding one hundred thousand. The key readership group for this semi-avant-garde testament to industry and infrastructure was the new Soviet elite of party officials, government functionaries, and industrial managers, who no doubt took a proprietary pride in the accomplishments of the new society, at the top of which they sat, much like American industrialists enjoyed the celebration of industry in *Fortune* and the photographs of Margaret Bourke-White. For the 1935 Seventh Congress of the Soviets, Lissitzky and Küppers produced a lavish seven-volume documentation of "Heavy Industry," complete with accordion foldouts, overlays, special papers, collages, and incorporated fabric.[62]

The documentary magazines and books and the many posters celebrating industry drew much of their splendor from the very high quality of Soviet photography. With most of the country illiterate during the first years after the revolution, the Bolsheviks saw photography, film, posters, and heavily illustrated magazines as more effective vehicles of propaganda and enlightenment than the written word. The first issue of *USSR in Construction* carried a notice that "The State Publishing House has chosen the photo as the method of illustrating socialist construction for the photograph speaks much more convincingly in many cases than even the most brilliantly written article." The camera itself became a mark of modernity; "Every progressive comrade," Anatoly Lunacharsky, the Commissar of Enlightenment, wrote, "must not only have a watch but also a camera."[63]

Soviet photographers engaged in fierce debates over the style and to a lesser extent the content of their images, forming rival organizations, but most shared a commitment to the socialist project and willingly followed government injunctions to document the giant projects of the Five-Year Plans. Even sites distant from Moscow and Leningrad attracted top photographers. Dmitri Debabov, Max Albert, and Georgy Petrusov all took extraordinary photographs at Magnitogorsk, with the latter spending two years there as the head of information for the factory. While there were some commonalities between the documentary approach of the Soviets and leading American industrial photographers, like Bourke-White and Charles Sheeler, there were important differences, too. Soviet photographers more quickly adopted the small, light 35-mm camera, introduced by Leica in 1924 (with a Soviet version beginning production in 1932), than the Americans, who largely stuck with their big, heavy, large-format equipment. The 35-mm camera made it easier to shoot from odd angles and unusual vantage points. Uncommon framing, diagonal positioning, unfamiliar angles, and shots taken from very low down or very high up characterized early Soviet photography, including the industrial work. Although by the early 1930s government authorities began to criticize unconventional artistic modes, moving toward an embrace of socialist realism, photography, as art historian Susan Tumarkin Goodman wrote, remained "the last bastion of a radical visual culture," imparting excitement to the documentation of the giant factory and associating it with modernist trends in the arts.[64]

Soviet filmmakers, who shared many stylistic approaches with Soviet photographers, also embraced the factory as a subject. In the 1931 film *Entuziazm (Simfoniya Donbassa)* ("Enthusiasm: Symphony of the Donbas"), the avant-garde newsreel and documentary filmmaker Dziga Vertov portrayed the transformation of Ukrainian towns, wracked by religion and alcoholism, through the development of coal mines and a giant steel mill. Dramatic images of steelmaking contribute to the visual inventiveness and frantic montage typical of Vertov's films, in this case complimented by the innovative use of sound, then just being introduced. Charlie Chaplin declared *Enthusiasm* the best picture of the year.[65]

While the Soviets favored visual imagery in their celebration of the factory, they did not ignore the written word. In 1931, Maxim Gorky proposed a massive project to document the "History of Factories and

Plants," both older facilities and the new giants of the First Five-Year Plan. Reflecting how important Soviet leaders viewed the representation of factories, the highest levels of the Communist Party became involved with the series, which was discussed by both the Central Committee and the Politburo. Bukharin (by then already beginning to fall out of favor) and Kaganovich, one of Stalin's closest colleagues, composed separate lists of possible editors and Stalin himself crossed off names and added others. Thirty volumes were published before the series was discontinued in early 1938 amid the height of the purges. Some documentary volumes were published in English as well as Russian. An abridged version of *Those Who Built Stalingrad, As Told by Themselves*, an oral history of the Soviet and foreign workers who built the tractor factory and started up production, with a foreword by Gorky, was published in New York in 1934, an innovative work that had some of the quality of the books Studs Terkel would assemble decades later in the United States, stressing the cultural and political transformation of the workers as much as the operations of the plant itself. A booklet about Magnitogorsk, with an image of a blast furnace embossed in copper on its cover, was sold at the Soviet pavilion at the New York World's Fair.[66]

Many American journalists, economists, and academic experts on the Soviet Union also were swept up by what one of the best known of their number, George Frost Kennan, called "the romance of economic development." Foreign correspondents like Walter Duranty and William Henry Chamberlin from the *Christian Science Monitor* regularly filed stories about industrial projects and wrote about them in books. Economists and social critics influenced by Thorstein Veblen's technocratic outlook, like Stuart Chase and George Soule, were particularly enthusiastic. Sharing the equation of progress with economic growth and industrialization, they admired the Soviet embrace of large-scale planning and thought the United States could learn much from it. Though the journalists and academics were well aware of the great sacrifices that were being made by the Soviet people to finance the crash industrial drive, most thought it was a price worth paying.

Louis Fischer, the Moscow correspondent for *The Nation*, later wrote that before World War II he had been "glorifying steel and kilowatts and forgetting the human being."[67]

Europeans flocked to the Soviet Union in even greater numbers than Americans. Many brought back positive reports from sites of industry. Dutch filmmaker Joris Ivens, who later became internationally famous for directing *The Spanish Earth*, a pro-Republican documentary made during the Spanish Civil War, spent three months in 1932 camped out in a barracks in Magnitogorsk, filming workers erecting huge blast furnaces. Left-wing Austrian composer Hanns Eisler, who agreed to create the soundtrack for his film, joined him, recording industrial sounds to use, much as Vertov had done in his recently released *Enthusiasm*. Ivens centered what became *Song of Heroes* on the transformation of a Kirghiz peasant into a Soviet worker. Complex political and artistic debates swirled around the film at a moment when cultural experimentation was being reined in across the Soviet Union. Premiered in early 1933, it soon disappeared from view. Meanwhile, Eisler and Soviet writer Sergei Tretyakov planned an opera about Magnitogorsk, which was scheduled to premiere at the Bolshoi Theater, but never did, perhaps for political reasons.[68]

At least in the United States, more than mainstream journalists or avant-garde leftists, perhaps the person most responsible for getting out the story of Soviet industrial behemoths was Margaret Bourke-White, in effect reprising the role she had played for American factories. She was drawn to the Soviet Union, where few foreign photographers had ventured, not out of any particular sympathy for the socialist experiment but by a desire to document rapid industrialization and the transformation of the peasantry into a working class. "I was eager to see what a factory would be like that had been plunged suddenly into being," she later wrote.

Getting into the Soviet Union proved quite a challenge. In spite of letters of introduction from Sergei Eisenstein, whom she met in New York, it took her unyielding persistence and a considerable wait to finally get a visa. Then came an arduous train trip from Germany carrying her bulky equipment. But when she finally got to Moscow, her portfolio of photographs of blast furnaces, oil derricks, locomotives, and coal freighters worked like a magic wand, opening all doors. "I had come to a country where an industrial photographer is accorded the rank of artist and prophet," she discovered. In short order Soviet officials organized a five-thousand-mile tour of key sites of the First Five-Year Plan, a kind of Stations of the Cross on the road to socialism, including a textile mill, the Dnieprostroi Dam, a collective farm, a Black Sea cement plant like the one fictionalized in Fyodo Vasilievich Gladkov's popular novel *Cement*, and the Stalingrad tractor factory.

Bourke-White published a book, *Eyes on Russia*, documenting her journey, the first time she complemented her photographs with substantial text. One picture she took at the Stalingrad plant appeared in both *Fortune* and, in a slightly different version, in *USSR in Construction* (which published several other of her photographs as well). In the summer of 1931, she returned to the Soviet Union at the invitation of the government, extensively photographing Magnitogorsk. The *New York Times Sunday Magazine* published six articles by her, accompanied by photographs, based on her trip. The next summer she returned to the Soviet Union yet again, in her first and last effort at filmmaking, a largely failed endeavor that nonetheless yielded two short films distributed in theaters when the United States recognized the U.S.S.R. in late 1933.[69]

Bourke-White's photographs of Soviet industry resembled her work in the United States: pictures of machinery that highlighted symmetry and repetition; large-scale equipment and installations set against dramatic skies; molten steel flowing within dark sheds. Her photographs of a textile mill on the outskirts of Moscow were not so different from those she had taken in Amoskeag. The main dissimilarity between her Soviet photographs and her early industrial work is her greater attention to workers, both in industrial settings and in portraits.

Compared to Soviet photographers, with their more unusual

compositions, Bourke-White's Soviet photographs seem a touch old-fashioned: staid, dignified, a bit static. As was the case in the United States, she often staged photographs to capture what she saw as the essence of the processes before her. For her photograph of a tractor at the end of the Stalingrad assembly line, she scoured the plant to find the right figure for the "exultant picture" of industry triumphant.

Bourke-White would become increasingly committed to the political left during the 1930s and 1940s, in part as a result of her experiences in the Soviet Union. But politics, in the usual sense, did not shape her images of Soviet industry. Rather, it was the physical machinery of industry and the people building and operating it that captivated her. In the Soviet Union, as in the United States, Bourke-White saw in large-scale industry beauty, progress, and modernity. It was the giant factory, not its social context, that she documented, and in doing so implicitly suggested a fundamental similarity of the factory as an institution in the communist world and the capitalist one.[70]

Paying the Bill

The heavy involvement of foreign workers and professionals in Soviet industrialization proved short-lived. By 1933, the net flow had reversed, with more foreigners headed out of the Soviet Union than into it. Within a few years, very few remained.

To a great extent, it was a matter of money. Foreign contracts, with workers and companies, generally called for payments in dollars, European currencies, or gold. To finance purchases of foreign equipment and expertise, the Soviets exported grain, gold, artwork, and raw materials, including timber, oil, flax, and fur. But the value of these goods was falling. A global drop in commodity prices began in the late 1920s, even before the American stock market crash, forcing the Soviets to increase foreign sales to stop the depletion of their gold and hard currency reserves. In August 1930, Stalin called for more than doubling grain exports, "Otherwise we risk being left without our new iron and steel and machine-building factories. . . . We must push grain exports furiously." That meant ever greater and more brutal requisitioning of grain in the countryside and hunger throughout the country.

By the time first new factories were ready to begin operating, the Great Depression had hit the United States and Western Europe, so that there were plenty of skilled workers out of jobs willing to head to the Soviet Union simply to get work, if not out of political sympathies. But the Depression also led to a further drop in prices for grain and raw materials, accelerating the decline of Soviet foreign reserves. In mid-1931, Soviet leaders began curtailing the importation of foreign equipment and the use of foreign experts. The Americans were hit particularly hard, as firms in other countries offered better terms and credit arrangements not available from the United States.

When Albert Kahn's consulting contract ended in 1932, the Soviets offered to renew it only if the company agreed to be paid in rubles, which were not convertible into dollars. Kahn traveled to Moscow in an effort to save the partnership, but in the end his company brought home its personnel. In 1934, after purchasing the number of vehicles and parts specified in their contract with Ford, the Soviets terminated that agreement, too.[71]

But it was not just a matter of money. Inflated expectations of what foreign equipment, methods, and experts could accomplish led to something of a disillusionment with Americanism as the new factories, erected at enormous cost, had difficulty meeting their targeted output. The effects of the Depression in the West also dampened enthusiasm for the United States and its representatives, as a traditional Marxist critique of the internal contradictions that crippled capitalist progress reemerged. Vassily Ivanov reported after his trip to the United States early in the Depression, "I saw with my own eyes how the productive forces were outgrowing their narrow capitalist framework. The factories were working at one third of capacity, repressing their powers, crushing and constricting their exuberance of technical potentialities." The United States might not continue to be a leader in industrial innovation.

At the same time, a new generation of Soviet managers and specialists was beginning to come into its own, with up-and-coming leaders increasingly confident they did not need foreign tutors, whom in some cases they came to resent. A general turn toward suspicion of outsiders and even xenophobia emerged in the mid and late 1930s, accompanied by ever more grandiose claims of Soviet superiority in all realms. In accounts of the industrialization drive, the role of foreigners became downplayed or erased. After the mid-1930s, nonnatives who decided to stay in Russia were viewed with suspicion and had difficulty finding work or even staying out of prison.[72]

While the Soviets began moving away from directly copying methods, plants, products, and processes developed under capitalism, the basic thrust of their industrial drive remained unchanged. The Second Five-Year Plan and a Third Five-Year Plan, begun in 1938 but cut short by World War II, continued to prioritize heavy industry, though they gave greater attention to consumer goods as well. After World War II, the Soviet Union resumed the prewar model of multiyear economic plans and concentrated investment in giant production and research complexes, often with accompanying new cities. The crash industrialization of the First Five-Year Plan created a template that would be used in the Soviet Union until its demise and in many of its satellites and allies as well.[73]

But many of the key figures in the industrialization drive did not live to see the spread of the model they helped create. The Great Terror of the late 1930s wiped out the pioneers who brought the "Amerikansky Temp" to the Soviet Union and built the first giant factories. Many of the participants in the industrialization debate of the 1920s, including Bukharin and other top Bolsheviks, were arrested on patently absurd charges, convicted in show trials, and executed, or, in Trotsky's case, murdered in exile. Sergo Orjonikidje committed suicide in 1937. Alexei

Gastev, the Soviet Taylorist, survived repeated internal party battles, only to be arrested in September 1938 on charges of "counterrevolutionary terrorist activity" and executed the following spring. Saul Bron, who as head of Amtorg had signed agreements with American companies worth tens of millions of dollars, including the pacts with Kahn and Ford, was arrested in October 1937, tried in secret, and shot in April 1938. Others executed at roughly the same time included Vassily Ivanov, the first directors of the tractor plants in Kharkov and Chelyabinsk, the first director of the Gorky auto factory, and the head of the auto trust who signed the contract with the Austin Company. Such leaders of the big industrialization projects, who generally had political but not technical credentials (making them dependent on foreign or old-regime expertise), having done their jobs and built potentially threatening local power bases in the process, were simply wiped out, replaced by a generation of newly trained managers, usually from peasant or working-class backgrounds, who had no ties to the early days of the Bolsheviks and the ideological and organizational resources they provided.[74]

Did It Work?

Did the giant factory succeed in the Soviet Union? The question carries a different meaning and weight than if asked about earlier incarnations of industrial giantism. Elsewhere, big factories had been built by individuals or corporations for a narrow purpose, their own economic reward. Sometimes they also had philanthropic or social goals, but those were almost always secondary and often instrumental to the economic success of the factory and its payoff to its creators and investors. By contrast, in the Soviet Union, giant factories were seen as a means to very large social and political ends: industrialization, modernization, national defense, and the creation of socialism. While earlier big factories were conceived of as a way to expand production, in the

Soviet Union they were seen as a way to transform society, culture, and, ultimately, world history.

By the measure of aggregate output and economic growth, the Soviet industrialization drive of the 1930s succeeded. The infrastructure and industrialization efforts under the Five-Year Plans accelerated the growth of industry and the overall national economy to rates that surpassed those in the West, where the Depression left the leading industrial countries in stagnation. In sector after sector, Soviet industrial output zoomed up, in many cases with industrial giants playing a critical role.

Economists have debated if the same kind of growth could have been achieved through a more balanced program of development, less focused on concentrated investment in landmark gigants. As the Soviets discovered, there were diseconomies of scale in creating islands of industrial giantism in a vast, undeveloped nation. Expensive, advanced equipment went unused, unmaintained, or prematurely worn down through overuse. Skilled labor shortages proved endemic and supply chains immensely difficult to create and sustain, given the thinness of the national industrial base and the difficulties of coordination through centralized planning structures rather than markets. Unable to depend on reliable flows of quality material through official channels, industrial managers built their own off-the-books networks of suppliers, using barter, favors, and other methods, creating shortages and difficulties elsewhere, while they themselves often passed defective goods up the chain.[75]

But the success of the giant factory cannot be fully judged using only economic measures. The scale of the great Soviet industrial projects, more than the scale of projects in the capitalist world, served an important ideological function. Giantism contributed to the massive social mobilization required for the industrialization drive, which became the moral equivalent of revolution and civil war. The world-historic scale of Soviet factories and infrastructure contributed to a cultural revolution in which modernity and progress were linked to Soviet power and mechanization. And it worked, as millions of Soviet citizens made heroic efforts to construct new facilities, a new economy, a new society.

At a price. The industrialization drive was linked, by design, to squeezing as much as possible out of the peasantry, even to the point, at times, of famine. The brutal collectivization of agriculture pushed millions of peasants away from their homes to industrial employment. Conditions during the First Five-Year Plan were worst in the countryside, but real wages and living standards for workers fell, too. Circumstances at the new plants were harsh and shortages widespread.

But *comparatively* the situation did not look quite so bad. Housing was very crowded, but lack of private space was nothing new to most peasants or even to most urban workers. What was new, for many, was electricity, clean running water, and central heating. Furthermore, by the standards of the early phases of industrialization in England and the United States, working hours in the Soviet Union were short, in the early 1930s generally seven hours a day (not counting dinner breaks) and six hours in dangerous occupations. By the late 1930s, material conditions for workers notably improved.[76]

Isolating the giant factory from everything else going on in the Soviet Union during the 1930s—including the collectivization of agriculture and the Great Terror—is impossible, so judgments on the efficacy of industrial giantism as a developmental strategy are difficult to make. But in one realm, the record seems clear. The creation of the metallurgy, automotive, and tractor industries, especially the plants located deep in the Soviet interior, proved critical to Soviet survival and ultimate victory during World War II. One reason the Soviets sited so many industrial behemoths in the Urals was to distance them from any invasion, positioning them beyond not only land attack but also aerial bombardment. Many key industrial facilities were designed to be quickly convertible to armaments production. While the Reuther brothers were working in the Gorky tool room, army specialists would show up regularly to supervise the construction of dies and fixtures for making military equipment, which would be tested and stored for possible later use. During World War II, the factory produced cars, trucks, jeeps, ambulances, armored cars, light tanks, self-propelled guns, and ammunition for the military. The Stalingrad tractor factory also poured out light tanks, until the Germans destroyed the factory during the epic battle for the city. The Chelyabinsk tractor plant proved even more important, before the war producing self-propelled artillery pieces, howitzers, and light and heavy tanks. After the Germans invaded, machinery and personnel from other factories, including the Kharkov tractor plant and diesel engine factory, were moved to Chelyabinsk. Over the course of the war, the expanded complex produced 18,000 tanks and self-propelled guns, 48,500 tank engines, and over seventeen million pieces of ammunition. As John Scott wrote in early 1942, the Magnitogorsk plant and the broader Ural industrial district it was part of were "Russia's number one guarantee against defeat at the hands of Hitler," which, of course, also helped ensure the victory of Britain and the United States.[77]

But if the giant Soviet factory contributed to industrialization, modernization, and national defense, its role in the creation of socialism depends on what is meant by the term. As state-owned endeavors, the Soviet giants were part of an economic and social system built around government and—to a lesser extent—cooperative ownership of the means of production. But whether this made the Soviet Union a socialist society, a state capitalist one, or something else entirely was a subject of fierce debate in the 1940s and 1950s and is still a matter of controversy in the much-shrunken universe of people who care about such things.[78]

Did socialism, or state ownership, change internal relationships within the factory? A bit, but not much. Even in the years of purges and terror, Soviet workers felt free to criticize managers and government officials about how plants operated, probably more so than workers at, say, Ford or U.S. Steel before they unionized. But ironically, at the same time that American unions began to grow, Soviet unions, which once gave workers something of an autonomous base, were defanged of independence and real power (though their role in providing social benefits expanded). During the late 1930s, workers sometimes used the atmosphere of suspicion and secret police power to bring down disliked officials. After the terror abated, harsh new labor laws criminalized absenteeism, lateness, and quitting without permission (a throwback to English law at the time of the first factories). More fundamentally, social relations inside the factory remained hierarchical, in much the same way they were in the West. As one journalist wrote, describing the Stalingrad tractor factory, the assembly line was "no longer an issue of disagreement between capitalists and socialists."[79]

When in 1931 H. J. Freyn, who had spent four years in the Soviet Union as a leading consultant to its metallurgy industry, gave a speech to a meeting of the Taylor Society—disciples of the father of scientific management—about the First Five-Year Plan, he described the Soviet Union as a dictatorship, but he felt that at its current stage dictatorship was "essential for the welfare of the people." And in any case, "a modern business enterprise can scarcely be operated or managed by applying the principles of democracy." Like Kahn, Freyn barely mentioned communism when he discussed Soviet industrial development.[80]

The giant factory shaped the path along which the Soviet Union developed, and became a mainstay of ideas of economic growth and modernity in the country for decades to come. But as an institution unto itself, it proved remarkably impervious to its surroundings.

CHAPTER 6

"COMMON REQUIREMENTS OF INDUSTRIALIZATION"

Cold War Mass Production

FROM THE EARLY 1940S THROUGH THE 1960S, IT BECAME common among political intellectuals and academics, especially in the United States, to argue that the United States and the Soviet Union were becoming more similar, that ultimately they would come to closely resemble one another. James Burnham first made this case to a broad public in his 1941 book, *The Managerial Revolution*. Burnham, an American backer of Leon Trotsky, initially accepted the exiled Russian leader's characterization of the Soviet Union as a "workers' state," even if degenerated by Stalinism and the rise of a "Bonapartist bureaucracy." But in late 1939, he broke with Trotsky, coming to see the U.S.S.R. as neither socialist nor capitalist but as a new type of social organism, in which a managerial elite ruled through control of state-owned property. Burnham contended that bureaucratic collectivism, or what he called "managerial society," represented a universal phase of historical development, the actual successor to capitalism, rather than socialism, which had been posited as such by leftists for a century. The Soviet Union, he argued, represented the advance guard of a form of social organization that the United States and European capitalist nations inevitably would come to adopt.[1] A few years later, Friedrich von Hayek, coming from the political right, made a similar claim, seeing a growth of collectivism in capitalist societies pushing them toward the "serfdom" to which he believed socialism was headed.

The idea that the Soviet Union and the United States were converging soon gained traction among American social scientists. The leading sociologist of the post–World War II era, Talcott Parsons, was an early adopter of "convergence theory," which came to be embraced, in one form or another, by such luminaries as C. Wright Mills, Alex Inkeles, Herbert Marcuse, and Walt Rostow. Leftists like Mills and Marcuse fretted that the stifling bureaucracy of Soviet life was being re-created in the West, while Parsons and other liberal proponents of modernization theories thought that the Soviet Union would inevitably become more like the United States.

What these theories shared was the belief that economic development was behind convergence. As Marcuse put it in 1958, both the Soviet Union and the United States were shaped by the "common requirements of industrialization," which pushed them toward bureaucracy, centralization, and regimentation. In effect, these authors believed, modern industry existed as a social and cultural system independent of the economic arrangements in which it was embedded. Ultimately it would mold the larger society. They adopted "industrial society" and "industrial civilization" as descriptive terms and analytic categories that bridged the Iron Curtain, capturing the central features of life in "developed" or "advanced" nations. By contrast, "capitalism" and "communism" were seen in sophisticated academic circles as atavistic slogans, of little explanatory value in understanding modern life.[2]

Ironically, at the very moment when some of the leading minds of the left, right, and center were declaring that industrial development was resulting in a convergence of the capitalist and communist blocs, their actual industrial practices were diverging. Through World War II, in both realms, industrial giantism was adopted as a road to economic development, social progress, and modernity, a heroic effort celebrated in art, literature, and politics. But after the war, American corporations moved away from ever-upping the scale of industry, deciding that the industrial behemoth had reached its limits of profitability and control. Rather than continuing to concentrate production in industrial colossuses, they began to decentralize manufacturing in smaller, scattered plants. By contrast, leaders in the Soviet Bloc—and in other parts of the world—retained a belief in the efficacy of gigantic industrial projects as means for rapid economic growth and as symbols of national prowess and social progress. Though there were multiple reasons for the diverging fate of the giant factory in the United States, the Soviet Bloc, Western Europe, and what came to be called the Third World, the course of labor organization was critical. The intensity of class conflict in the United States brought unprecedented benefits to workers in large-scale industry, making what retrospectively has come to be called the "American Dream" come true, at least for a while. But it also contributed to the demise of the giant factory. Elsewhere, with labor less volatile, industrial giantism continued to be seen as a viable path to the future.

Military Giantism

The downsizing of American factories came after a final wave of industrial giantism during World War II, devoted to making military goods. Some armaments production took place at government facilities, which swelled during the war. The Brooklyn Navy Yard doubled its size, taking over adjacent land to build the world's largest dry docks and the world's largest crane, with its employment roll hitting seventy thousand. But most defense production took place in corporate-run factories, mills, and shipyards, facilities converted to war production or newly built for the purpose.[3]

Albert Kahn designed some of the largest war plants in a last burst of activity before his death in December 1942. They included the Chrysler Tank Arsenal in Warren, Michigan; the East Chicago cast armor plant for American Steel Foundries Company; Amertorp Corporation's torpedo plant in Chicago; the Curtis-Wright Corporation plant in St. Louis; the Wright Aeronautical plant in Cincinnati; and the Dodge Chicago plant, which made aircraft engines (the last three of these were huge structures). But Kahn's largest war plant, the best known of all the wartime defense facilities, was the Ford Willow Run aircraft factory, an effort to bring Fordism to an industry even more complex than the auto industry.[4]

As World War II loomed, in a dash to build up American air warfare capacity, defense officials—and Walter Reuther, by this time a top UAW leader—pressed for the partial conversion of the automobile industry to airplane production. Officials at Ford, which previously had manufactured small aircraft, with limited success, proposed to use assembly-line methods to produce the newly designed B-24 heavy bomber. When defense officials agreed, a crash effort began to build a massive factory and adjacent airport on Ford-owned land in Ypsilanti, Michigan, twenty-five miles west of Detroit. The main building—which covered sixty-seven acres, making it at the time the largest factory structure in the world—went up quickly, but getting production going was a whole other matter. The federal government and Ford proved not much better than the Soviets in starting up such a massive endeavor and faced similar problems in assembling a workforce in an area distant from existing pools of skilled labor (which, in any case, were too small to meet wartime demand).

Part of the blame for Willow Run repeatedly falling behind production schedules—which became a political hot potato—came from applying mass-production techniques to the manufacture of bombers. Creating specialized tools and fixtures delayed the start of parts making, usually done in the aircraft industry using standard machine tools. Repeated design changes from the Army impaired a manufacturing approach predicated on long runs of standardized parts. As in the Soviet Union, slow delivery of materials contributed to delays. So did repeated reorganizations and personnel changes in the federal defense agencies and managerial chaos at Ford. (Contrary to its rationalist public image,

Figure 6.1 The B-24 Liberator assembly line at the Willow Run
bomber plant in Michigan, circa 1944.

Ford suffered from personal fiefdoms, fierce competition among executives, and a lack of clear lines of responsibility.) But an inability to find
and retain enough workers presented the biggest problem.

Across America, defense industries scrambled to find workers, especially with industrial skills. The location of Willow Run added a burden. As construction and production workers began flooding into the
sparsely populated rural area, they found virtually no homes they could
buy or rent, forcing them to room with local residents or live in trailers,
tents, or jerry-built structures, shades of Gorky and Magnitogorsk.

The UAW proposed the construction of a ten-thousand-unit
"Defense City," a permanent new settlement to house plant workers.
It commissioned Oscar Stonorov, a German-born modernist architect
who had designed a union-sponsored housing complex in Philadelphia,
to lay it out. (In 1931, Stonorov, with a partner, had taken second place
in an international competition to design the Palace of the Soviets in
Moscow, besting such celebrities as Le Corbusier and Walter Gropius.)

Defense City, and a similar plan by federal officials for a "Bomber City," proposed multifamily structures and extensive communal facilities, social housing of the sort that had been pioneered in interwar Europe and tried at Gorky and elsewhere in the Soviet Union. Stonorov and his partner at the time, Louis I. Kahn (later famed for his modernist structures, and no relationship to Albert), produced striking designs for a variety of types of dwelling units. But nothing came to pass in the face of fierce opposition from local real estate interests, Ford, and even some union members who—like their Russian counterparts—preferred individual living (in this case single-family homes) to the communalism promoted by left-wing planners. Reversing gears, federal authorities quickly threw up prefabricated temporary dwellings, including—again shades of the U.S.S.R.—worker dormitories.

With living conditions difficult and jobs elsewhere easy to find, workers flowed out of Willow Run almost as fast as they flowed in. Most had little industrial experience, requiring considerable training before they could begin efficient work. Though at one point Ford projected a plant workforce of 100,000, in practice it peaked at 42,506, massive but not massive enough to meet production schedules. Reluctantly giving up the idea of Rouge-style total integration, Ford began moving some B-24 parts production to other plants and even did some subcontracting.

Eventually tools were finished, production methods perfected, and a large enough workforce trained to achieve high-volume output. By 1944, the plant turned out one plane every sixty-three minutes. When production ended in June 1945, the plant had manufactured 8,685 B-24s. Some were shipped out as kits for final assembly elsewhere, but 6,792 were put together on site and flown off, many almost immediately into action.[5]

No other aircraft plant tried as thoroughly to apply mass production methods, but industrial giantism characterized the wartime aviation industry as a whole. North of Baltimore, at Middle River, Glenn L. Martin employed even more workers than Willow Run, 45,000, making B-26 bombers and PBM Mariner flying boats at a complex that included a huge Kahn-designed assembly building with the longest flat-span trusses ever used and massive lift-up doors that allowed airplanes to move in and out. On Long Island, Republic Aircraft Corporation swelled from a few hundred workers to more than 24,000 and Grumman Aircraft from 1,000 to more than 25,000. In the Seattle area, Boeing employed 50,000 workers, nearly half of them women.[6]

Wartime shipbuilding also depended on huge facilities and assembly-line methods. Until the war, ships had been custom-built by highly skilled workers, a practice that continued for naval vessels at facilities like the Bethlehem Steel shipyard at Sparrows Point, Maryland, which employed eight thousand workers. But for cargo ships, needed in massive numbers for the war effort, assembly line techniques were developed, including the standardization of design, extensive prefabrication of parts, the use of welding instead of riveting, and a highly developed division of labor. At Bethlehem's newly constructed Fairfield yard in Baltimore harbor, 45,000 workers, 90 percent of whom had never worked in a shipyard, produced over four hundred vessels during the war. On the West Coast, Henry J. Kaiser, a construction company owner new to shipbuilding, threw up a series of huge yards to produce Liberty ships and other vessels using mass-production methods. His Richmond, California, shipyard employed some 90,000 workers, making it one of the most populous industrial worksites in American history. To support his operations, Kaiser built the first integrated steel mill on the West Coast, in Fontana, east of Los Angeles; constructed new cities for his workers, like Vanport in Portland, Oregon, with homes for nearly ten thousand families; and expanded his prepaid comprehensive medical program, which he renamed Kaiser Permanente—altogether an American *Kombinat*. After the war, Kaiser leased the Willow Run plant from the federal government to produce automobiles for the newly established Kaiser-Frazer Corporation, which remained in the car business until 1955.[7]

Defense production—especially in huge factory complexes—

elevated the social prestige of the blue-collar worker, already raised by the substance and imagery of the New Deal and the great union organizing drives. Political, military, and labor leaders repeatedly stressed the importance of the industrial home front to victory, overlaying patriotism on the Promethean heroism already associated with the giant factory and the workers within it. Flags, bond sale rallies, blood drives, and collection points for British, Soviet, Greek, and Chinese relief made factories, mills, and shipyards into arenas of patriotic expression. Newsreels, billboards, and magazines celebrated war workers—female and male—for their skill and dedication, their ease in operating giant machines and building huge objects, their role in the defense of the nation. Workers responded to such publicity, higher income brought by steady work, unionization, and the tight labor market with a confidence evident in the many short wartime strikes, held in defiance of the union movement's no-strike pledge, and in a jauntiness that characterized the industrial workforce across the country. It could be seen in wartime photographs of industrial workers, like those Dorothea Lange took at the Kaiser Richmond shipyard. Though few realized it at the time, the war brought the giant factory and the blue-collar worker to their apogee in American life.[8]

The Bounty of Unionized Industry

The end of World War II led to a rapid shrinkage of employment at defense plants, fears of mass unemployment, and a tectonic crash between workers and their employers. The immediate issue was the desire by workers for wage boosts to catch up with inflation and compensate for diminished hours once war production ended. But the larger question was the place of organized labor in the postwar world, the desire by unions to solidify their New Deal and wartime gains and by employers to check or roll them back. In the year following the end of the war, five million workers went on strike in the largest strike wave in American history. At its height, in January 1946, two million workers were off the job, including 750,000 steelworkers, 175,000 GM autoworkers, 200,000 electrical manufacturing workers, and more than 200,000 meatpacking workers. Left-wing reporter Art Preis wrote from Pittsburgh of steel plants "sprawled lifeless," while fires to warm pickets formed "a mighty chain up and down the valley and the river banks."

A similar clash had taken place at the end of World War I. Unions had won some battles and lost others (including the steel strike), but, in the face of repression, an economic downturn, and a conservative political turn, the net result was a sharp decline in the size and power of the labor movement. The post–World War II strike wave proved a different story. Generally peaceful, with widespread public support, the big strikes ended with an eighteen-and-a-half-cent per hour wage increase (the equivalent of $2.46 in 2017) or something close to it, a huge boost. For the only time, the United States effectively had a national wage settlement. Price hikes soon cut deeply into the wage gains, but the strikes marked just the beginning of a quarter century of dramatic improvements in pay and benefits for industrial workers.[9]

Before World War II, the newly formed industrial unions had not stressed wage rates, in part because in a deflationary period steady wages meant rising real income. Instead, they fought to check the power of management on the shop floor through union recognition, increasingly detailed contracts, shop stewards, grievance procedures, and the use of seniority in layoffs and jobs assignments. After the war, unions successfully pressed for wage increases and a growing array of employer-provided benefits, including health insurance, pensions to supplement social security, and supplementary unemployment insurance.

The cumulative result was a revolution in the daily lives of workers in large-scale industry, and for their families and communities. Steelworkers' union president Philip Murray once said that for working people a union meant "pictures on the wall, carpets on the floor and music in the home." A quarter century after World War II, workers in heavily capitalized, unionized industry had achieved that and much more.

Things once unusual or unknown among workers—home ownership, modern appliances, vacations, cars and second cars, children sent to college, retirement while still healthy—became common. Unionism grew so established that in 1949 a critic in a left-wing newspaper could write that "In revealing the beauty of factory architecture, [Charles] Sheeler has become the Raphael of the Fords. Who is it that will be the Giotto of the U.A.W.?"

Higher income and welfare programs provided by the government and employers, including pensions, unemployment insurance, disability insurance, and health insurance, gave workers an unfamiliar sense of security and well-being. Many resented the high price they paid for their improved way of life, especially the continued, if diminished, authoritarianism of Fordist production, the monotony of assembly-line work, and the physical toll of manufacturing labor. Still, as Jack Metzgar, the son of a Johnstown, Pennsylvania, steelworker wrote of his family's experience, "If what we lived through in the 1950s was not liberation, then liberation never happens in real human lives."[10]

Dispersion and Downsizing

While for workers the 1945–46 strike wave launched a trajectory of material improvement and union power, for industrialists it brought home a lesson some had begun to glean during the strikes of the 1930s, the danger of extreme industrial concentration in large-scale facilities. Even before the burst of labor militancy in the mid-1930s, a few large corporations had begun to hedge their bets, building smaller plants to supplement their main production facilities. By the late 1920s, the big three tire makers, Goodyear, Goodrich, and Firestone, in addition to their giant plants in Akron, all had factories in Los Angeles to meet the demands of the West Coast market. In 1928, Goodyear built another tire factory, this time in Gadsden, Alabama, a low-wage, antiunion center far from any major tire market. The purpose seemed primarily to lower labor costs and gain a threat to use against Akron workers. After the 1936 strike at the main Goodyear plant, the company expanded the Alabama factory. Other Akron firms also started decentralizing production. By 1938 the Firestone workforce in Akron had fallen from 10,500 to about 6,000, as the company shifted work to a factory it built in Memphis and to other outlying plants. The Goodyear Akron plant shed a fifth of its workforce.

Labor was not the only reason tire companies began dispersing production. Technological innovations and the increasing standardization of tire sizes made it possible to build mass-production plants that oper ated efficiently at smaller scales than the Akron monsters. As car ownership spread and population redistributed, siting plants near growing markets meant lower shipping costs.

But the biggest factor seemed to be a desire to stop being held hostage by small groups of workers. The sequential nature of tire production meant that if one department went on strike, a whole plant might be shut down. And that happened repeatedly in Akron, where sit-downs and other strikes, often begun without official union involvement, became endemic, as a volatile worker culture of direct action developed. In an October 1944 strike at the Goodyear factory, just four striking workers idled five thousand others.

When siting new factories, companies looked for locations where labor costs were lower and unionism was less likely to succeed, or at least be of a less militant sort. Repeated prewar efforts by the United Rubber Workers to unionize Goodyear's Gadsden plant and Firestone's Memphis plant failed, with a reign of terror in Alabama that included severe beatings of union organizers by company thugs and antiunion workers in cahoots with local law enforcement.[11]

The Radio Corporation of America (RCA) also reacted quickly to labor militancy. In 1936, a month-long strike, overcoming imported strikebreakers and police violence, led to the unionization of the company's two-million-square-foot complex in Camden, New Jersey, just across the Delaware River from Philadelphia, where 9,700 workers (75 percent female) produced nearly all of its products. Almost immediately, RCA began moving operations elsewhere, between 1936 and 1947 setting up a component plant in Indianapolis, a radio plant in Bloomington, Indiana, tube plants in Lancaster, Pennsylvania, and Marion, Indiana, a record plant in Hollywood, and a cabinet shop in Pulaski, Virginia. By 1953, only three hundred consumer-electronics jobs remained in Camden. The original complex continued to be an important center for the company, primarily for research and development and manufacturing military equipment, but all mass production of consumer goods had been scattered to smaller plants.[12]

General Motors likewise realized early the threat labor militancy presented to its integrated production system. A 1935 strike at its Toledo transmission factory forced the shutdown of every Chevrolet plant in North America. Soon after, the company launched a $50 million program to expand and modernize its manufacturing, which included building new plants so that the interruption of production at one factory would not halt operations elsewhere. Most of the new factories, which included a plant in Muncie, Indiana, to duplicate the output of the Toledo factory, were in small towns or cities with weak union movements.[13]

The GM program came too late to block the UAW's triumph in 1937. The Flint sit-down and the strikes that followed reinforced the message about industrial concentration. While there might be economies of scale in producing every Chevy engine in a single plant or bodies for all GM cars of a particular body type in a single factory, when workers grew militant it brought danger, too.

No company, even giants like General Motors with huge financial resources, could quickly build factories to duplicate all the production of their most centralized facilities—plants like the Rouge or Dodge Main or the Chevy and Buick complexes in Flint. But World War II provided an opportunity to begin or further the process. As in the Soviet Union, national security dictated the siting of defense plants in the interior of the country, safe from bombardment. Warm weather and vast empty expanses made the Southwest especially attractive to military planners. With government financing, the rubber companies built new tire plants to meet war needs in Iowa, Texas, Pennsylvania, Alabama, Oklahoma, and Kansas. After the war, Washington sold off the plants at bargain prices to the corporations that operated them. Other big wartime defense factories were sold, too, and converted to civilian production, like the North American Aviation bomber plant in Kansas City, Kansas (which had twenty-six thousand workers), taken over by General Motors to assemble cars and, briefly, jet fighters, and the Louisville, Kentucky, war plant that became the nucleus for General Electric's "Appliance Park."[14]

The postwar strike wave provided further impetus for industrial relocation and more but smaller plants. The country had never seen anything like it before. Not only were the strikes huge, they were highly disciplined, with very few workers breaking ranks, even as some of the walkouts dragged on and on, GM for 113 days, textile workers for 133 days, glass workers for 102 days. Corporate leaders found deeply disturbing the support the strikers won in industrial centers. In steel towns, where for a century local officials, newspapers, and businesses had backed the companies in their clashes with labor, now they stayed neutral or supported the strikers. Striking electrical workers won support from college students, the mayors of Cleveland and Pittsburgh, and fifty-five members of Congress. Veterans played a conspicuous role in many of the postwar walkouts, lending them moral capital earned on the battlefields. In Bloomfield, New Jersey, which housed both GE and Westinghouse factories, the local branch of the American Legion, a notoriously conservative group with a history of antiunionism, backed the strikers, even though leftists led their union. In Chicago, pharmacies and grocery stores extended credit to striking packinghouse workers, while priests joined their picket lines. The Truman administration vacillated in its handling of the walkouts, but it took the legitimacy of unionism for granted and ultimately used federal power to force the major corporations to grant large wage hikes.[15]

Figure 6.2 Pittsburgh Mayor David L. Lawrence addressing a crowd of Westinghouse strikers in April 1946.

The strikes made painfully clear to manufacturing companies that they no longer controlled the physical, social, and political environments in which their largest factories operated. GE president Charles Wilson bitterly complained in congressional testimony that strikers had kept even nonunionists—managers, scientists, and office workers— from entering struck facilities. "I don't think a corporation should have to go with its hat in hand to a union and ask permission to bring its engineers and so on into a plant." Politics and daily life in industrial communities changed as prounion politicians got elected to local and state office, small businesses allied themselves with their working-class customers, and unions injected themselves into all aspects of civil life, from the Community Chest to recreational sports to cultural activities. In Yonkers, New York, manufacturing companies like Otis Elevator and Alexander Smith, which, with a peak workforce of seven thousand workers at its massive mill, was the premier carpet manufacturer in the United States, had effectively controlled the town. But after the war, decisions about taxes and public policies became subjects for debate, with a well-organized, ambitious local labor movement throwing around its weight. Giant industrial complexes, once fortresses of corporate power, had become hostages to communities of workers in dense urban centers, where working-class solidarity developed in ethnic organizations, veterans groups, churches, bars, bowling alleys, and social venues, as well as within factory gates.[16]

GE had the most multifaceted response to the upsurge of union power in and around its leading factories. After the 1946 strike, the company named a public relations expert, Lemuel R. Boulware, as vice president of employee and community relations. Boulware took a hard line toward unions, in negotiations presenting the company's offer as a take-it-or-leave-it proposition, while arguing its reasonableness through newspaper advertisements and other media to employees and residents in the towns where GE plants were located. In addition to promoting the virtues of the company, Boulware worked to educate GE workers and the general public about the merits of free-market capitalism, hiring Ronald Reagan to be a spokesperson for the firm in its ideological offensive. GE's efforts, though unusually extensive, were part of a broad corporate campaign to reshape public thinking about the economy, an extended drive to counter the ideological and political impact of the New Deal.[17]

GE and other electrical equipment manufacturers also started transferring operations out of their large factories to smaller plants located in the South, the border states, the West Coast, rural New England, the Midwest, the mid-Atlantic region, and Puerto Rico. The resulting drop in employment in older factories could be very substantial. When GE transferred some of the production of small home appliances from its Bridgeport, Connecticut, plant to new factories in Brockport and Syracuse, New York; Allentown, Pennsylvania; and Asheboro, North Carolina; the workforce shrank from 6,500 to less than 3,000. At the historic GE Schenectady factory, which produced heavy-current products and at its height during World War II employed 40,000 men and women, the workforce plummeted from 20,000 in 1954 to 8,500 in 1965, as the company shifted work to plants in Virginia, Indiana, Maryland, New York, Vermont, and California.[18]

Multiple reasons figured in the dispersals. In the case of GE, building geographically distributed plants was linked to a corporate reorganization, which created decentralized product divisions. As had begun before the war, many companies built plants to be near growing markets, especially in the South and West, facilitated by improvements in transportation, communication, and air conditioning. Modernization sometimes necessitated relocation. In cities like Detroit, few large empty tracts of land with good railroad connections (necessary for producers of large products, like automobiles) remained. As manufacturers sought to replace old, multistory plants with single-story facilities, with room for truck-loading docks and employee parking, they often turned to suburban sites, small or medium-size cities, or even rural areas, where large tracts were readily available. Government incentives also came into play, including tax breaks, tax-free industrial development bonds, and labor training programs, all widely used by Southern states to attract Northern industry.[19]

In the large, theoretical literature on industrial location, labor rarely gets much attention. Differential wage rates are sometimes considered, but the presence or absence of militant workers and unions almost always is ignored.[20] However, in practice, labor often was a key factor in corporate decision-making. One guidebook "for executives charged with evaluating the placement of a company's productive capacity" frankly and matter-of-factly noted an "informal decision rule that some corporations follow is no plant which is unionized will be expanded on-site," a dictum "grounded in management's concern for maintaining productivity and flexibility at its facilities." When companies embarked on major expansions, rather than enlarging unionized plants they generally built new ones, "often new locations in right-to-work states." GE publicly justified its downsizing of older plants and job relocations as an effort to remain competitive with companies using low-wage Southern labor, but privately Boulware discussed it, along with speedup, as a way to discipline the workforce.[21]

Some large corporations with national union contracts faced opposition when they began moving production to areas hostile to organized labor. In 1960, striking workers sought a contractual measure limiting the ability of GE to shift work from Northern plants to the South, but the company rejected the idea and the walkout proved a dismal failure. A decade later, the UAW took on the same issue when it accused GM of a "southern strategy" in building parts plants in Louisiana, Alabama, Georgia, and Mississippi and an assembly plant in Oklahoma City. Ultimately, all the GM plants were unionized, but many companies, like RCA, found that in moving out of established factories to new communities they might end up with unions, but weaker and less militant unions than they were leaving behind.[22]

Not all new plants were smaller than the ones they replaced or partially supplanted, but most were. Sometimes this reflected a desire to multisource intermediate or final products, building plants for just some of the production previously done at a larger factory. Automation also led to downsizing. Many manufacturers embraced new technologies after World War II that allowed machines to be self-regulating and perform tasks that previously required human labor. Motives included greater precision and speed and the elimination of physically onerous tasks. But a desire to lower labor costs and reduce the power of workers contributed significantly to the automation drive.

In the automobile industry, Ford took the lead. Setting up an "Automation Department," the company began shifting work out of the Rouge, which had one of the most militant UAW locals in the country and where wildcat strikes and slowdowns remained common. The labor savings proved considerable. In the mid-1950s, the company transferred production of Ford and Mercury engines from the Rouge to a newly automated plant in Cleveland. It also built a plant in Dearborn to make Lincoln engines. At the Rouge, it had taken 950 workers to make piston connecting rods, but at the Cleveland and Lincoln plants it required only a combined workforce of 292. During the 1950s, Ford transferred many other operations out of the Rouge to more automated plants, including stamping, machine casting, forging, steel production, and glassmaking. As a result, employment at the Rouge shrank from 85,000 in 1945 to 54,000 in 1954 to 30,000 in 1960, making it still one of the largest factories in the United States though only a shadow of what it had been in its heyday.[23]

Dodge Main underwent a similar metamorphosis, as the Chrysler Corporation deintegrated, decentralized, and automated production. From a peak of 40,000 workers during World War II, the plant production workforce shrank to 8,300 in 1963. With parts production moved elsewhere, the sprawling plant housed little more than assembly operations. When, in 1980, the company shuttered it completely, only 5,000 men and women remained.[24]

Automation and mechanization contributed to an impressive rise in productivity. During the quarter century after World War II, employment in the automobile industry plateaued at three-quarters of a million, while output roughly doubled. Between 1947 and 1967, total employment by manufacturing enterprises rose 27 percent, while value added (adjusted for inflation) jumped 157 percent. More efficient management and speedup accounted for some of the boost, but new plants and equipment figured heavily.

Large factories continued to be built; in 1967 there were 574 factories in the United States with 2,500 or more workers, compared to 504 twenty years earlier.[25] But companies rarely erected the kind of giant, showcase plants that had sprung up across the manufacturing belt in the late nineteenth and early twentieth centuries. GE's Appliance Park in Louisville—where the company manufactured refrigerators, washers, driers, electric stoves, dishwashers, disposals, and later air conditioners—was something of an exception. Begun in 1951 on a 700-acre site (eventually expanded to 920 acres), the heavily landscaped complex included six factory buildings, a research and development center, a warehouse, and its own powerhouse. It even had its own zip code. With 16,000 workers in 1955 and 23,000 at its peak in 1972 (15,000 union represented), the complex was large by any standard. But it never reached the size of the workforce at the company's Schenectady complex during its heyday and was only a fraction of the one-time size of such giants as the Rouge and Dodge Main.[26]

The Disappearing Worker

With the shrinkage of the giant factory and broad social changes, the industrial worker faded in popular culture and political saliency. For a brief period after World War II, the media still paid attention. In 1946, *Fortune* sent Walker Evans to photograph the Rouge for a story on "The Rebirth of Ford."[27] One early television show, *The Life of Riley*, featured a Los Angeles airplane worker, occasionally showing the lead character, played first by Jackie Gleason and then by William Bendix, in a factory, riveting wings and complaining about work and the pretensions of the rich (though most episodes revolved around domestic doings). The show lasted until 1958. Blue-collar workers would not again appear regularly on television screens until the 1970s.[28]

With white-collar workers beginning to outnumber blue-collar workers in the mid-1950s, and unions increasingly integrated into established economic and political relationships, intellectuals, too, largely lost interest in the men and women working inside the biggest industrial plants, or at least no longer saw them as key to the future. Left-wing scholars like Mills and Marcuse and many of their followers in the New Left abandoned the idea that the industrial proletariat would act as an agent for progressive social change. While in 1972 there were 13.5 million manufacturing production workers in the United States (more than two million of them working in facilities with 2,500 or more workers), one-time socialist Daniel Bell, a leading sociologist, announced in a book the next year, *The Coming of Post-Industrial Society*. For Bell and many others, "knowledge workers" or "symbolic analysts" had elbowed aside blue-collar workers to constitute the key economic group.

In the late 1960s and early 1970s, there was a brief flurry of political and cultural interest in the discontent of industrial workers—the so-called "blue-collar blues"—but an economic downturn quickly put an end to that. The next time factory workers captured public attention, they did so as a result of deindustrialization and the massive social crisis it brought to the "rust belt." Between 1978 and 1982, employment in the automobile industry fell by a third, with more than three dozen factories shuttered in the Detroit area alone. During those same years, the steel industry shed more than 150,000 jobs. Bethlehem cut ten thousand jobs at Sparrows Point and phased out operations in Lackawanna, New York, and Johnstown, Pennsylvania. U.S. Steel eliminated twenty thousand jobs in Gary, devastating the city, and in 1986 shut down the historic Homestead mill. The worker in the giant factory, once a heroic figure, mastering volcanic forces and massive machines, at least in the United States came to be seen as an atavism, a problem, a sad relic of a passing age.[29]

Soviet Giantism Marches On

As American companies downsized and dispersed their factories, in much of the rest of the world giant industrial complexes continued to be built and celebrated. After World War II, the Soviet Union revived the model of the outsized production facility with an accompanying worker city. Under Soviet influence, the *gigant* model spread to Eastern Europe and China. On the other side of the Cold War divide, the giant factory remained alive and well, too, in parts of Western Europe and some developing countries. As before the war, very large scale industrial complexes were seen as a quick means of economic advance and an efficient investment strategy, especially in countries with centralized planning mechanisms. They also continued to serve important ideological and cultural functions, as carriers of ideas about modernity and the good life and a means of asserting national pride. While in the United States the industrial behemoth was becoming associated with the past—a receding era pictured in black and white—in much of the rest of the world the giant factory remained associated with the future.

The Soviet Union, after being devastated by World War II, initially concentrated on reconstruction. Giant factories like the Stalingrad tractor plant were rebuilt, in many cases continuing to produce military equipment while also resuming the manufacture of civilian goods. Unlike their counterparts in the United States, Soviet managers did not worry about worker militancy or the risk of workers using industrial chokepoints to assert their power.

Magnitogorsk, after playing a vital role in the war effort, doubled in size during the 1950s and 1960s. By the late 1980s, it was the largest steelmaking complex in the world, with 63,000 employees, 54,000 directly connected to steel production, annually putting out almost as much steel as Great Britain. New large-scale infrastructure projects were launched as well, of the sort associated with the First Five-Year Plan—canals, dams, power stations, and irrigation systems—"the giant construction projects of communism."[30]

In the late 1940s and 1950s, the U.S.S.R. also built a series of new cities, variants of the industrial *gigant* model, as centers for scientific research and nuclear weapons production, like Ozersk in the Urals, which housed the huge Maiak plutonium plant. The scientific and atomic cities, in many cases constructed in part by prison labor, like Magnitogorsk were self-contained settlements, with schools, cultural institutions, and housing estates linked to large employers. Many were closed cities, with no access for nonresidents and sometimes no exit for residents, secret places that literally did not exist on maps or in directories.[31]

When the Soviet Union sought to up the production of civilian goods, belatedly embracing the idea of consumer society, its leaders, many of whom had begun their careers with technical training and as factory managers, turned to the giant factory for that as well. For their generation, the First Five-Year Plan had been a formative experience. During his 1959 tour of the United States, Premier Nikita Khrushchev recalled—probably to blank looks from the Americans around him—"when you helped us build our first tractor plant, it took us two years to get it going properly," an episode still vivid in his mind a quarter century later.[32]

In the mid-1960s, the automobile industry once again took the forefront in Soviet industrialization. Vehicle production had languished in the Soviet Union as the military and other industries ranked higher for investment. Also, some communist leaders, most notably Khrushchev, favored mass transit over private car ownership. In 1965, the country manufactured only 617,000 vehicles, mostly trucks and buses, paling before the 9.3 million cars that poured out of U.S. factories. Following Khrushchev's ouster, Soviet leaders set out to jump-start the vehicle industry by returning to the methods of their youth, in 1966 signing an agreement with FIAT for technical assistance and training for a huge new factory to mass produce a version of a current FIAT model. It was the most important foreign commercial contract the country had signed since the deal with Ford decades earlier (which in monetary terms it surpassed).

The Soviets located the plant in Togliatti, a small city on the Volga River that had recently been renamed for the deceased Italian communist leader. Though the site was not selected primarily because of the link to Italy, both sides made the most of the connection, portraying the new plant as an exemplar of Italian-Soviet friendship. The vertically integrated plant, which included its own smelter, eventually covered more than a thousand acres. When it began operation in 1970, it had over 42,000 employees, including nearly 35,000 production workers, with a majority under the age of thirty. The workforce kept growing, reaching an astounding 112,231 (46 percent female) in 1981.

To house the workers and their families, the Soviets created what amounted to a new city, Avtograd. In something of a reprise of the 1930s, young workers from all over the Soviet Union came to build the plant and city (in this case without prison labor). Like other Soviet factory cities, extensive club and sports facilities, schools, libraries, and day-care centers were provided, with the factory taking charge of everything from the local hockey team to a military museum. What made the city unusual, though, were the extensive accommodations made for cars, a novelty in a country where individual automobile ownership always had been rare.[33]

The Soviet government launched a second giant vehicle factory, KamAZ, to build heavy duty trucks in Naberezhnye Chelny, along the Kama River in Tatarstan. One hundred thousand workers were mobilized to build the plant. The Soviets purchased much of the equipment to make a projected 150,000 trucks and 250,000 engines annually from foreign firms. Later, the factory added minicar production. The adjacent city grew to a population of a half million.[34]

The latter-day Soviet vehicle-making giants lasted until the end of the U.S.S.R. itself and beyond. At the start of the twenty-first century, the Togliatti auto company, renamed AvtoVaz, still employed some 100,000 workers (some outside the city). After the company was privatized and looted by managers, oligarchs, and criminal gangs to the point of near collapse, Renault and Nissan eventually obtained majority control. When they began cutting the workforce and reorganizing the plant in 2014, it still had 66,000 employees, far more workers than at any U.S. factory and, with the exception of the Rouge, more workers than had ever been employed at an American auto plant. In a deeply troubled economy, excess staffing served a social-welfare function difficult to disrupt. KamAZ (with Daimler AG buying a minority stake in 2008) kept going, too, producing its two-millionth truck in 2012.[35] Stalinist giantism lived on in Russia long after the statues of Stalin, and the country he helped build, disappeared.

First Cities of Socialism

In the late 1940s, as the Soviet Union helped the communist parties of Eastern Europe consolidate their control, it fostered on the region its template of model socialist cities built around large-scale industrial projects. As had been the case in the U.S.S.R., the motive was partly economic, to promote accelerated growth through concentrated investment in heavy industry. Most of Eastern Europe never had much industry, except parts of East Germany and Czechoslovakia, and much of what there had been had been destroyed during the war or, in the case of Germany, taken by the Soviet Union as reparations. But showplace industrial-urban complexes served important political and ideological functions as well. The Eastern European communist parties were very small when World War II ended, able to achieve power only because of the presence of the Red Army. Communist leaders faced a huge challenge in establishing their legitimacy, mobilizing the population for reconstruction (Germany and Poland, in particular, had suffered massive destruction), and winning popular favor for their Soviet protectors. Model industrial cities, forerunners of new socialist societies, were meant to serve all these functions.[36]

Several of the cities supported new steel plants: Stalinstadt in East Germany, Sztálinváros in Hungary, Nowa Huta in Poland, and Nová Ostrava in Czechoslovakia, part of what one historian dubbed a "cult of steel" linked to the cult of Stalin (whose adopted name meant "man of steel"). Communist leaders saw steel as key to industrial development and arms production, a priority as the Cold War settled in. Breaking the pattern, Bulgaria built its model city, Dimitrovgrad, around a large chemical plant (named after Stalin) and a big power plant. Dimitrovgrad and Stalinstadt also had cement plants, supplying a favored construction material in the Soviet Bloc.[37]

Launched with great fanfare, the new factories and cities were presented as the first living embodiments of what socialism would be, part and parcel of a valorization of industry and workers seen in the iconography and rituals of the new people's democracies. The 100-zloty note issued in Poland in 1948 featured a picture of a miner on one side and an industrial landscape, with rather old-fashioned factory buildings and belching smokestacks, on the other (quite a contrast to the American one-hundred-dollar bill, with Benjamin Franklin on one side and a pastoral view of Independence Hall on the other). Governments called for heroic efforts to rapidly build the industrial settlements. Youth brigades were organized for short-term labor and full-time workers recruited mostly from rural areas. Most workers were young, their presence offered as evidence of the promise of the new societies.

Though each model city had distinct features and a distinct history, reflecting national circumstances, they shared many characteristics. Their planners and architects all consulted with Soviet specialists about overall layouts and even individual buildings. The most striking thing about the new cities was not their socialism but their urbanism. Initially, some of the plans envisioned dispersed housing, eliminating a hard boundary between countryside and city and providing green space and areas for growing food. But the planners quickly switched gears, moving toward higher density, with a concentrated population and no garden plots within city boundaries.

Several factors explain the shift. First, cost. Building apartment blocks, often of standardized design and in many cases using prefabricated materials, was cheaper than constructing many small dwelling units, an important consideration for countries with vast housing needs. Second, compact, dense cities made it easier to provide extensive social and cultural services, important features of cities meant to prefigure what socialist life would be like. Third, the urbanism of the industrial cities constituted an explicit rejection of the postwar vogue in the capitalist West for dispersion: British new towns, Scandinavian satellite towns, American suburban sprawl. (Divided Berlin became a showplace for competing planning visions: density and continuous streetwall in the East; greenswards, lower density, and dispersed buildings in the West.) Grand boulevards and large squares were featured as sites for parades and rallies, but there were smaller-scale urbanist gestures, too, like arcades. The industrial cities were meant to represent modernity, newness, gateways to the future. Anything that smacked of the old rural village, with its individual ramshackle houses and garden plots, seemed a reactionary repudiation of the very spirit of the enterprise.

Though they owed their existence to the Soviet Union, the Eastern European showcase cities served as centers of nationalism. Ritualist expressions of friendship with the Soviet Union abounded, in monuments, buildings donated by the Soviets, statues of Stalin, and the naming of some of the cities and factories for Soviet leaders. But the settlements were projected as vehicles of nation-building, albeit socialist nation-building, not of an abstract, generic socialist revolution. Socialist realism, a forced import from the Soviet Union, ironically furthered this by promoting the somewhat vague idea that buildings should be socialist in content but national in form. Accordingly, many of the buildings at the new factory sites and accompanying cities incorporated motifs and styles associated with national pasts. Building socialism, figuratively and literally, was portrayed as a *national* drama.

Most of the industrial showcases were never finished, at least as originally planned. Stalin's death in 1953 loosened the Soviet reins in its satellite bloc and ended the need for ritualistic deference to the Soviet dictator. Building large industrial facilities and new cities at breakneck speed proved very expensive. What once seemed like economies of scale in concentrating investment on large-scale projects, which were meant to stimulate broader economic development, no longer looked so favorable, as the distorting effects of putting so much financial and political capital into just a few sites became evident. A few years after they were begun, plans for the industrial centers were cut back or abandoned, and what growth did occur generally was improvised and haphazard. Most of the "first cities of socialism" quickly faded into obscurity, renamed and largely forgotten, except as kitschy remnants of the Stalin years.[38]

But not Nowa Huta ("New Mill"), site of the largest and most important of the new factories, arguably the last Stalinist utopia. The idea for a steelworks in central Poland predated World War II and the communist regime. In 1947, the Polish government ordered plans for a large mill from Freyn Engineering, the same U.S. firm that had done work in the Soviet Union, including at Magnitogorsk. But the intensification of the Cold War led to the cancellation of the contract. A 1948 economic agreement with the U.S.S.R. and the creation, the following year, of the Council for Mutual Economic Assistance, linking the Soviet Union and the Eastern European states, provided the framework for a new start. This time the Poles worked with the Soviets, who pressed for a very large facility that would serve the whole communist bloc, much larger than the steel plants around which other model cities in the region arose. The U.S.S.R. lent Poland $450 million to build the plant (a substitute for funds that might have been lent by the United States if the Soviet Union had allowed the Eastern European nations to participate in the Marshall Plan), picked a site six miles east of Kraków, designed the equipment and built much of it, trained 1,300 Polish engineers in Soviet steel plants, and sent skilled workers and specialists to help get the factory going, taking on many of the roles foreign companies had played in the Soviet Union two decades earlier.

In the Stalinist spirit, the government made a crash effort to rapidly build the Nowa Huta plant (later named the Vladimir Lenin Steelworks), the lead project in the Polish Six-Year Plan (1950–55). The sprawling enterprise, on a 2,500-acre site, ultimately encompassing five hundred buildings (including its own power and heating plants), grew in stages over several decades. It began operations with its first blast furnace in 1954. More blast furnaces, coke ovens, a sintering plant, and open-hearth and electrical steel converters followed. By the time the cold-rolling mill went on line in 1958, the plant had 17,929 employees, producing 1.6 million tons of steel a year (half of what twenty-three Polish steel mills had produced before the war), much of it exported to the Soviet Union. And the complex kept growing, with more coke ovens and open-hearth furnaces, a pipe-welding operation, a galvanizing mill, and a basic oxygen steel mill (by this time with some of the equip-

Figure 6.3 Uneven and combined development in Poland, as shown in Henryk Makarewicz's 1965 photograph of the Lenin Steelworks.

ment imported from the West). In 1967, a fifth blast furnace opened, one of the largest in the world and bigger than anything in the Soviet Union, and the plant's workforce reached 29,110. One Polish account argues that the continued expansion of the plant was "clear evidence of the authorities' love of grandeur—motivated more by politics than by economy," with the giant blast furnace, which required anthracite coal, a poor investment. New slabbing and rolling mills followed. Annual output peaked in 1978 at 6.5 million tons of steel and employment a year later at 38,674 (a larger workforce than ever seen at an American steel plant, though smaller than at Magnitogorsk).[39]

Though like the mill, the city of Nowa Huta stood as a national priority, its construction proved a long, difficult haul. While heavy equipment was used in building the steelworks, limited funds meant that the residential and commercial area was largely built by hand, with shovels, wheelbarrows, and occasional cranes. Material shortages and mismanagement slowed construction, while the poor quality of building supplies led to later problems. Authorities used agitation campaigns, labor competitions (which pitted workers against one another), and extra voluntary labor to push the pace of construction at what was dubbed the "great building site of socialism." Women were hired in large numbers, at both the mill and in the construction effort, to promote sexual equality and help meet the demand for labor. Many held blue-collar jobs traditionally reserved for men, like the all-female casting crews in the mill and the bricklayers and plasterers in the city. With housing construction lagging behind the growth of the steel mill and the flood of arriving workers, for years most people in Nowa Huta had to live in crude, cold, single-sex barracks, sometimes with over a dozen men or women sharing a single room, lacking basic sanitary provisions. Magnitogorsk redux.[40]

But by the mid-1950s, the housing shortage and generally miserable living conditions began to ease. Between 1949 and 1958, workers built 14,885 apartments in Nowa Huta, with the original plan essentially completed two years later, as the population reached 100,000. Many residents came to view the city quite favorably.[41]

The pre-1960 part of Nowa Huta forms half an octagon, with major boulevards radiating out from a central square on one edge (in 2004 renamed after Ronald Reagan). The steel mill gates are a half mile away, far enough so that the plant is barely visible from the center of the city, though, no doubt, in its heyday smoke from the mill, a notorious polluter, could have been seen. A tramline connects the mill and the original housing and commercial district.

A distinct urbanism characterizes the city center, reinforced by the appropriation of elements of Renaissance design, like galleries and squares, a marked contrast to contemporary residential developments in the United States of roughly the same size, like Levittown, New York, and Lakeland, California, with their small, single-family, detached houses and automobile-based design. Apartment buildings line the main avenues and fill the areas between them, organized into clusters designed for five to six thousand residents. From along the avenues, the long facades of the housing blocks, ranging from two to seven stories high, feel ponderous, but behind them are enclosures, quiet and

Figure 6.4 An aerial view of Nowa Huta.

humanly scaled, with little traffic. Lawns, playgrounds, schools, day-care centers, garages, and clotheslines fill the space. Each neighborhood unit was meant to be largely self-sufficient, with stores on the ground floors, health centers, libraries, and other services. Cinemas, a theater, a department store, restaurants, and public institutions generally were within walking distance from the residences, while a tramline provided a connection to Kraków proper (which in 1951 administratively absorbed Nowa Huta). The social organization in effect constituted a more fully realized, if less radical, embrace of communal life along the lines of the early worker housing in Gorky.

Plans for Nowa Huta kept being changed, in some ways to the benefit of the city. The first housing units were quite basic, but, keeping with the idea of Nowa Huta prefiguring a new socialist society, many of the estates that followed were built to standards far above the norm for ordinary Poles, with more space, private bathrooms, built-in radios, shared telephones in every entryway, cooling cupboards, and balconies. The blocks completed in the first half of the 1950s had a generic, socialist realist stodginess, but their lower height and smaller scale compared to similar housing elsewhere, like along the Stalinallee (now Karl-Marx-Allee) in East Berlin, avoided the monumentality sometimes wrongly attributed to the city. Contributing to the human scale was the abandonment of plans for an unattractive, towering city hall and a monumental theater, meant to bookend the central axis. Efforts to incorporate traditional Polish elements ranged from the charming, like the octagonal cupolas on the small Ludowy Theater (which housed one of the most innovative theater companies in the country) to the absurd, like one of the two factory administration buildings, built to resemble a Renaissance palace with a "Polish parapet."

With Stalin's death, greater variety crept into Nowa Huta housing, including the modernist "Swedish house" apartment block, derivative of Le Corbusier. Cost-cutting, however, led to the elimination of such features as elevators and parquet floors. As the city population grew to exceed original expectations, new housing estates were built on the outskirts of town. Many of these were modernist in appearance but of poor-quality construction, with low- and high-rise buildings separated by green space, with few nearby stores or amenities, the sort of "tower in the park" developments that became the vogue for urban housing in both the communist and capitalist blocs.[42]

Meant to be a showcase for socialist Poland, Nowa Huta garnered national and even international attention. Over the years, visitors included Khrushchev, Charles de Gaulle, Haile Selassie, Kwame Nkrumah, and Fidel Castro. The steelworks and town figured in numerous novels, journalistic accounts, films, and even musical compositions. The mill appeared on postage stamps in 1951 and 1964. Generally, propaganda and artistic renderings presented Nowa Huta extremely positively, as the start of a socialist future, "the pride of the nation," "the forge of our prosperity." But having been elevated by communist authorities to a prominent place in the national narrative, it also became a pole for critiques of the socialist project. Adam Ważyk's sensational 1955 "Poem for Adults," openly critically of Polish socialism (by a writer until then known as a communist hardliner), painted an ugly portrait of Nowa Huta ("a new Eldorado") and its residents ("A great migration, carrying confused ambitions, . . . A stack of curses, feather pillows, a gallon of vodka, a lust for girls"). Andrzej Wajda's acclaimed film *Man of Marble,* released in 1977, used Nowa Huta for a wrenching, clear-eyed look back at the history and mythology of Polish communism, prefiguring the revolution that would soon come to the steelmaking city, the nation, and the whole communist bloc.[43]

Socialist Citizens

Like their Soviet predecessors, the showcase industrial cities of Eastern Europe were meant to not only produce steel, concrete, and other vital supplies, they also were to produce new men and women, templates for the socialist citizens of the future. One youth brigade in Bulgaria chose as its motto "We build Dimitrovgrad, and the town builds us." But the lived reality proved far more complex.

Some workers did move to Nowa Huta and the other showcase cities out of genuine enthusiasm for the socialist project and the new people's democracies. And some found the experience of helping build and launch new factories and cities intoxicating, something they would look back on fondly. But many workers joined the construction efforts and took jobs at the new plants not out of any particular ideological identification but from necessity.

As in the Soviet Union, the recruitment of construction and industrial workforces was intimately connected to miserable conditions in the countryside, the result of increased taxes, dictated crop sales, collectivization, long-standing poverty, and the impact of years of war. Many rural Hungarians who moved to Sztálinváros were hostile to the communist government because of policies they saw as attacks on their home villages and way of life. The lack of any church in Sztálinváros added to their alienation. For at least some, Sztálinváros came to be seen not as a beacon to a brighter future but as a symbol of everything that was wrong with the socialist state. Experienced industrial workers who came to the pioneer Hungarian city had a more positive view, appreciating the better housing and higher wages available than elsewhere, but nonetheless they often resented the autocratic management in the plant, the intensity of the labor, and the ongoing shortages of food and other goods.[44]

Poland, with the tacit approval of the Soviet Union, did not attempt to collectivize agriculture, so there was no direct link between forced displacement and worker recruitment for Nowa Huta. Nonetheless, the bulk of the construction force and city population came from the countryside, mostly people under the age of thirty. Even in the steel mill, where many jobs required industrial skills, in 1954, 47 percent of the workers came from peasant backgrounds. Many were landless peasants from the immediate area. "Looking into the future," historian Katherine Lebow wrote, "they saw a life of relentless drudgery and cultural marginalization and found the prospect intolerable." More pushed out of their old life than drawn to a vision of a new one, they hoped that Nowa Huta would provide an opportunity to gain skills and money, escape the boredom of rural life, and achieve a brighter individual future. As later remembered by trade unionists, the attraction was not any pride in the idea of working in the country's leading industrial establishment but the desire to enjoy the superior wages, housing, and privileges offered in Nowa Huta once it got past its start-up difficulties.[45]

For many newcomers, Nowa Huta, especially in the early years, proved a disappointment, with its challenging living and working conditions, including high rates of industrial accidents. Many simply left, creating a serious problem of labor turnover (also the case in other showcase cities). Rather than Nowa Huta forging socialist men and women out of peasant stock, the opposite seemed to be happening, as what the communists saw as ills of rural backwardness infected the city. Same-sex barracks, a very large cohort of young men but far fewer women, and the paucity of entertainment, recreation, or religious opportunities led to boredom and rowdiness. Alcoholism became epidemic, despite drastic efforts by authorities to control it. With it came a great deal of brawling and sexual assault, lumped by communist officials into the category of "hooliganism." With civil and familial authority thin and religious authority absent, sexual freedom (and venereal disease) flourished, to the dismay of government officials. And when former villagers did adopt a kind of modernity, it was not necessarily the kind authorities wanted. Some young men became *bikiniarstwo* ("Bikini boys," named after the bomb site, not the bathing suit), who adopted dress and hairstyles modeled after American youth culture.

Similar problems arose elsewhere. In Dimitrovgrad, former peasants took over public parks and courtyards to plant vegetables and raised goats, chickens, and rabbits in the cellars of apartment blocks, until communist authorities finally managed, during the 1960s, to stop the urban farming. In Sztálinváros, young factory workers from urban backgrounds brawled with construction workers from the countryside.[46]

Communist authorities wrung their hands over the behavior and attitudes of the working class they were creating and intensified efforts to inculcate socialist urbanity. In private and sometimes even in public, they acknowledged that the leap to socialist personhood was not taking place as planned. But as long as misbehavior remained outside the political realm, they took no drastic action.

Serious political trouble first occurred in Sztálinváros, not as a reaction to conditions specific to the steel mill but as part of the 1956 Hungarian Revolution. Sztálinváros became a center of revolutionary action, with a workers' council challenging government authority. After troops fired on a demonstration, killing eight, workers fought back, forcing the soldiers to retreat and seizing the local radio station. Later, when the Soviet army arrived to pacify the city, workers joined defecting Hungarian soldiers and officers to defend what its citizens had renamed Dunapetele, the name of the village that had preceded the steelworks. The factory and city that in their very appellation were to be testaments to Soviet-Hungarian friendship turned into the opposite. Ironically, workers finally seemed to embrace an identity linked to the showcase project when they declared that they would defend from Soviet troops what they themselves had built, a form of nationalist expression the planners of Sztálinváros had not anticipated. After 1956, an effort by the new communist leadership, installed by the Soviets, to woo worker support through improved wages and social benefits ultimately shifted opinion in what was once again called Sztálinváros, as a local socialist patriotism developed in the late 1950s and early 1960s, a sense of shared class experience and pride.[47]

Trouble came later in Nowa Huta, following a different course. Steelworkers helped lead a challenge to the ruling powers, at first not over work issues but in assertion of their Catholicism. Like Magnitogorsk and Sztálinváros, Nowa Huta was designed without any church, forcing residents to worship in nearby villages. Requests from the Kraków diocese to build a church in the city were repeatedly turned down until the fall of 1956, when, in response to widespread protests, the Polish Communist Party brought back as its first secretary the once-imprisoned Władysław Gomułka. Attempting to improve relations with the Church, Gomułka gave the OK. A year later a site was chosen and a cross erected there. Then authorities began stalling, and in 1960 reassigned the site to a school, ordering the cross removed. But the crew sent to take it down was blocked, first by a group of neighborhood women and then by a crowd swelled by workers finishing their shift at the mill. The defenders of the cross sang both "The Internationale" and hymns, a sign of their multiple allegiances. The day ended with a full-scale battle between four thousand residents and militia troops, who used water cannons, tear gas, and bullets, while the crowd threw stones, vandalized stores, and torched a building. Nearly five hundred people were arrested, some given substantial prison terms. The authorities, belatedly realizing the explosive symbolism, let the cross remain.

Within a few years, Catholic leaders resumed their campaign for a church, with the backing of the new archbishop, Karol Wojtyła, the future Pope John Paul II. In 1965 the government gave approval for a church near a new housing development. It took an extended campaign to raise money for the building and erect it (with no cooperation from the government), culminating in the consecration of what was called the Lord's Ark by the then-cardinal Wojtyła in May 1977, with seventy thousand people in attendance.[48]

The defense of the cross and building the church helped forge a culture of resistance and networks of mobilization that soon would be used for a more profound challenge to the establishment. But the politics of Nowa Huta were by no means simple. In 1968, when student protests broke out across Poland, authorities had to move vigorously to keep secondary and technical school students in Nowa Huta from joining demonstrations in Kraków. At the same time, workers from the steel mill were bussed into the nearby city, where they beat up students from Jagiellonian University, perhaps reflecting class and cultural antagonisms as much as political differences (as in the hard-hat demonstrations in the United States two years later, when construction workers beat up student antiwar protesters). As late as 1980, about a quarter of the workers in the mill belonged to the ruling Polish United Workers' Party.

By then, intellectual and worker opponents of the Polish regime had become increasingly vocal and well organized. In Nowa Huta, in April 1979, a group drawing on Catholic social teaching, the Christian Community of Working People, formed just months before Pope John Paul II spoke at a monastery on the outskirts of the city, after being denied government permission to visit the Lord's Ark. "The cross cannot be separated from man's work," he declared. "Christ cannot be separated from man's work. This has been confirmed at Nowa Huta."[49]

Both national and local developments undermined steelworker support for the regime. Price hikes in 1970 and 1976 led to widespread worker protests across the country, while in Nowa Huta the construction of a large steel mill in Katowice and a growing environmental movement criticizing pollution by the Lenin Steelworks raised fears about the future.[50] When in July 1980 yet another price hike led to a new wave of strikes, workers in Nowa Huta joined in, winning concessions from management. The following month, they began forming units of the independent Solidarity trade union, founded at the Lenin Shipyard in Gdańsk. The Nowa Huta steelworkers had long had a union, but it had little authority; workers wanting something often went straight to the party, the real power in the shop. When an alternative appeared, workers flocked to it.

By the fall of 1980, with 90 percent of the workforce signed up, the steel mill unit became the largest workplace Solidarity branch in the country, second in importance only to Gdańsk. In a measure of their new confidence to assert their own values, workers began bringing crosses, consecrated at the Lord's Ark church, into the mill (along with Solidarity banners), reversing the flow of culture creation from civic society to the workplace rather than the other way around as communist planners had envisioned. Nowa Huta Solidarity activists also joined in creating the "Network," linking together the largest industrial workplaces in Poland, acknowledging their vanguard role.[51]

The declaration of martial law on December 13, 1981, began a prolonged "state of war" in Nowa Huta (and elsewhere) between Solidarity, now driven underground, and the government. Workers occupied the Lenin Steelworks for three days before militia units with tanks regained control. By the next year, workers had begun building a clandestine Solidarity structure in the mill. The size and resources of the showcase enterprise facilitated organizing. Solidarity activists used mill supplies and printing presses to produce underground newspapers and propaganda in large quantities, for circulation both within and without the complex. Mill technicians helped set up and maintain a clandestine radio network that served the southern part of the country. Supplies lifted from the factory were distributed to Solidarity activists elsewhere. Overseas backers sent aid to the Nowa Huta unionists, who eventually obtained a computer before the mill itself had one.

With so many workers toiling and living together, norms and networks of resistance spread inside and outside the plant, as Nowa Huta became one of the most militant centers of opposition to the government. In 1982, regular protest marches began, first led by workers but over time increasingly consisting of youths. Often the protesters assembled in churches before setting out for the center of the city, inevitably to be confronted by police and militia. In the regular running battles, at least three protesters were killed. Solidarity was less successful in its efforts to hold protest strikes in the mill itself.

In 1988, Nowa Huta helped push the country to a radical resolution of what had become a permanent economic and political crisis. Once again, price hikes led to protest. On April 26, workers at the Lenin Steelworks, still the largest industrial enterprise in the country, launched a sit-down strike demanding an increase in wages and the legalization of Solidarity. Taking control of the complex, workers' spouses and children, sympathetic priests, and outside Solidarity leaders came into the plant to support the protest. On May 4, soldiers took back control of the mill and arrested the strike leaders. But by then, the strike had sparked strikes elsewhere, most importantly at the Lenin Shipyard in Gdańsk. In an effort to end the protests, the government reached out to Lech Walesa, who had helped launch Solidarity, ultimately leading to the "Round Table" negotiations with the group, the legalization of independent unions, and, in 1989, open elections for the national senate. The massive victory by Solidarity candidates brought an end to communist rule in Poland and hastened the end of communist control in all of Eastern Europe.[52]

The rise and ultimate victory of Solidarity demonstrated—too late—to Polish authorities the dangers of factory giantism and industrial urbanism. Nowa Huta, intended, among other things, to create a politicized working class largely out of children of the peasantry, succeeded, but in a way its planners had not anticipated. By the account of Solidarity unionists, Nowa Huta workers came to have a shared pride in working in the plant not because of its role in creating a socialist Poland but because of its role in fighting it.[53] As Goodyear, GM, Ford, GE, and other American corporations had learned decades earlier, large assemblages of workers who work together, live together, pray together, drink together, and die together can turn the largest, most important factories from models of efficiency into weapons of labor power.

The aftermath of victory proved ironic for Polish workers. Giant fortresses of industry, built to lead the transition to socialism, stood little chance of surviving intact the transition back to capitalism. Most of the outsized Polish industrial complexes suffered from underinvestment, low productivity, and overstaffing, lacking advanced machinery found in the West. As government subsidies were lessened, captive markets lost, and privatization begun, they could not compete. What had been the Lenin Shipyard in Gdańsk underwent repeated reorganizations, layoffs, and privatization, until its workforce, 17,000 in 1980 when it gave birth to Solidarity, shrank to fewer than 2,000 in 2014.[54]

In Nowa Huta, one Solidarity unionist, soon after the first noncommunist government took power, estimated that a mill in the West with the same output as the Lenin Steelworks would have 7,000 workers, not 30,000, a measure of more modern equipment, more intense work, and no obligation to keep aging, ill, or alcoholic workers. With production in Nowa Huta plummeting, in 1991 the government, after negotiating with various unions (Solidarity, at that point, represented only about a third of the workforce), began a program of deintegration, spinning off various support functions, like the internal railroad network and slag recycling, and some finishing operations to twenty new enterprises, which together employed about 60 percent of the old workforce. The original company focused only on basic steel operations. To reduce pollution, large parts of the plant were simply shut down, including two blast furnaces, the open-hearth furnace, the sintering plant, and some coke ovens. The broad social mandate for the mill was reduced, too; over the years it had taken on many functions for the workforce and the city, including running a farm, canteen, medical center, vacation facilities, and a football club. These, too, were spun off or downsized.

In 2001 the Nowa Huta steelworks (by then renamed for Polish engineer Tadeusz Sendzimir) were merged with the other major steel mills in the country. Following privatization and a later merger, it became part of the largest steel company in the world, ArcelorMittal. The new owner invested some money in modernization, with an advanced hot rolling mill opening in 2007. But in 2015 only 3,300 employees remained on the payroll, with another 12,000 workers at separate companies linked to the mill. Wages, once considerably above the norm, now were comparable to those at other area businesses. The great heroic days of socialist construction and the fight for faith and freedom were over. The mill had become ordinary, like many others across Europe and the United States, employing a modest-sized workforce, providing only a small percentage of the output of its parent company, and facing the challenge of a worldwide glut of steelmaking capacity—the result of many countries, especially China, still seeing steelmaking as a prerequisite to national greatness and modernity.[55]

Global Giantism

During the era when American companies moved to smaller, dispersed factories and the Soviet Union stuck to the giant factory model, spreading it to Eastern Europe, very large factories continued to be built and acclaimed in other parts of the world, too. Some giant factories operated in Western Europe, most notably in Germany. There also were some very large factories in the developing world.

Today, the largest automobile factory in the world is in Wolfsburg, Germany, where 72,000 workers at a 1,600-acre industrial complex turn out 830,000 Volkswagens a year. With nearly 600,000 employees worldwide, including 270,000 in Germany, the Wolfsburg workforce represents only 12 percent of the company total.[56] Still, no other company in Europe or North America concentrates so many workers at one site.

Germany had an industrial history somewhat different than the United States or Britain. In the nineteenth century, the Krupp steelworks in Essen was one of the largest factories in the world. But in the first half of the twentieth century, small and midsize firms dominated German industry, often working in collaboration with one another, as the country's industrial strength lay in the production of diversified, high-quality goods rather than standardized, low-cost products. There were some very large plants making producer goods—most notably steel and chemicals—but consumer-products plants remained smaller. Though Fordism attracted a great deal of attention, in practice German companies were slow to adopt its production techniques and the very large scale factories that came with them due to capital shortages, trade barriers that limited the scale of the market, and a highly skilled labor force that would be underutilized using American methods.[57]

German auto companies began experimenting with the assembly line in the early 1920s, but only slowly moved toward integrated, mass production. When the National Socialists took power, Adolf Hitler, a great admirer of Ford, pressed the companies to join together to mass produce a German equivalent of the Model T, a "people's car" or Volkswagen. When they declined, the government itself took charge. In 1938, Hitler laid the cornerstone for a Volkswagen factory at what was originally called Stadt des KdF-Wagens bei Fallersleben or the City of the Strength Through Joy Car at Fallersleban (the nearest village, later to be renamed Wolfsburg). Like the Soviets, the Nazis turned to the United Sates for specialized, single-purpose machinery. But the war intervened before the people's car could go into mass production; instead, the factory engaged in war production using forced labor, mostly conscripted in Eastern Europe.

German manufacturers gained experience with mass production making armaments during the war. By the early 1950s, conditions in West Germany facilitated its application to civilian production, as domestic spending power and trade increased. The Wolfsburg factory, which survived the war with little damage, converted back to its original purpose. In a throwback to the early days of Ford, for years it produced only one model, the Volkswagen Beetle, later adding a closely related van. The company resisted building plants overseas to keep up volume and make extensive automation profitable. The German model of codetermination, which gave an extensive role to unions in corporate management, and high wages and generous social benefits (including large profit-sharing payments) helped ensure peaceful labor relations. Unlike contemporary American manufacturers, Volkswagen did not fear that workers might take advantage of concentration to disrupt production and force their will on the company.[58]

Though the Mittelstand of small and medium-sized enterprises continued to dominate the West German and later unified German economy, there were, besides Volkswagen, some manufacturers with very large plants. The chemical giant BASF, once part of IG Farben but reformed as a separate entity after World War II, concentrated production at its long-established complex along the Rhine in Ludwigsafen. In 1963, its managing board acknowledged "that a company whose production volume is concentrated in one geographical spot is especially vulnerable in many respects (e.g. to strikes, earthquakes, and other forces beyond one's control)." Nonetheless, it decided to continue investing and expanding its historic main plant, while later adding others to increase capacity. In 2016, some 39,000 employees worked at the four-square-mile site, which had some 2,000 buildings.[59]

But Volkswagen remained the showcase of German industry and Wolfsburg a temple to factory giantism. Like Henry Ford, aware that a factory could be a merchandising tool, Volkswagen's management built an automobile theme park, Autostadt, next to the main plant, which in 2014 had 2.2 million visitors. Many purchasers arranged to pick up their newly manufactured vehicles there. After German unification, the company built an extraordinary new plant in Dresden to make its highest-priced models. Glass walls make the production process completely visible, with finished cars displayed in a twelve-story glass tower, a Crystal Palace for the twenty-first century.[60]

If Volkswagen exemplified postwar Western European industrial giantism, dependent on stable labor relations through firm-level and national social democratic policies, the Misr Spinning and Weaving Company in Mahalla el-Kubra, Egypt, in the heart of the Nile delta, demonstrated again the explosive potential when giant factories brought together masses of workers and treated them poorly. Year after year, regime after regime, the Mahalla workers have been at the forefront of the Egyptian labor movement, defending their immediate economic interests and increasingly intervening in national political events as well.

The Misr company was founded in 1927 by Bank Misr, an explicitly nationalist enterprise created to fund Egyptian-owned businesses during an era when Britain still occupied the country and controlled much of its economy. Despite the long history of the Egyptian cotton industry, Misr was the first modern mechanized textile plant to be owned by Muslim Egyptians. At the end of World War II, the integrated mill, which did spinning, weaving, and dyeing, employed twenty-five thousand workers, making it the largest industrial establishment in the Middle East.

Egyptian authorities and company officials projected mechanized textile mills as "citadels of modernity, national progress, and economic development." But the workforce, largely recruited from the peasantry, did not accept the elite notion of the mill as a shared nationalist project, repeatedly protesting harsh work conditions and low pay. In 1938, the first large strike at the mill demanded higher piece rates and a switch from twelve-hour to eight-hour shifts. A brief strike in 1946 was followed the next year by a massive walkout protesting layoffs and autocratic management. Tanks entered the plant to crush the strike, and three workers were killed in the confrontation. When in 1952 army officers led by Gamal Abdel Nasser seized power, overthrowing the Egyptian monarchy, workers at the mill expected improved conditions, but when they struck, the army once again smashed their walkout.

In a measure of the symbolic and practical importance of Misr, when in 1960 Nasser took a left turn to embrace "Arab Socialism," the mill was the first industrial enterprise to be nationalized. Under government ownership, the tradition of worker militancy continued, including participation in a three-day strike in 1975 that led to substantial wage increases for industrial workers employed by the state. In 1986, workers struck again, winning a wage hike, and two years later struck yet again, this time explicitly criticizing President Hosni Mubarak. A strike at the mill in late 2006, when the government reneged on promised bonus payments, set off a wave of worker protests at other textile mills and was the prelude to an even bigger strike the following year that won a big boost in the bonuses.

An April 2008 protest by Mahalla workers, broken up by thousands of police, leaving at least three dead, helped spark open opposition to Mubarak, culminating in his tumbling in 2011 during the Arab Spring. In February 2014 workers at the mill struck, demanding the removal of Mubarak-era officials still in company management. Even after yet another quasimilitary regime took power, led by Abdel Fattah el-Sisi, the textile workers kept up their militancy, striking in another conflict over bonuses, to protest a government decision to end cotton subsidies, and to call for the ouster of corrupt company officials. As had happened elsewhere, the launching of a giant factory in the name of nationalism and modernity created a workforce with its own views of what that meant, in a strategic position to make their ideas about the past, present, and future matter.[61]

CHAPTER 7

"FOXCONN CITY"

Giant Factories in China and Vietnam

I N MID-2010, A RASH OF WORKER SUICIDES BROUGHT global attention to a company that three years earlier the *Wall Street Journal* had dubbed "the biggest exporter you've never heard of," Hon Hai Precision Industry Co., operating under the name Foxconn. That year, eighteen workers between the ages of seventeen and twenty-five attempted suicide at Foxconn factories in China, fourteen successfully, all but one by jumping off a company building. Though startling by themselves, what made the suicides a big story around the world was that they occurred at factories that assembled iPads and iPhones, among the hottest consumer goods on the market, symbols of modernity and good living. The juxtaposition of workers feeling so oppressed and alienated by their jobs that they took their lives with the elegantly designed Apple products—seamless, luxurious, futuristic—for a moment raised uncomfortable questions about the sausage factory in which the meat of modernity was being produced, and the human cost of stylish and convenient gadgets.[1]

The corporate reaction to the suicides proved almost as disturbing as the deaths. The companies that used Foxconn to produce their products, including Apple, Dell, and Hewlett-Packard, took a low-key approach, expressing concern and saying that they were investigating. Apple CEO

Steve Jobs called the suicides "very troubling," adding, "We're all over this." In 2012, after more bad publicity about Foxconn, Apple contracted with the nonprofit Fair Labor Association to inspect Foxconn plants and their compliance with the monitoring group's workplace code of conduct. But none of Foxconn's major clients, including Apple, stopped using its services.

Foxconn founder and chairman Terry Gou at first dismissed the suicides as insignificant, given the size of his workforce. But as the deaths and bad publicity mounted and Foxconn's share price fell, the company began to act. In June 2010, Foxconn raised basic wages at its Shenzhen plants, where most of the suicides occurred, from the legally mandated minimum of 900 renminbi a month ($132) to 1,200 renminbi ($176) and in October raised wages again. It also set up a twenty-four-hour counseling center for its workers and put on an elaborate celebration at its largest plant, complete with a parade, floats, cheerleaders, Spider-Men, acrobats, fireworks, and chants of "treasure your life" and "care for each other to build a wonderful future."[2]

But there was a darker side to the Foxconn reaction, too. The company tried to limit its liability for future deaths by requiring employees to sign disclaimers saying "Should any injury or death arise for which Foxconn cannot be held accountable (including suicides and self-mutilation), I hereby agree to hand over the case to the company's legal and regulatory procedures. I myself and my family members will not seek extra compensation above that required by the law so that the company's reputation would not be ruined and its operation remains stable." Worker outrage soon led it to abandon the effort. The company also began moving production from Shenzhen to new factories in the interior of China, largely to lower wages but also believing that if its migrant workers—the vast majority of its employees—were closer to home they would be less likely to kill themselves. Finally, the company began putting wire mesh around the balconies and outdoor staircases at its dormitories and latches on upper-story windows to keep workers from jumping, while surrounding all its factory and dormitory buildings with netting twenty feet above the ground, so that if a worker did manage to leap they would not die. It used more than three million square meters of yellow netting in the process, almost enough to cover all of New York's Central Park if a latter-day Christo got really ambitious. Foxconn's Swiftian response—not to change a production regime that was leading young men and women to jump off buildings, but instead to try to catch them before they hit the ground—seemed a return to the warped utilitarianism of Charles Dickens's Thomas Gradgrind, applied to factories so large that they made Manchester's textile mills look like mom-and-pop shops.[3]

Although some of the stories about the Foxconn suicides noted the very large size of the factories involved, none mentioned that the company's Longhua Science and Technology Park in Shenzhen, better known as "Foxconn City," was, as far as can be determined, the largest factory, in number of employees, in history. Foxconn's extreme secretiveness makes it impossible to be sure about even such basic information as the numbers of workers at its plants, but journalistic and scholarly accounts have reported that at the time Longhua had more than 300,000 employees, and by some accounts more than 400,000, dwarfing even such monuments to giantism as River Rouge and Magnitogorsk, which combined had far fewer workers than the Foxconn plant. A visiting Apple executive, finding his car stuck in a mass of Longhua employees during a shift change, declared, "The scale is unimaginable."[4]

Though no other factory equaled Longhua in its number of workers, there are plenty of other supersized factories in East Asia. Foxconn itself owns many of them. In 2016, the company employed 1.4 million employees in thirty countries, over a million of whom worked in factories in China that ranged in size from eighty thousand to several hundred thousand workers. A second Foxconn factory in Shenzhen, run in close coordination with Longhua, employed 130,000 workers in 2010. One hundred and sixty-five thousand workers produced iPads at a Chengdu factory, a ten-square-kilometer complex several times larger than the Longhua campus. And at peak moments during 2016 an astounding 350,000 workers made iPhones at a Foxconn complex in Zhengzhou, one of the most populous factories in history.[5]

Other electronics firms also have very large factories in China. In 2011, after the trouble at Foxconn, Apple began shifting some iPad and iPhone production to Pegatron, like Foxconn a Taiwanese-owned firm. In late 2013, Pegatron had more than 100,000 workers at its Shanghai plant, including 80,000 living in overcrowded dormitories.[6] Electronics plants with 10,000, 20,000, or even 40,000 workers are not unusual in China. Though small by Foxconn standards, they have more employees than almost any factory in the United States. The 2006 film *Manufactured Landscapes*, about Canadian photographer Edward Burtynsky, begins with a tracking shot moving slowly down the aisle of a factory in Xiamen City, Fujian Province, which housed some 20,000 workers making electric coffeepots, irons, and other small appliances. The shot goes on for nearly eight full minutes, giving a sense of the immensity of a factory with even just 20,000 workers.[7]

A few other industries in Asia besides electronics have very large factories. The Huafang Group, a leading Chinese textile producer, had one factory complex with more than a hundred buildings and 30,000 employees. A few toy factories also are very large.[8] And there are some truly gigantic factories making sneakers and casual shoes.

The Foxconn of footwear is Yue Yuen Industrial (Holdings) Limited, a subsidiary of the Taiwanese firm Pou Chen Corporation, founded in 1969. A little more than an hour's drive north of Foxconn City sits a Yue Yuen factory in Dongguan that in the mid-2000s had 110,000 workers, making it the largest shoe factory in history. Workers produced nearly a million pairs of shoes a month for international brands like Nike (which had offices inside the plant) as well as for Yue Yuen's own YYSports brand, sold through a chain of company-owned retail stores in China. Like Foxconn City and many other Chinese factories, the plant included dormitories and dining rooms for its workers, as well as a reading room and disco built by Nike. Yue Yuen had five other factories in China, including three more in Guangdong Province. Pou Chen, which in 2015 had revenues of \$8.4 billion, controlled shoe factories in Taiwan, Indonesia, Vietnam, the United States, Mexico, Bangladesh, Cambodia, and Myanmar as well. In June 2011, more than 90,000 workers went on strike at a Yue Yuen plant in Vietnam, probably the largest single-site strike anywhere in the world in decades.[9]

Two developments led to the latest chapter of factory giantism. First was the opening up, starting in the 1980s, of China and Vietnam to private and foreign capital, part of national efforts to boost living standards and embrace a modernity increasingly measured by global, largely capitalist, standards. Second was a revolution in retailing in the United States and Western Europe, as in many product lines merchants, rather than manufacturers, became the key players in design, marketing, and logistics. The convergence of these changes resulted in the construction of the biggest factories in history.

Twenty-first-century factory giantism in many ways resembles earlier moments of outsized industrialism, almost eerily so. But in some respects it is quite dissimilar, representing a new form of the factory behemoth. While contemporary Asian factory giants have built on past experience in their organization, management, labor relations, and technology, they play different economic, political, and cultural roles than earlier giant factories. Like the largest and most advanced factories in the past, today's industrial giants embody the possibilities and horrors of large-scale industry. But they do so largely out of the spotlight, hidden rather than celebrated as factories once were.

Maoist Giantism

The giant factories built in China and Vietnam over the past two decades came after one of the last substantial efforts to reconceive the factory as a social institution. In the years following the victory of the communist forces in China in 1949, a complicated story of scale and struggle unfolded in the effort to modernize the country through industrialization. Fitfully, the Chinese communists experimented with new ways of organizing production, not content to simply transplant the factory as it developed under capitalism and Stalinism to revolutionary China. The attempts proved deeply controversial, contributing to divisions that nearly split the country apart and ultimately led to a radical political and economic reorientation.

At first, the factory story in communist China seemed like a rerun of the Soviet experience, much like what was occurring in Eastern Europe. After a period of economic recovery following the end of the civil war, in 1953 the communist government, with Soviet advice, launched a Five-Year Plan. Following the Soviet precedent, China's plan placed heavy emphasis on industry, which accounted for more than half the planned investment in the overwhelmingly agricultural country. Producer goods, especially the iron and steel, machine-building, electric-power, coal, petroleum, and chemical industries, had priority. Six hundred and ninety-four large-scale, capital-intensive projects were to be the driving force for economic growth, a quarter of which were to be built with Soviet assistance. China imported much of the machinery and equipment from the Soviet Union using short-term loans. Like Eastern Europe, China became an heir to an industrial tradition that had traveled from the United States through the Soviet Union, with a stress on specialized tasks and equipment, high-volume output, hierarchical management, and incentive pay.[10]

But even before their Five-Year Plan ended, Chinese leaders began edging away from the Soviet model. First they rejected "one-man management" of factories, seeking broader party and worker involvement, and began abandoning individual incentive pay. Then, in the preliminary planning for a Second Five-Year Plan, priority shifted from huge, capital-intensive projects to smaller-scale, more widely distributed plants, seen as more appropriate for China's limited financial capacity.

The Second Five-Year Plan never was completed because of a more radical departure, the Great Leap Forward, launched in 1958 in an effort to accelerate economic growth through mass mobilization and decentralized innovation. The Great Leap Forward had a deeply disruptive, antibureaucratic thrust. In industry, the new policy embraced "walking on two legs," continuing capital-intensive, large-scale, modern factory development while also promoting small-scale, labor-intensive, technologically simple industry that used local resources. Microindustry was meant to take advantage of underutilized rural labor and materials, serve agriculture, and provide inputs to large-scale industrial concerns. Most famous were the several hundred thousand very small "backyard" blast furnaces built across the country, which, along with small mines to feed them, at one point employed sixty million workers. Local initiatives took on a more prominent role in industrial development, while the importance of central directives diminished.

In addition to experimenting with factory scale, supporters of the Great Leap Forward also tried to break down the division between management and labor within the factory and the unequal distribution of power and privilege between them. In May 1957, the Central Committee of the Communist Party directed that all managerial, administrative, and technical personnel in factories spend part of their time directly engaged in productive activities, exposing them to the conditions, concerns, and views of workers. At the same time, workers were given greater opportunities to participate in the management of factories, or at least to have some say over the behavior of managers. Periodic congresses of workers evaluated managerial action while wall newspapers provided a more immediate outlet for criticism. Some administrative tasks, including accounting, scheduling, quality control, job assignments, and discipline, were shifted from managers to teams of workers. To enable workers to engage technical and administrative issues in an informed way, the country launched a massive program of technical education, reminiscent of the Soviet Union during the 1930s.

The efforts to create small-scale rural industry and give workers greater say over factory management reflected a Maoist belief in the centrality of popular mobilization to economic development and building socialism. But the Great Leap Forward, including its radical experiment with industrial scale, proved a disaster. Output of some goods soared, but they were of such low quality and often in unneeded varieties that they proved virtually useless. Meanwhile, pulling labor out of agriculture to local industry, along with the chaos that came with a weakening of central planning and wild misestimates of upcoming harvests, led to a severe famine. Even the strongest backers of the Great Leap Forward, including Mao, had to acknowledge that economic growth could not be achieved simply through mass mobilization.

Yet even as the Chinese leadership shut down most of the backyard iron furnaces, reasserted central control, and put experts back in charge of industry, experimentation continued, promoted particularly by Mao, in an effort to avoid what were seen as the flaws in the Soviet model and the hardening of hierarchy and bureaucracy at the expense of communist ideals. While again embracing industrial giantism as a path of national development, Mao hoped to grant large enterprises considerable autonomy in order to diminish the complexities and rigidities of central planning and create an environment for greater worker involvement in management.

The Anshan Iron and Steel Company, along with the Daqing Oil Field, became a model for the leftist approach to industrial management promoted by Mao. Anshan, located in the northeast, had been one of the two largest steelmakers in precommunist China, expanded with Soviet help during the First Five-Year Plan. In 1960, Mao approved a "constitution" for the management of the mill, supposedly written by its workers. Though its details were not published, its general principles stressed putting politics in command, relying on mass mobilization, bringing workers into management, avoiding irrational rules and regulations, and creating work teams that joined together technicians, workers, and managers. The "Anshan Constitution" was presented explicitly as a counter to the management approach at Magnitogorsk, which subordinated workers through restrictive rules and regulations.[11]

Giant industrial enterprises, Mao believed, could become anchors for new social arrangements. Rather than simply pouring out a narrow range of goods, a steel plant could also operate machinery, chemical, construction, and other enterprises, in effect becoming an all-purpose commercial, social, educational, and even agricultural and military organization. The factory would be the core of an all-encompassing community, going beyond even the expansive role of large factories in the Soviet Union and Eastern Europe. The Daqing Oil Field, like Magnitogorsk, developed in what had been a sparsely settled area, presented an opportunity to conceive a new type of settlement to break down the urban-rural divide. Unlike at Magnitogorsk, where the Soviets built a new city along conventional lines, at Daqing the Chinese developed dispersed residential areas, while providing support for agricultural production and a range of social and educational services.[12]

Mao believed that the key to the advance to a socialist society, with both greater equality and more rapid growth, lay in the relations of production, not simply in the level of material development. Who ruled the factory made all the difference. But there were plenty of critics among Chinese leaders as a debate unfolded in the late 1950s and early 1960s—somewhat reminiscent of the debate in the Soviet Union during the 1920s—over economic policies and industrial practices. Many Chinese leaders, in the wake of the Great Leap Forward, rather than promoting enterprise self-sufficiency and worker self-rule, called for greater specialization of enterprises and workers and greater use of material incentives.

Minister of Labor Ma Wen-jui represented one side of the debate when in 1964 he argued—much like Trotsky four decades earlier—that modern industry, with its complex machinery and coordinated activity of large numbers of workers, required a particular form of organization, regardless of whether it operated in a capitalist society or a socialist one. Maximizing output "to satisfy the needs of society" remained the "basic task" of state-owned enterprises. Socialism eliminated the inherent class conflict within the factory under capitalism because all output was for the benefit of society as a whole—workers and managers no longer had different interests. But the actual internal organization of the factory need not differ significantly from capitalist models. Ma endorsed worker involvement in overseeing managers but did not anticipate eliminating the distinction between them.

For others, though, a change in ownership constituted only the first step in the transformation of the factory and the larger society. Politics, they argued, needed to take command inside the factory as well as outside of it, promoting not only greater equality but also "the revolutionization of man." Socialism should lessen the distinctions between mental work and manual work and between manager and worker. Practically, that meant requiring everyone associated with the factory to do some physical labor, bringing workers into administrative and leadership bodies, and having the Communist Party oversee factory management. Workers might continue to engage in highly specialized activities within a detailed division of labor, but that would not be all they would do. With their colleagues, technical personnel, and political cadre they would join with managers in determining all aspects of plant operation.[13]

The Cultural Revolution that began in 1966 intensified the struggle over who should run the factory and what it should be doing. The factory, though slow to be drawn into the escalating political strife, eventually became a center of battle as the turbulent political climate encouraged attacks on entrenched factory leaders and the powers and privileges they enjoyed. Worker critics and their allies challenged what they saw as bloated bureaucracies, full of officials doing little of real use, while workers were locked out of participation in such key areas as technical innovation. More radically, supporters of the upsurge questioned the notion that the factory should be understood simply as an economic unit responsible for maximum production. Harking back to Mao's view during the Great Leap Forward, they argued that the factory should be a social institution, serving the multiple needs of its workers and the surrounding community, even at the cost of diminished production and profit. Some pushed for the despecialization of factories, especially in rural areas, so that their equipment and expertise could be used to serve local needs and make varied products for local consumption, rather just a narrow range of products for the national market.

The period of radical experimentation proved short-lived. As political conflict in schools, government agencies, and factories intensified and threatened to spin completely out of control, top communist leaders moved to reassert their authority using the army as their agent, as local Communist Party units were hopelessly sundered. As order was restored, so was hierarchy, though with great variation from factory to factory, as some degree of worker participation in management and experimentation with organizational forms continued. Still, the shift in the tide was clear.[14]

"Feeling the Stones"

The Cultural Revolution led to a break between the first Chinese industrial revolution, based on capital-intensive, state-owned enterprises making producer goods like steel and petrochemicals, and a second, based on labor-intensive consumer-goods manufacturing by privately owned enterprises. The chaos of the Cultural Revolution, followed by Mao's death in 1976, left an opening for reformers, led by Deng Xiaoping, who sought to revive the stagnant Chinese economy and improve Chinese life. In many cases themselves victims of the Cultural Revolution, the reform leaders rejected basic Maoist tenets, including the centrality of mass mobilization and the need to reject all capitalist forms of organization. By the late 1970s, many communists came to believe that China's continuing poverty, and its lag behind not only developed Western countries but also rapidly developing Asian nations like Singapore, stemmed from the country's lack of markets.

To stimulate growth, the reformers sought at least the limited introduction of markets. They also pressed for a shift away from state investment in heavy industry. Somewhat like Bukharin and others in the Soviet Union a half century earlier, they argued that labor-intensive production of consumer goods would provide a more effective path to economic growth and rising living standards in a country lacking in capital but with plenty of underutilized labor. Over time, funds generated by light manufacturing could be channeled into more advanced, capital-intensive endeavors.[15]

Deng and his allies sought foreign capital and expertise to help expand industry without having a long-term blueprint. Instead, Deng called for "crossing the river by feeling the stones." As an experiment, in 1979 the government established "special economic zones" in Guangdong and Fujian provinces, designed to attract foreign businesses. Within these zones, firms would be taxed at lower rates than elsewhere in the country. Additionally, companies could obtain tax holidays of up to five years; repatriate corporate profits and, after a contracted period, capital investments; import duty-free raw materials and intermediate products going into export products; and pay no export taxes. Local authorities within the zones were granted considerable autonomy and generally aligned themselves with the privately owned businesses being courted. Seen as a success, additional special zones were established over the course of the 1980s in other coastal areas and, in 1990, in the Pudong New Area of Shanghai. Two years later came a new set of zones in other parts of the country.[16]

During the 1980s, Chinese leaders came to share the cultlike faith in the power and efficacy of markets associated in the West with Margaret Thatcher, Ronald Reagan, and their followers. The dream of modernity in China, wrote Hong Kong–based social scientist Pun Ngai, became associated with "the great belief in capital and the market," a one-hundred-and-eighty-degree shift from the prior belief that socialism represented a more advanced phase of history. "Search for modernity" and "quest for globability" became catchphrases as the marketization of a once almost completely socialist economy began.[17]

A similar swing took place in Vietnam. The long war with the United States, the subsequent wars with Cambodia and China, and the international boycott after the Cambodian conflict had severely drained the Vietnamese economy. Communist leaders had great difficulty integrating the capitalist economy in what had been South Vietnam with the socialized economy in the North. Measured by per capita income, Vietnam was one of the poorest countries in the world.

In an attempt to revive the southern economy, in 1981 and 1982 local authorities allowed Chinese merchants in Saigon to resume their activities, leading to a burst of prosperity. By 1986, the communists who had led the Saigon effort had won national-leadership positions, promoting pro-market reforms. The Doi Moi ("renovation") policy, meant to move Vietnam toward a "socialist-oriented market economy," included reforms in the state sector and opening up the country to foreign investment, market activity, and export industry. As in China, ideological change accompanied the shift in practical policies, with the Communist Party speaking of the objective laws of the market with a certainty once reserved for the virtues of central planning. Membership in the World Trade Organization (WTO) in 2007 deepened Vietnam's integration into global markets and further facilitated export manufacturing.[18]

In China, the new market-oriented policies rapidly transformed the Pearl River Delta region in Guangdong. The region was selected as one of the first special economic zones because of its relative isolation from the major population and power centers of the country and its proximity to Hong Kong and Macao, and that proved critical to its success. At the time, the economy of Hong Kong (still under British control) depended heavily on manufacturing, trade, and transportation. With land and labor costs rising, the opening up of the adjacent part of the People's Republic provided an opportunity to shift manufacturing to a much lower-cost area with which many Hong Kong businesspeople had family ties. At first, Hong Kong–run businesses largely aimed their operations within China at its domestic market, but by the middle and late 1980s, as the Chinese government eased restrictions on direct foreign investments, export-oriented manufacturing became increasingly prevalent, first in the garment industry, then in footwear and plastics, and finally in electronics.

The Hong Kong–Guangdong combination proved a remarkable profit machine, reflecting the advantages for capitalists of uneven global development. Hong Kong businesses, in many cases with extensive experience in international trade, initially moved their simplest, most labor-intensive operations to the People's Republic, taking advantage of far lower labor and land costs and the free reign they were given in managing labor relations. They kept their administrative, design, and marketing operations in Hong Kong and used the territory's advanced infrastructure, including the world's busiest container port and extensive airfreight capacity, for exporting Chinese-made goods. As the authors of a study of the Pearl River Delta put it, "Third World level costs are combined with First World caliber management, infrastructure, and market knowledge."[19]

As the initial Hong Kong–based forays into manufacturing in China proved successful and the Chinese government further loosened regulations and spent heavily on infrastructure serving the special economic zones, more investment flowed in. Hong Kong firms began shifting more complex manufacturing processes, logistics, quality control, sourcing, and packing to China. At the same time, companies based in Taiwan began manufacturing in mainland China, too, soon followed by companies from Japan and Korea, at first almost always operating through Hong Kong or Macao middlemen. Many of the Taiwanese firms were headed by executives with family ties to the mainland. Terry Gou, head of Foxconn, which built its first Chinese plant in Shenzhen in 1988, was a charismatic army veteran whose family came from north-central China and whose father fought with the Kuomintang before fleeing with Chiang Kai-shek to Taiwan in 1949. Once the United States granted China permanent normal trade relations in 2000 and China joined the WTO the following year, American companies began shifting manufacturing operations to China as well.[20]

Dagongmei and Dagongzai

A measure of the explosive growth of export-oriented Chinese manufacturing can be seen in the dizzying rise of Shenzhen's population, which shot up from 321,000 in 1980 to more than seven million in 2000, one of the most rapid urban growths in history. Most of the new residents were migrants from elsewhere in China who came to work in the factories that were popping up all over.[21] With the local labor pool quickly exhausted, a system of migrant labor developed that has been central to China's second industrial revolution and which has made possible the hyper-giantism of twenty-first-century Chinese manufacturing.

Soviet and Eastern European factories recruited peasants displaced by the collectivization of agriculture. In China, it was the *decollectivization* of agriculture that freed up a workforce no longer ensconced in the benefits and obligations of the collective farm. After Mao's death, communal farms were broken up, with small parcels of land leased to individual farmers under the "household responsibility system," which allowed them to sell produce exceeding quotas on the open market. Initially, the new system brought a rapid boost to the rural standard of living. But further changes, including opening the country to food imports and rising costs for health care, education, and other social benefits, left the countryside far poorer than the cities. Many children from farm families, seeing limited economic and social opportunities at home, moved to the new export-oriented manufacturing centers to take factory jobs.

But usually only temporarily. Unlike those in England or the Soviet Union, Chinese peasants were not dispossessed of their property; though the state continued to own all agricultural land, thirty-year leases gave families effective control. Workers could and did move back and forth between farms and factories, knowing that they had something to return to in their home villages.[22]

In most cases they had to return home, like it or not, because of the Chinese *hukou* system of residency permits, instituted in the 1950s.

Chinese citizens need a permit to live in particular areas and most social benefits, including health care and public schooling, are linked to the specific *hukou* they possess. Migrant workers received temporary residency permits arranged by their employers, which expired when they left their jobs. Obtaining a permanent shift of residency permit to a city was all but impossible. For the first generation of migrant workers, factory jobs (and urban construction work and service jobs) were necessarily interludes, usually lasting a few years, often between the time of finishing or dropping out of school and beginning a family, much as had been the case for New England mill workers.[23]

Migrant factory workers had a different and inferior social status than workers in state-owned enterprises. Until reforms that began in the late 1980s, state-owned and collective employers in China provided a broad range of benefits, including permanent job tenure, training, housing, lifetime medical care, pensions, and other welfare provisions, even subsidized haircuts. Generally, the intensity of work was light and managerial discipline minimal.[24] This was not the case in the privately owned factories that blossomed as the state-owned enterprises began to shrink. Job turnover in the special economic zones was astoundingly high. Many companies provided dormitory housing to migrant workers for free or a fee but otherwise took no responsibility for their welfare. Whatever benefits workers were eligible to receive—including educational opportunities for their children and pensions—came from their home area, where they were registered under the *hukou* system. Private employers were legally required to contribute to social benefit funds for their workers, but like minimum wage and overtime regulations the requirement was frequently ignored. The intensity of work in private-sector factories was high and discipline harsh.[25]

In effect, China developed two quite different systems of factory production, one state or collectively owned, the other privately owned, with different informing ideologies, laws, customs, standards of living, and workforces. Even the terminology for workers differed. Employees at state enterprises were *gongren* ("workers"), holding, at least in theory, the highest social status in China during its communist heyday. Rural migrant workers, by contrast, were often called by newly coined terms *dagongmei* or *dagongzai* ("laboring girl") or ("laboring boy") with the connotation of hired hand, a low-status appellation.[26]

The migrant labor system provided employers with a vast workforce, expandable and shrinkable at will. The pool of rural young men and women was so large that it took nearly two generations before labor shortages began. And it was a pool of cheap labor. Most factories paid migrant workers the legal minimum wage (which in China is set by local governments) or less, as enforcement generally was minimal. Recruiting out of a rural labor market, where living standards and wages were far below urban norms, the coastal export factories did not have to match wage levels for local workers or what state-owned enterprises paid, able to attract workers because the low wages they offered were substantial by village standards. Furthermore, because the factories did not pay for most social welfare benefits for their workers, they were effectively being subsidized by the rural governments that did, allowing their labor costs to be below the cost of social reproduction in the areas they were located. Like Stalinist industrialization, Chinese industrialization has depended on squeezing wealth out of the countryside.[27]

Lodging workers in company dormitories was both a necessity and an advantage for big export factories. Migrant workers, because of housing shortages in the factory boomtowns and their lack of permanent resident status, often had difficulty finding lodging. To attract workers, factories provided it themselves, just as the Lowell mills and the Soviet industrial giants had done. Doing so allowed them to pay workers less than they would have to if those workers had to obtain housing on the open market.

In the early years of private factory growth, most of the migrant workers were young women, so lodging them in company dorms also had an element of providing a chaste environment. One large electronic firm required as a condition of employment that all young, unmarried, single women live in dormitories within the factory complex. Even after men began to be hired for production jobs, dormitories generally remained sexually segregated.

The dormitory system gave companies extraordinary control over their workers. As in the Lowell-style mills, many Chinese factories had (and have) detailed rules for behavior, imposing fines not only for being late to work, poor-quality work, or talking on the job but also for littering or leaving dormitory rooms untidy. Foxconn forbids workers of the opposite sex from visiting one another in their rooms, bans drinking and gambling, and imposes a curfew.

Having workers in company housing allows factories to mobilize large numbers of workers rapidly when rush jobs come in and makes it easier to have large numbers of young women working night shifts. Extremely long working hours—sometimes twelve hours at a time or more, a common practice, especially during busy seasons—are easier to demand if workers live right at the factory.[28]

In the mid-1990s, there were an estimated 50 million to 70 million migrant Chinese workers. In 2008, 120 million. By 2014, more than 270 million, nearly double the number of employed civilian workers of any kind in the United States, an oceanic movement of population from farms to factories and back.

Hometown networks play an important role in the movement, as migrant workers tell sisters and brothers and neighbors about the opportunities and city life and help them find jobs. Provincial and local governments have facilitated the flow. Interior provinces helped recruit workers for factory labor elsewhere, prizing the remittances they sent back home. Some local governments set up offices in Shenzhen to connect workers from their region to foreign-owned factories. Without active state support, the whole system would not have been possible.

The urban employment of rural workers has turned the Spring Festival week around Chinese New Year into an epic of logistics, emotion, and labor recruitment. Each year, millions upon millions of migrant workers return home for the holiday, to be reunited with parents, children, and village friends, in what has become the world's largest, regular human migration. In 2009, the Chinese railway system expected to carry about 188 million passengers during the holiday period. Huge crowds fill stations and spill over into neighboring streets. Ticket systems crash under the weight of demand. Trains and buses are crammed and overcrammed with people and baggage (though the recent expansion of the Chinese railroad system has somewhat eased the chaos). When the holiday ends, not everyone goes back. Each year, millions of migrant workers decide to stay home, forcing factories and other employers to scramble to find replacements.[29]

Why So Big?

Migrant labor made possible the rapid expansion of export-oriented manufacturing in China—and also Vietnam—but it does not explain the creation of factories larger than any ever seen before.[30] For the most part, their size is not a result of technical requirements of production. Look at a photograph of a large sneaker factory in, say, Vietnam and what you will most likely see will be rows of workers sitting at individual workstations assembling precut pieces. (Sneakers and casual footwear are made by gluing and stitching together pieces of rubber, synthetic fabrics, synthetic leather, and sometimes actual leather.) Masses of workers may be under the same roof, but for the most part their labor is individual or in small groups, doing work identical to other individuals or groups nearby, without interacting with them.[31] In this respect, these plants are less like River Rouge or Magnitogorsk and more like the early English textile mills, where weavers or spinners stood side by side doing individual tasks.

Even when products require more complex assembly, there is often no clear relationship between the number of workers needed to make a particular product and the size of a factory. In the EUPA factory, the Taiwanese-owned small appliance plant featured in *Manufacturing Landscapes* and one of Edward Burtynsky's best-known photographs,

Figure 7.1 Workers making Reebok shoes in a factory in
Ho Chi Minh City, Vietnam, 1997.

assembly workers are housed in a vast, modern, single-story shed. But
each assembly line within it is short and relatively simple. Thirty lines
made electric grills, but each had an average of only twenty-eight workers, not the hundreds found on integrated assembly lines in automobile or tractor plants. Rows of assembly workers face each other across
a slowly moving belt. For the most part they use simple hand tools,
without mechanical pacing of production, taking pieces on and off the
belt rather than working on moving components, as in an auto plant.

Electronics firms are notoriously secretive, so it is difficult to get
a full sense of their manufacturing processes. But one account of an
Apple production area within the Foxconn Longhua complex described
assembly lines ranging from dozens to more than a hundred workers
each, larger than the lines in shoe or small appliance factories but still
very modest in size compared to the overall size of the factory, with its
several hundred thousand workers.

Vertical integration adds to plant size. Some footwear plants make
the synthetic materials that go into sneakers and shoes, mold and cut

pieces, and embroider logos. EUPA manufactures most of the parts used in the goods it produces. Foxconn makes some of the components that go into the devices it assembles, though most of the high-end elements come from elsewhere.

Still, even adding in parts manufacturing, technological requirements do not explain giant plant size. Rather it is like Alfred Marshall's comment about cotton spinning and weaving, that "a large factory is only several parallel smaller factories under one roof." At Foxconn City, that was almost literally the case, with separate buildings used to assemble similar products for different companies.

Beyond some point, economies of scale in production diminish or disappear. In his classic study *Scale and Scope* Alfred D. Chandler, Jr., after noting that at one point close to a quarter of the world's production of kerosene came from just three Standard Oil refineries, wrote: "Imagine the *diseconomies* of scale that would result from placing close to one-fourth of the world's production of shoes, textiles or lumber into three factories or mills! In those instances the administrative coordination of the operation of miles and miles of machines and the huge concentration of labor needed to operate those machines would make neither economic nor social sense." Yet something close to that has happened in the production of electronic devices and some types of footwear. In the case of Apple, production concentration has gone beyond what Chandler imagined as absurd; every iPad is assembled in a single factory and most iPhone models in just one or two.[32]

Why are the factories so large? The answer seems to lie in economies of scale and competitive advantages, not for manufacturers, but for the retailers that sell the products they make. This reflects a fundamental shift in relations between the two parties. Until fairly recently, the design, manufacture, and marketing of consumer products generally occurred within the confines of one company. But since the 1970s they have been delinked. And, as sociologist Richard P. Appelbaum has argued, in contemporary global supply chains it is retailers and branders (designers and marketers that depend on others for manufacturing) who have the most power to establish the arrangements and terms of production, not factory owners. Factory giantism serves their interests.[33]

Early in the history of factory production, some of the most successful manufacturers established their dominance by selling their products under brand names and controlling distribution networks. In the United States, the Lowell mills pioneered this approach, which was adopted by such iconic companies as the McCormick Harvesting Machine Company. The Singer Manufacturing Company extended the model to a global scale, as its salesmen and distribution agents sold sewing machines across Europe and the Americas, largely produced in just two factories. The big automobile manufacturers used the model as well, selling cars that they branded—Fords and Chevys, Chryslers and Cadillacs—through independently owned dealerships that they effectively controlled. General Electric, IBM, and RCA likewise sold or leased their products under their own names and exerted considerable influence, if not total control, over distribution networks.

The manufacturer-dominated system of branded products stayed in place in Europe and the United States through the 1970s. Goods producers like Volkswagen, GM, Siemens, Sony, Ford, Whirlpool, Levi Strauss, and Clarks shoes (which first garnered wide attention when its products won awards at the 1851 Crystal Palace exhibition) persisted as household names. The companies, their products, and the factories that produced them remained tightly bound to one another in reality and image.[34]

The severe global recession of the 1970s and a series of subsequent developments unraveled the ties. With profit rates declining as a result of increased international competition, rising energy and labor costs, tight credit, and inflation, many American corporations, under pressure from corporate raiders, sought to reduce costs and shed less profitable operations. To become leaner and more flexible and show a rapid drop in spending, they began outsourcing to other firms functions they had traditionally performed themselves. They tended to start with support services, such as data processing and communications. But over time, companies began outsourcing core functions, too, including manufacturing.[35]

Take sneakers. From their introduction in the nineteenth century through the 1960s, sneakers generally were designed and made by the same companies, mostly large, stodgy rubber firms like United States Rubber Company (Keds) and BF Goodrich (PF Flyers). But then dominance shifted to companies like Adidas, Puma, Reebok, and Nike that were built around athletic footwear and clothing rather than rubber and focused on technological innovation, fashionable design, and marketing. While into the 1980s most of the industry leaders, including Nike, did at least some of their own manufacturing, increasingly they contracted out production, until they became essentially just branders.[36]

In the electronics and computer industries as well, leading corporations began contracting out some of their manufacturing. Sun and Cisco, two Silicon Valley success stories, worked with specialized contract manufacturers, like Solectron and Flextronics (before the rise of Foxconn, the largest such firm), to manufacture advanced products, sold under their brand names. Some companies, including IBM, Texas Instruments, and Ericsson (a large Swedish telecommunications manufacturer), sold off individual factories or even whole manufacturing divisions to smaller firms, with which they then contracted to do their manufacturing. Over time, contract manufacturers became increasingly sophisticated in their design and logistics capacities, partnering with their clients in integrated, multifirm production systems, stitched together by electronic data communication.[37]

During the same years, a revolution in selling took place as well. It had two facets, the rise of new, giant, low-price retailers and the burgeoning of global brand companies that did little or no manufacturing themselves.

In the United States, the new mass retailers had their origins in the 1960s, when a series of discount store chains, including Wal-Mart and

Target, were founded. But it was not until the 1980s that they really took off. Wal-Mart, using a combination of low-wage labor, low prices, advanced technology, and highly efficient logistics, grew into the largest retailer in the world. In 2007 it had 4,000 stores in the United States and 2,800 elsewhere. Though no company came even close to Wal-Mart in size, other retailers based in Europe and the United States, like Carrefour, Tesco, and Home Depot, ballooned through expansion and acquisitions.

With their massive purchasing power, giant retailers won an edge over their suppliers, whether well-known companies like Levi Strauss or obscure firms that made products sold under the retailers' house labels. New communications and logistics technology, including bar codes, computer tracking systems, and the internet, allowed retailers to monitor, communicate with, and direct suppliers on an almost instantaneous basis. Faced with the possibility of the loss of massive orders, companies that made goods for megaretailers were at their mercy and often restructured their operations to meet their needs and desires.[38]

A parallel process developed in the growth of branded product companies like Apple, Disney, and Nike. Such firms achieved massive global sales by concentrating on product design and, above all else, marketing, making their products symbols of hipness, worldliness, modernity, and fun. Some of the big brands at one point or another did some of their own manufacturing, but typically they eventually outsourced most or all of the production of the goods they sold. Koichi Nishimura, the CEO of Solectron, in 1998 said of his customers that "The more sophisticated companies work on wealth creation and demand creation. And they let somebody else do everything in between." Apple initially manufactured its own products, some in factories near its Silicon Valley headquarters. But in the mid-1990s it began selling and shutting down plants, contracting out almost all of its physical production. In 2016 Apple made only one major product, a high-end desktop computer, in the United States. Similarly, in the 1990s Adidas, which had made most of its footwear in factories in Germany, began getting out of the manufacturing business, closing down all of its plants except for one small operation it used as a technology center.[39]

One advantage of contracting out manufacturing was that it distanced brand companies from the work conditions under which their products were made. Seeking lower labor costs usually meant relocating manufacturing to low-wage regions, often with autocratic or corrupt governments; avoiding unions; and paying less attention to worker health, safety, and well-being. If child labor, excessive hours, use of toxic chemicals, repression of unionists, and the like took place within the facilities of a brand company, its image—its most important asset—might well be damaged. But if the problems could be blamed on a contractor down the supply chain, the damage would be less costly and more easily contained. Nike and Apple were both able to survive with remarkably little long-term harm revelations about work conditions and worker treatment in the plants that made their products by blaming contractors, promising better oversight and more transparency, and issuing new codes of conduct.[40]

The location and size of the contract factories serving large retailers and brand companies varied greatly and changed over time. Early on, many American electronic companies contracted with local firms, some in or near Silicon Valley, to build their products. But logistical and political changes made it ever easier to locate manufacturing plants at great distances from contracting firms. Container shipping and expanded airfreight capacity increased the speed and lowered the cost of shipping. Cheap international telephone rates, satellite connections, and the internet improved communications. Lower tariffs reduced the surcharge on manufacturing across borders.

As retailers and brand-name firms like Wal-Mart and Apple relentlessly pressured their suppliers and subcontractors to lower their prices, firms scouted the world for low-wage regions to locate their factories. Mexico was one favored site. So, following the collapse of Soviet communism, were Eastern European countries. Textile and garment manufacturers built plants in Central America, the Caribbean, South Asia, and Africa. Malaysia, Singapore, and Thailand attracted contract electronics manufacturers. And, more and more, manufacturers looked to China to locate their plants, with its vast, cheap labor pool and cooperative government authorities.[41]

The staggering size of orders from transnational corporations like Hewlett-Packard, Adidas, and Wal-Mart made it convenient for them to depend on concentrated production centers, minimizing the administrative and logistical tasks that would result from using many widely scattered suppliers. The changed economics of shipping made it possible for them to concentrate manufacturing in a single small region or just a single factory. In the nineteenth and twentieth centuries, even companies known for centralized, vertically integrated production, like Ford, set up branch plants to assemble products for markets distant from their main factories. But the radical reduction in shipping costs and increase in shipping speed, largely as a result of container shipping and highly efficient port logistics, meant that companies like Apple could supply a particular product to retail stores and internet customers around the world from just one or two locations.[42]

Concentrated production did not necessarily mean big factories. Sometimes it meant industrial districts or centers where many small plants and ancillary services clustered together. In the mid-2000s, over a third of the world's socks—nine billion pairs a year—were produced in Datang, China, not by one company but many, supplying retail giants, including Wal-Mart. Production of neckties began in Shengzhou, China, in 1985 when a Hong Kong company moved its production there. Soon various managers left to start their own companies and tie production grew until the city became the global leader, able to meet orders of hundreds of thousands of units at a time. At one point Yiwu, China, had six hundred factories where workers, who in many cases did not know what Christmas was, produced over 60 percent of the world's Christmas decorations and accessories.[43]

But sometimes scaling up meant just one giant factory. For some products, including footwear and electronics, big buyers, especially brand marketers, have preferred very large factories, which can consistently provide the vast quantity of goods they sell and quickly gear up to make new products or meet rush orders. Apple represents this tendency taken to the extreme. It produces only a very limited number of products but in mind-boggling quantities. Its marketing strategy depends on carefully choreographed, highly publicized annual or semiannual product introductions, stimulating global stampedes by consumers eager to get the newest product and demonstrate their position on the leading edge of technology, style, and modernity. In June 2010, Apple sold 1.7 million iPhone 4s in the three days following its introduction. In September 2012, it sold five million iPhone 5s on the first weekend of sales. Three years later, the company sold more than thirteen million iPhone 6 and 6 Plus units during the first three days after launch. With final product design often locked up only shortly before sales begin, Apple needs to mobilize a vast amount of labor in a very short time to produce inventory for the sales rush to come. Factory giantism has been the solution Apple has adopted, though the giant factories are not its own.

Using giant contract manufacturers, like Foxconn and Yue Yuen, has allowed Apple, Nike, and their ilk to operate without large standing inventories of products that tie up capital and run up warehouse expenses. Even more important, just-in-time production avoids the possibility of being stuck with piles of outdated cell phones, laptops, or sneakers in what are essentially fashion industries. Tim Cook—the Apple executive who masterminded the company's shift from in-house production to contracting out before succeeding Steve Jobs as CEO— once called inventory "fundamentally evil." "You kind of want to manage it like you're in the dairy business. If it gets past its freshness date, you have a problem."[44]

Foxconn and Pegatron keep Apple's milk fresh by rapidly mobilizing hundreds of thousands of young, poorly paid Chinese workers, often under harsh conditions (perhaps closer to evil than inventory). In 2007, just weeks before the scheduled unveiling of the first iPhone, Jobs decided to switch from a plastic to a glass screen. When the first shipment of glass screens arrived at the Foxconn Longhua plant at midnight, eight thousand workers were awoken in the dormitories, given a biscuit and a cup of tea, and sent off to begin a twelve-hour shift fitting the screens into their frames. Working around the clock, the plant was soon pouring out ten thousand iPhones a day. On occasion, to fulfill an order, Foxconn moved large groups of workers from one factory to another in an entirely different part of the country. Meeting surges of demand requires not only a vast army of labor but also a large corps of junior officers, thousands of industrial engineers to set up assembly lines and oversee them, something that China, with its massive program of technical education, can provide. It is this ability to quickly scale up (and, when the rush is over, quickly scale down) production that Apple and other customers prize in the giant contract manufacturing plants that have sprung up in East Asia.[45]

A combination of Fordism and Taylorism facilitates the rapid mobilization of unskilled workers. Apple is ideal for this approach, because it makes a very limited number of highly standardized products, just as Henry Ford did. Some of the final assembly procedures for Apple's computers and mobile devices are highly automated, but most are not. Rather, they involve an extreme division of labor, very simple tasks repeated over, and over, and over again. Workers can be taught them in virtually no time—critical given the very high turnover of workers at factories employing Chinese migrant workers (who have no reason to be loyal to their employers and frequently switch jobs) and the need to bring on fleets of new employees rapidly when big orders come in. The orientation for new hires at Foxconn involves lectures about company culture and rules, but no training in actual production tasks.[46]

Many large contract manufacturing firms cope with big rush orders by subcontracting some of the work to small companies with which they have relationships. Rather than either/or, large and small factories often work in symbiotic relationships, with the bigger companies helping small ones, sometimes just family workshops, to set up as parts suppliers or as subcontract assemblers or processors. Such networks enhance the ability of big firms to quickly scale up production without adding to their fixed costs.[47]

Some contract manufacturers have preferred large-scale factories for their own convenience or out of a kind of corporate vanity, separate from the preference of their customers. The head of a firm that made cases for PCs and game consoles related that he preferred to buy land in low-wage areas close to major markets, build a large factory, and set up suppliers right there. Rather than many small factories, his company runs six big industrial parks spread around the globe. Yue Yuan built gigantic factories in part simply as a strategy to quickly raise its capacity to produce a vast volume of shoes in its successful quest to become the world's largest footwear company. Foxconn's Longhua plant grew very large out of a rush to scale up production, as well as to serve as a show-case for the company and its CEO, Gou. The manager of the complex felt it far too big for efficient operation. Most subsequent Foxconn factories have been considerably smaller, though still very large.[48]

Asian industrial giantism requires state support. In recent decades, the Chinese government has maintained the Soviet and early Mao-era view that very large concentrations of productive capacity are the quickest route to industrial advance and economic growth (a policy Vietnam has followed as well), with distributed, small manufacturing no longer a major thrust. Concentration has not necessarily meant giant factories. The Chinese government actively encouraged the creation of the sprawling clusters of small and midsize firms making specialized products, providing big parcels of land for development, creating industrial parks, building infrastructure and transportation, and providing tax benefits. But often it has meant outsized plants. One manager in the Chinese automobile industry, which is partially owned and heavily guided by government entities, told sociologist Lu Zhang "the government wants big firms. To achieve large scales and high volumes in a short time, we rely not only on highly advanced machinery, but also on our hard-working workers—our comparative advantage." Provincial Chinese governments have embraced industrial giantism as a development strategy. Companies seeking to build large new plants have been offered land (sometimes for free), tax breaks, reduced-cost electricity, and help in recruiting a workforce (including student interns, an increasingly important source of cheap labor for manufacturers).[49]

Inside the Behemoth

What is it like to work in the industrial behemoths of modern Asia? In some ways, the experience is remarkably like that of factory workers generations and even centuries ago in England, the United States, and the Soviet Union. As was the case with nineteenth-century Lowell-style mills, many young women and men have been attracted to twentieth- and twenty-first-century Asian factories by the opportunity to earn money and help their families, build houses or pay for a sibling's education, or amass savings to start a business or bring to a marriage (providing women some protection in case it goes bad). Some women sought to escape arranged marriages, patriarchal control, or family disputes. Just as in the Lowell-style mills, most workers returned to their home village after a few years of factory work to settle down to marriage and family in the countryside, farming or sometimes setting up small businesses.

But factory work in China has not only been a means to make money but also a way to escape rural provincialism and experience city life and what is seen as modernity. The first generation of migrant workers, in the 1980s and 1990s, had little idea of what to expect. Returning migrants were living billboards for a different world. One teenaged woman from an ethnic minority in Guangxi Province recalled that when young people from her village came back for the New Year celebrations in their new clothes, she was envious, echoing the experience of New England teenagers nearly two centuries earlier. She soon left to take a job in an electronics factory. Later migrants were more sophisticated, having seen images of city life and modern factories on television and become at least superficially familiar with fashion and fashionable products through smartphones. One young female worker from Hunan Province, who took a job in an electronics factory near Guangzhou, recalled "When I saw factories on TV, they always seemed so nice: well-built buildings, tiling, and a clean environment, so I thought it would be fun."[50]

Going from the countryside to a factory hundreds of miles away, teeming with tens or even hundreds of thousands of workers, could be deeply disorienting. Recently industrialized Chinese cities do not look like modern equivalents of Manchester. Because so many low-paid workers live in company dormitories, there are not sprawling districts of slums. Some industrial centers, like Shenzhen, contain within them neighborhoods or villages filled with migrant workers and businesses serving them, which reproduce something of the feel of village life. But most new industrial regions are modern and large scale. Upriver from Shanghai, sociologist Andrew Ross reported "Spotless, newly laid highways reached out in all directions. Crowding out all the other buildings were the industrial newcomers—fat, squat warehouses with high-tech roofs, rows of factories as long as freight trains, and a multitude of postmodern boxes that carried the brand of their corporate owners but said nothing about what was done inside their walls." Driving across Dongguan, Nelson Lichtenstein and Richard Appelbaum saw "broad but heavily trafficked streets, continuously bordered by bustling stores, welding shops, warehouses, small manufacturers, and the occasional large factory complex. This is how the cities of the old American rust belt must have once looked, smelled, even vibrated."[51]

Simply finding the way around vast factory complexes like Foxconn City could be bewildering to teenagers who had rarely left their small villages, if ever. The Longhua plant covers over two square kilometers; it takes an hour to walk from one side to the other. Many signs at Foxconn plants were English acronyms, meaningless to newcomers. Frustration and anomie from sudden immersion in an alien world contributed to the rash of Foxconn suicides.

But there was excitement, too. Many migrant workers marveled at new sights and experiences. One worker from Hunan, assigned to a factory dormitory, recollected, "I had never lived in a multi-story building, so it felt exciting to climb stairs and be upstairs." Just as had been the case in the Soviet Union in the 1930s, something as simple—and taken for granted—as a staircase could be the divide between two universes.[52]

The factory giants in China and Vietnam are not sweatshops. Generally, they are recently built and modern looking, though undistinguished. Inside they are mostly clean, orderly, and well lit. Some are air conditioned. As a rule, conditions, pay, and benefits are better in foreign-owned large factories than in locally owned small plants and workshops. And large factories are less likely than small ones to cheat workers out of what they are owed, a big problem in China.[53]

Still, work inside large factories often is difficult and the atmosphere oppressive. Many Taiwanese-owned industrial giants use quasimilitary discipline to control their workforces, full of newcomers. Workers at EUPA, Foxconn, and other large plants wear company uniforms. Plant security is intense. The entire perimeter of Foxconn City is walled, with barbed wire topping some sections. Like River Rouge, entry is only though manned security gates. Identification cards are necessary to enter most large factory complexes and sometimes again to enter particular buildings. At Foxconn plants, surveillance cameras are ubiquitous.

Foxconn puts particular stress on following detailed rules and work instructions—a kind of hyper-Taylorism—enforced by a multilayered management hierarchy. Line leaders, themselves poorly paid workers, supervise individual production lines, in turn overseen by layers and layers of higher-level supervisors. Workers are forbidden from talking on the job (though in practice enforcement varies greatly) or moving about the plant. Slogans adorn banners and posters on factory walls, some reminiscent of Alexei Gastev: "Value efficiency every minute, every second," others more hyperbolic, "Achieve goals or the sun will no longer rise," and still others crudely threatening, "Work hard on the job today or work hard to find a job tomorrow."

At Foxconn and other foreign-owned factories in China, there is no echo of the experiments with worker involvement in management that took place in state-owned enterprises. The genealogy of the internal organization of modern Chinese factory giants lies in Western and Japanese systems of management, not in the earlier years of communist China. Hierarchy is unquestioned, rules and regulations extensive. Quality-control systems, imported from developed capitalist countries, further top-down organization.[54]

Assembly-plant jobs that require the rapid repetition of a series of motions for long periods of time are exhausting and even debilitating, reminiscent of early English textile mills where child workers suffered physical damage from doing the same tasks over and over again. At the Foxconn Chengdu plant, some workers' legs swelled so badly from standing all day that they had difficulty walking. Extremely long working hours compound the problem. Though Chinese laws stipulate a normal workweek of forty hours and limit overtime to nine hours a week, factories routinely ignore them, scheduling much longer workweeks. Schedules of well over sixty hours are not uncommon. At Foxconn, workdays of twelve hours (including overtime) are common, but there and elsewhere, when order deadlines approach, workdays can stretch even longer. Foxconn workers switch between day and night shifts once a month, much like American steelworkers used to rotate shifts every two weeks, leading to sleep loss and disorientation. Though workers like extensive overtime for the boost it gives to their earnings, they have fought to control their hours and raise wages so that huge amounts of overtime are not necessary to make a decent living. At one giant Yue Yuen factory, workers found the mandatory overtime so exhausting that they struck in protest. Just as in Marx's time, much of the struggle between labor and capital in today's megafactories revolves around the length of the working day.[55]

Discipline is another point of contention. In many giant Chinese factories, discipline is harsh and degrading. Firms commonly impose fines for negligent work and even minor rule violations, like talking or laughing on the job, recalling English textile factories, where, Marx noted, "punishments naturally resolve themselves into fines and deductions from wages, and the law-giving talent of the factory Lycurgus so arranges matters, that a violation of his laws is, if possible, more profitable to him than the keeping of them." (By contrast, fines as a form of labor discipline are illegal in Vietnam.) Some foreign managers believe that especially strict disciplinary measures are required in China because of a lax pace of work inherited from socialism, along with a culture of everyone "eating from the same rice bowl," collective rather than individual effort and reward.

At Foxconn, supervisors verbally abuse workers for breaking minor rules. In one instance, a supervisor forced a worker to copy quotations from CEO Gou three hundred times—a cross between schoolhouse punishment and the Cultural Revolution. Security guards sometimes beat up workers suspected of theft or simply violating a rule (shades of the Service Department at River Rouge). Some Chinese factories hire off-duty policemen as guards, giving them a sense of impunity.[56]

Xu Lizhi, a Foxconn worker who committed suicide in 2014, addressed factory discipline in a poem, "Workshop, My Youth Was Stranded Here," published in the company newspaper, *Foxconn People*:

> *Beside the assembly line, tens of thousands of workers line up like*
> *   words on a page,*
> *"Faster, hurry up!"*
> *Standing among them, I hear the supervisor bark.*

In "I Fall Asleep, Just Standing Like That," he wrote:

> *They've trained me to become docile*
> *Don't know how to shout or rebel*
> *How to complain or denounce*
> *Only how to silently suffer exhaustion.*[57]

Some giant Asian factories have had severe health and safety problems. In 1997, an internal report commissioned by Nike found serious problems with toxic chemicals in a large Korean-owned contracting plant in Vietnam. Levels of toluene in the air far exceeded both U.S. and Vietnamese standards. Pervasive dust and oppressive heat and noise added to the poor conditions. In China, too, exposure to toluene, along with benzene and xylene, created hazardous conditions in footwear factories. Chemical solvents used to clean screens are a hazard in electronics factories. Aluminum dust, from making and polishing cases for iPads, presents another danger; workers breathe it in and it can be highly explosive. A 2011 blast at the Foxconn Chengdu plant caused by the dust killed four workers and severely injured eighteen others.[58]

In Lowell, boardinghouses, centers of sociability and relaxation albeit strictly regulated by the companies, provided something of a respite from the monotony, fatigue, and regimentation of the factory. At many Chinese factories that is less the case. About a quarter of Foxconn's Shenzhen workers live in company housing, one of the thirty-three dormitories inside its factory complexes or the one hundred and twenty dorms it rents nearby. Foxconn dorm rooms typically house six to twelve workers, more than housed in Lowell boardinghouse rooms, though unlike in Lowell, each worker has her or his own bunk bed. (Many Taiwanese-owned factories also have higher-grade housing for managers.) Workers are assigned to rooms randomly, so that friends, relatives, workers from the same production area, or workers from the same region rarely bunk together. With some roommates working day shifts and others at night, disruptions come regularly and rooms cannot be used for socializing. As in Lowell, strict rules regulate dormitory behavior: curfews are enforced, visitors restricted, and cooking forbidden.[59]

But many industrial giants, including some though not all Foxconn plants, have extensive on-site social and recreational facilities that provide opportunities for relaxation, socializing, and entertainment. Foxconn City, in addition to dormitories, production buildings, and warehouses, includes a library, bookstores, a variety of cafeterias and restaurants, supermarkets, extensive sports facilities including swimming pools, basketball courts, soccer fields, and a stadium, a movie theater, electronic game rooms, cybercafés, a wedding-dress shop, banks, ATMs, two hospitals, a fire station, a post office, and huge LED screens that show announcements and cartoons. In 2012, a central kitchen used three tons of pork and thirteen tons of rice every day to feed workers. Another company's factory complex, where workers made small motors for electronic devices and automobile accessories, contained a skating rink, basketball courts, badminton fields, table-tennis courts, billiards, and a cybercafé (though workers complained about the lack of Wi-Fi in the dormitories).

At Foxconn City, the giant outdoor television screens and extensive shopping and recreational venues brought consumer modernity into the plant itself, offering workers a taste of the world they left their villages seeking. Migrant workers often quickly assimilate to it. Journalist James Fallows wrote after visiting Longhua in 2012, "At factories I'd previously seen across China, workers looked and acted like country people weathered by their rough upbringing. Most of the Foxconn employees looked like they could have come from a junior college." Many second-generation migrant workers own—or are saving to own—the products they themselves make that symbolize modernity, like smartphones and stylish footwear and apparel.[60]

Militant Workers

Fallows sees China as a feel-good story, a country rapidly moving from working-class living conditions like those in William Blake's England to those like in the United States in the 1920s, and continuing upward. Since China began allowing foreign entities to build and run factories, there has been an enormous decline in poverty, also the case in Vietnam. According to World Bank data, between 1981 and 2012 more than a half billion Chinese rose above a poverty line defined as living on the equivalent (in 2011 dollars) of $1.90 a day or less. Life expectancy at birth rose from sixty-seven in 1981 to seventy-six in 2014. Nonetheless, even at the most modern industrial giants in China and Vietnam, factories with pay and conditions above local norms, workers have repeatedly expressed their dissatisfaction through high turnover rates, strikes, and protests.

In recent decades, China has experienced a massive, if not well publicized, strike wave. The *China Labour Bulletin* details 180 strikes in 2014 and 2015 that involved a thousand or more workers, estimating that it has information on only 10 to 15 percent of all strikes that occurred. By contrast, during those same two years, there were only thirty-three strikes with a thousand or more workers in the United States.[61]

All kinds of factories in China have been hit by walkouts—large and small, state-owned and privately owned. Strikes have occurred at leading industrial giants in the electronic and footwear industries over pay, benefits, and working hours. Tactics, beyond stopping work, have included threatening suicides, blocking roads, and marching on government offices. With many strikers living in company dormitories, stoppages often become de facto occupations or sit-downs.

Even the largest contract manufacturers have been affected. In 2012, 150 workers at a Foxconn plant in Wuhan spent two days on a building roof threatening to jump off to protest a pay cut that accompanied their transfer from Shenzhen and conditions in the new plant. In the Spring of 2014, most of the forty thousand workers at a Yue Yuen factory in Guangdong Province struck to demand that the firm comply with a law obligating it to make pension contributions, one of the largest single-site strikes China has seen. Some protests have been violent. Workers at the Foxconn Chengdu plant rioted several times in fury over uninhabitable dormitory conditions and pay cuts. In one case, it took two hundred police officers to end the protest.

Chinese strikes occur in a legal gray zone. For years, workers had a right to strike, encoded in the constitutions of 1975 and 1978. But in 1982, as the government moved to attract foreign investment and reject the mass mobilizations of the Cultural Revolution, the right was removed from the fundamental law. Now workers cannot openly organize or publicize job actions. But they strike nonetheless. Most walkouts arise with little if any prior organization, no union involvement, and no clear leaders, and last a day or two at most. Often they end when the government intervenes to mediate.

As long as the stoppages are local, short, and nonpolitical, the government generally tolerates them. But if they get out of hand or last too long, physical force and arrests are used to break them up. Authorities want to make sure that labor turbulence does not drive away foreign investors or threaten the political status quo. For their part, foreign factory owners seem confident that the government will keep labor militancy under control, not hesitating to concentrate production in very large plants that if shut down would halt most or all production of particular goods.[62]

Strikes are even more common in Vietnam than China. Workers there have a legal right to strike, though in practice most walkouts have taken place without the elaborate steps necessary for authorization. Worker strikes hit large South Korean and Taiwanese-owned factories making shoes for Nike, Adidas, and other global brands in 2007, 2008, 2010, 2012, and 2015. The gigantic 2011 strike at the Yue Yuen factory, protesting low wages, captured international attention for its sheer size.

Even more startling were the riots three years later, which damaged or destroyed scores of foreign-owned factories outside of Ho Chi Minh City. The disturbances began with a rally of workers protesting China's deployment of an oil rig into waters claimed by Vietnam. But the protesters soon turned against nearby sneaker and clothing factories, many of which were Taiwanese, South Korean, Japanese, or Malaysian owned, angry about stagnating wages and foreign exploitation. A staff person at the Taiwanese Chutex Garment Factory reported that some eight thousand to ten thousand workers were involved in an attack on the plant, burning "everything, all of the materials, computers, machines."[63]

In China, worker militancy has pushed up wages and improved conditions, aided by pressure from international labor rights groups and brand companies afraid of their reputations being sullied by stories of worker abuse. Even so, by the 2010s large factories were having difficulty recruiting and retaining migrant workers. The rapid expansion of manufacturing, a shrinking rural population, a gender imbalance favoring men, and the growth of service-sector female employment meant that the pool of young women from the countryside that the factories preferred was effectively tapped out. Foxconn and other firms were forced to broaden their hiring practices, turning to men—who now constitute the majority of Foxconn employees—and older workers.[64]

Companies responded to rising wages and labor shortages by building new plants in lower-wage regions of central China. Many also turned to semicoercive measures to recruit and retain workers, echoes—though much attenuated—of practices from the earliest days of the factory. Some companies insisted that migrant workers make "deposits" to obtain their jobs, which would only be refunded if they left with permission of the firm. Similarly, companies withheld parts of workers' wages, promising to pay them at the end of the year.[65] Larger factories, under greater scrutiny and more attuned to international standards, were less likely to engage in such tactics. Instead, they turned to student interns as a new, cheap labor supply.

Chinese vocational schools require completion of a six-month or one-year internship before graduation. Foxconn and other firms have exploited this requirement by working with government and educational authorities to have large numbers of student interns sent to their factories, along with their teachers, who serve as de facto foremen and forewomen. In the summer of 2010, Foxconn had 150,000 interns, including more than 28,000 making Apple products at its Guanlan factory in Shenzhen. Generally, interns engage in basic production jobs that have no relationship to their field of study. Instead, the internships are simply enforced labor—students can leave, but doing so jeopardizes their ability to graduate. Interns receive basic entry-level wages but no benefits, making them cheaper than regular employees. Though not bound labor like parish apprentices in English textile mills, the students, who have become an increasingly important component of the Chinese factory labor force, are not exactly free workers hired through an open labor market, either. Rather, they are mobilized by state-company institutional arrangements that give them no real freedom of choice.[66]

Hiding in Plain Sight

The giant factory in China and Vietnam has not received the kind of notice it did in its earlier incarnations in England, the United States, the Soviet Union, and Eastern Europe. Considerable attention has been paid to the plight of migrant Chinese workers, particularly in film, but much less to the factories they work in.[67] Part of the reason is the secretiveness of factory owners, who for the most part see only a downside in allowing their facilities to be visited or documented. In the nineteenth and twentieth centuries, companies saw their factories as good advertising, symbols of their position at the cutting edge of industry and a way to get their products better known among consumers. Soviet and Eastern European authorities viewed their giant factories as showcases for socialism, also appealing, in a different way, to a broad public. By contrast, owners of giant Chinese and Vietnamese manufacturing enterprises do not want anything to do with the public. For the most part, their customers are not end users but other companies. And as far as those companies go, by and large the less known about the manufacturing processes the better.

For one thing, companies like Apple and Adidas want to keep secret proprietary methods and details about products about to be introduced. For another, they fear criticism of the working conditions under which their products are made, including by international social-justice groups adept at circulating images and information about worker abuse.

While ordinary tourists could visit River Rouge, and still can, the idea is unthinkable with Foxconn plants or most other giant factories in China. Scholars, journalists, and documentarians have great difficulty getting past factory gates, and when they do, they are closely guided by minders, not given full access. The leading media of their day were awash with images of English textile mills, Lowell, Homestead, the Stalingrad *Tractorstroi*, and Nowa Huta. By contrast, photographs of factories owned by Foxconn, Pegatron, and Yue Yuen are surprisingly uncommon, and pictures of what goes on inside them rarer still.[68]

Because the largest Asian factories do not serve as advertisements or symbols for the products made within them, there is no incentive to invest in distinctive or innovative architecture, as leading manufacturers did in the nineteenth and twentieth centuries. There is no Belper Round Mill or FIAT Lingotto in China. Instead, there are generic factory buildings, modern looking but utterly lacking ornamentation or distinguishing features, even the distinctive fenestration that once marked major manufacturing sites. Many Chinese plants look like they could be suburban office buildings. *Bloomberg Businessweek* described Foxconn City, with its multistory buildings faced in gray or white concrete, as "drab and utilitarian." In recent decades, China has been the leading world center for hiring celebrity architects to build unusual, large-scale, modernist structures, but they are office buildings, concert halls, stadiums, museums, libraries, shopping malls, and hotels, not factories.[69]

Recently built factories in China and Vietnam are not held up as sources of national pride, as steel mills in Braddock, Pennsylvania, and Nowa Huta once had been. Unlike the showcase factory giants of the past, the new massive factories in China and Vietnam are largely foreign owned, run by foreign managers, making goods largely for consumption out of the country. Rather than symbolizing how advanced their host countries are, they serve as reminders of how much catching up they have to do to match countries like South Korea, Taiwan, and Japan in technology, design, and management.

Many leaders in the developing world, including China, do not see having locally owned, large-scale manufacturing as their real target nor as a badge of entry to the club of First World nations. They are acutely aware that rich countries like the United States have been shedding mass-production manufacturing, concentrating instead on higher-end production of specialized goods, design, technological innovation, marketing, services, and finance. Basic manufacturing, for better or worse, seems like yesterday in much of the advanced world, especially the United States, an attitude picked up in less developed countries. Modernity does not mean the assembly line for Chinese policy makers and elites. Rather, they see mass manufacturing as a stage to go through and leave behind in achieving modernity. Chinese officials still see a role for mass production in raising living standards; they hope to hold on to lower-end, lower-paid manufacturing by moving it into poorer interior regions. But in wealthier parts of the country, including pioneer special economic zones, the push is to move beyond basic assembly-line production. In Shenzhen, the epicenter of the explosion of Chinese industrial giantism, older factories are being knocked down to build upscale residential and commercial buildings.[70]

Seen more as a necessity than a triumph, giant Chinese and Vietnamese factories are devoid of the heroic overtones associated with earlier large-scale industrial projects or with modern Chinese infrastructure projects like the Three Gorges Dam or the skyscrapers, bridges, and high-speed rail lines that have remade the landscape. In part, this is an issue of gender; modern apparel, footwear, and electronic plants are heavily staffed by women, unlike steel and automobile factories and big construction sites, where men have dominated the workforce, and largely still do. Heavily female industries sometimes have been associated with utopian dreams, like the early New England textile mills, but Promethean daring generally has been associated with brawny male workers, workers resembling the common portrayal of Prometheus himself.[71]

The nature of the products that pour out of Asian factory giants contributes to their banality. The twenty-first-century factories with the most employees typically churn out small things, like coffeepots, sneakers, or smartphones, which could fit into a small box or the palm of a hand, not the large, awe-inspiring cannons, beams, machines, vehicles, and aircraft produced by the largest nineteenth- and twentieth-century factories. Billions of people worldwide may want iPhones or Nike sneakers and see them as symbols of modernity, but these accessories lack the world-historic aura of the products that came out of the giant steel mills and auto plants of yore.

Rather than representing an enlargement of the human spirit, modern factory giants often seem to symbolize its diminishment. Images of Chinese factories typically do not celebrate machinery or man's mastery of nature but instead document bland, boring structures or portray repetitiveness—size as endless replication.[72] What makes Burtynsky's photographs of Chinese factories so extraordinary is not the extension of human power through the mastery of materials and machines or the beauty of the machinery itself, themes of so many earlier factory portrayals, but the shrunken scale of people, regimented in lines and grids within the vast confines of factory sheds. Burtynsky, like Andreas Gursky, also known for his spectacular photographs of factories and public spaces in Vietnam and China, generally takes large-format pictures from a distance, showing humans in almost abstract patterns, rarely focusing on any individual, the way earlier factory photographers like Margaret Bourke-White and Walker Evans did, at least occasionally.[73]

Foxconn, Yue Yuen, and the other modern giants of Asian manufacturing represent a culmination of the history of industrial giantism. They build on the past, incorporating all the lessons about assembling and coordinating masses of workers, the detailed division of labor, externally powered equipment, mechanical transfer of components and pacing of production, economies of scale, and shaping every aspect of workers' lives. All of the past lives in the present. But the future does not, except in the most limited, technical way. The giant factory no longer represents a vision of a new and different world a-coming, of a utopian future or a new kind of nightmare existence. Modernity, Foxconn-style, may be associated with higher living standards and innovative technology but not with a new phase of human history, as giant factories once were, whether it be the coming of a new type of class society in England and the United States or a new type of classless society in the Soviet Union and Poland. The future has already arrived, and we seem to be stuck with it.

CONCLUSION

OUTSIZED FACTORIES HAVE BEEN WITH US FOR THREE centuries. But no individual factory has lasted anywhere nearly that long. The Lombes' Derby Silk Mill, the first modern, large factory, also proved one of the longest lasting; with ups and downs, workers continued to produce silk thread in the mill until 1890, a run of 169 years. By contrast, Awkright's first cotton mill in Cromford all but shut down within seventy years. The first mill complex in Lowell, Massachusetts, built by the Merrimack Manufacturing Company, outlasted its successors, producing textiles for 134 years. Amoskeag, once the largest textile complex in the world, closed after barely a century. The pioneering Cambria Iron Works in Johnstown, Pennsylvania, operated until 1992, a total of 140 years, thirty-five years longer than the Homestead Steel Works, the scene of epic labor battles.[1]

Some landmark factory giants have kept going. Though Dodge Main, the Chevy complex in Flint, and FIAT's Lingotto plant all shut down decades ago, River Rouge remains part of Ford's now decentralized production system, with some six thousand workers pouring out F-150 trucks, the best-selling vehicle in the United States.[2] The Stalingrad *Tractorstroi*, Magnitogorsk, and Nowa Huta likewise continue to operate.

One hundred or one hundred-and-fifty years might seem like a long time, but many other institutions routinely function in their original buildings for far longer: parliaments, prisons, hospitals, churches, mosques, colleges, prep schools, even opera companies, to name a few. Seen from a distance of time, giant factories lose the solidity, the air of permanence, that so impressed contemporaries. Few last more than a lifetime or two.

The very dynamism of modernity that creates the giant factory leads to its demise. Giant factories have a natural life cycle. They arrive with explosive force, transforming not only production methods but whole societies. Their success generally rests at least in part on the exploitation of workers previously outside the labor market—children and adolescents, small farmers and peasants, nomads, prisoners, and wards of the state. During a period of primitive accumulation, workers could be exploited, sometimes brutally, through long hours, low pay, and harsh conditions, because they lacked the freedom of movement, legal rights, or ready alternatives.

Radical factory innovations have been followed by periods of incremental improvement or stagnation. The vast amounts of capital tied up in existing buildings and machinery promotes institutional conservatism, allowing new competitors using more advanced methods and technology to become more efficient producers. Meanwhile, worker protest and pressure from reformers pushes up labor costs. Some companies succeed in extending the high profits of primitive accumulation by repeatedly recruiting new workforces, new waves of young workers or immigrants from afar. But at some point, a combination of archaic technology, aging buildings, and rising labor costs forces a decision about whether to modernize, start over elsewhere, or milk a property and then shut it down.

The owners of the Boott mills in Lowell were typical. In 1902, a consultant they hired reported, "Your old buildings have perhaps served well their purpose in the past, but they were long ago out of date, and are of no value now.... I therefore recommend the entire demolition of the present structures, or at least so much of them as are dangerous to work in, or would in any way interfere with the best arrangement and construction of a first-class new mill." But at a time when many New England mill owners were investing in factories in the lower-wage South, Boott's owners decided, instead of building a "first-class new mill," to limp along using their dangerous old facility, where workers kept making textiles and profits for investors for another half century.[3]

In the socialist world, cost calculations played out differently, because workers had greater ideological and political standing and big factories were more central to social welfare systems. Closing factories, or even downsizing them, presented huge social and political risks that states shied away from, instead keeping bloated workforces at increasingly uncompetitive plants. Even today, the Chinese government moves gingerly in its prolonged effort to shut down unneeded or inefficient state-owned factory giants. But the collapse of the Soviet Bloc and the move of most of what remains of the communist world toward market economies has brought a convergence in industrial practices between the once-socialist and always-capitalist spheres.

Taken together, the overlapping cycles of giant factory development, spread across time and space, represents continuity and progress, with ever bigger and more efficient manufacturing operations appearing, which nonetheless retain a clear genetic heritage going back to the Derwent Valley three hundred years ago. Imitation, licensing, and theft have allowed each wave of factory developers to incorporate past innovations, as industrial giantism leapt across oceans and political divides. But if resilient and durable as a totality, industrial giantism in any given place has proven unsustainable. Specific communities have experienced the giant factory not as a continuity of progress but as an arc of disruptive innovation, growth, decline, and abandonment. As historians Jefferson Cowie and Joseph Heathcott wrote in a book about American deindustrialization, "the industrial culture forged in the furnace of fixed capital investment was itself a temporary condition. What millions of working men and women might have experienced as solid, dependable, decently waged work really only lasted for a brief moment."[4] Especially in areas dominated by a single industry, when the cycle of factory giantism moved on, prolonged devastation resulted, even as somewhere else on the globe giant factories were creating new possibilities and wealth.

The current cycle of industrial giantism shares much with earlier iterations, but it also differs in important ways. For one thing, the arc from development to decline has become shorter. Just thirty years after Foxconn built its first factory in Shenzhen, the region has passed its peak as a center of large-scale manufacturing, as companies, including Foxconn, move on to other areas of China and elsewhere for cheaper land and labor. For another thing, many workers in the new giant factories do not share the illusion of permanence that workers in earlier plants might have had, expecting to spend most of their lives elsewhere, back where they came from or in some other, better situation.

Though still seen as vehicles for profit and national economic development, giant factories are far less likely these days to be celebrated or held up as models for the larger society than they once were. Often, they are all but unknown to the purchasers of the goods they make, who are likely to be many miles and national borders away. Once, a buyer of a Singer sewing machine or a Ford Model T knew precisely where it was made. Today, the purchaser of a sneaker or a refrigerator or even an automobile probably has no idea what country it was manufactured in, let alone what factory. Production—work—once proudly associated with the physical goods we need and cherish is now largely hidden away.

As a global phenomenon, the giant factory may have reached its apogee. While very big factories continue to be built, many manufacturers have moved in other directions, seeking to lower labor costs and avoid the possibility that—as has happened in the past—their workers will take advantage of the concentration of production to assert their power. Continuing mechanization and automation is one path, the cause in the United States of much greater loss of factory employment than the movement of plants abroad. Even Foxconn, the world's largest employer of factory workers, is experimenting with greater automation. At its smartphone factory in Kunshan, China, not far from Shanghai, the company has invested heavily in robots, allowing it to reduce its number of employees from 110,000 to 50,000, still a very large workforce but no longer near the top of the list of the world's largest factories.[5] Other companies have turned to many small and midsize factories in very low wage regions, like Bangladesh, seemingly turning back the clock, as young women, freshly arrived from rural villages, make goods for global giants like Wal-Mart and H&M in low-tech, crowded, and often extraordinarily dangerous factories that more resemble late-nineteenth-century American sweatshops than modern Chinese factory giants.[6]

But if the giant factory has lost some of its allure, there still are entrepreneurs eager to begin the cycle again in fresh territory, unsullied by a history of labor activism or environmental spoilage. Huajian Shoes, a Chinese firm that manufactures footwear for international brands like Guess, in 2012 opened a factory in Ethiopia, where in 2014 the basic after-tax minimum wage was $30 a month, compared to an average manufacturing wage in China of $560. Within two years it had 3,500 workers. But the company had far grander plans, centered on a proposed new complex near Addis Ababa that would employ 30,000 workers and include worker dormitories, a resort, a technical university, a hotel, and a hospital, all on a site ringed by a replica of the Great Wall of China and shaped like a woman's shoe. In October 2016, Huajian announced plans to move its production of shoes for Ivanka Trump's line from its factory in Dongguan to Ethiopia.[7]

Whatever the future of the giant factory might prove to be, it already has left behind a transformed world. In some ways, industrial giants have fulfilled the dreams of their promoters, having been part and parcel of an extraordinarily rapid and large improvement in social well-being, comfort, longevity, material possession, and security, one without precedent in human history. The Industrial Revolution, which the giant factory pushed forward, contributed to not only higher living standards but also the creation of the modern state, urbanized society, and a transformed face of the planet.

It also helped create a "new man." Perhaps not exactly a new man at one with the automatic machinery and industrial processes of the giant factory as envisioned, in their own ways, by Henry Ford, Alexei Gastev, and Antonio Gramsci. But a new man and a new woman nonetheless, with a time sense dictated by the needs of mass, coordinated activity and the rhythms of machinery; with a commitment to the idea of progress through technical innovation and increasing efficiency; worshiping factory products and an industrial aesthetic; and taking for granted the idea of sacrifice for future gain.

In short, the giant factory helped produce modernity, the now we inhabit still, even if it no longer has the awe-inspiring novelty it once had. And it is a modernity that transcends particular political and economic systems. Usually, the large-scale factory is portrayed as a product of capitalism, a stage in its historic development. Yet, as this study shows, to portray the giant factory as strictly a capitalist institution requires eliding much of its history, including some of the largest factories ever built. The giant factory was central to both capitalist and socialist development, not only economically but socially, culturally, and politically as well. The factory is never just the same within different cultures and social systems, but its essential features have proved remarkably stable and enduring, as it has roamed the world, setting down in places seemingly utterly different from one another. The giant factory, rather than a feature of capitalism, turns out to be a feature of modernity, in all its variations.[8]

The giant factory has made dreams reality, but it has rendered nightmares real, too. In every society, the great productivity of the giant factory has rested on great sacrifices, almost always unevenly shared. In the capitalist world, it was the workers in the factory itself who most obviously suffered, exploited to produce rivers of products and profits. But the workers who produced the raw materials for the factory, including, at various times, the slave growers of cotton, coal miners, iron miners, rubber harvesters, and today miners of rare-earth elements needed for electronic components, suffered, too. So did workers using older methods, forced to compete with factory production.

In the socialist world, factory workers toiled hard, but often held a relatively privileged place in society, with better housing, food, and benefits than other citizens. The greatest sacrifices in the Soviet Union and Maoist China came distant from the factory itself, in the countryside, where peasants were squeezed hard, sometimes to the point of death, to generate the resources needed to build industrial giants.

For a few decades after World War II, in Europe and the United States, the giant factory became a vehicle for an extraordinary improvement in the pay, benefits, and security of workers (though the actual work remained physically draining, monotonous, and alienating). Largely because of unionization, workers shared the great productivity gains of large-scale industry, a moment of relative equality and democracy in the long history of capitalist society. In the light of the past four decades of stagnating working-class income and growing insecurity, the post–World War II era in retrospect looks golden, and the critiques of the factory during it all but forgotten. But the ugly residue of the giant factory is difficult to ignore.

Just as the costs of creating and operating the giant factory were unevenly shared, so has been the environmental and social wreckage it has left behind. No place better symbolizes the nightmarish afterlife of industrial giantism than Flint, Michigan, once the center of the great General Motors empire, now a shrunken, deeply impoverished community, given so little respect that state and local officials poisoned the population with lead-contaminated water to save money during a state-imposed administratorship. The toxic fate of Flint can be found with variations around the globe, in the depressed former industrial centers of the American Midwest, northern England, northern France, Eastern Europe, Russia, Ukraine, and northern China, with their plagues of unemployment, poverty, contaminated earth and water, drug and alcohol use, and despair.

What comes next? It is too soon to declare the giant factory over as a global institution. But in many cities, areas, and nations, its import has vastly shrunk or all but disappeared. Cities deserted by industry have tried to reinvent themselves, frequently hoping to use depressed land prices and abandoned industrial structures as the basis for reemergence as cultural and entrepreneurial centers, a strategy that has yielded, at best, modest returns.[9] On a national level, capital in Great Britain, the United States, and other pioneer industrial countries has increasingly shifted from production to finance, enabling continued economic gains from the factory system, but now through financing the system and its many ancillary activities rather than through actual operation, a strategy that has yielded great wealth for a few, growing economic inequality, and deep social fissures.

If the coming of the giant factory was associated with visions of utopia (along with dystopian fears), its passing has been associated with social malaise and shriveled imagination. The Industrial Revolution and the giant factory have left in their wake a continuing belief in a teleology of progress and techno-determinism. But for many, the future has already come and gone, perhaps leaving them with sneakers and a smartphone, but with little hope or belief in their ability to create a new world, a post-factory world that builds on the extraordinary advances of the giant factory to forge a new and different kind of modernity, one more democratic and more sustainable, socially, economically, and, perhaps most important, ecologically.

We are all in this, all implicated. In 2016, the second-largest holder of stock in Hon Hai Precision Industry, the major owner of Foxconn Technology Company, was The Vanguard Group, Inc., a mutual fund company with a benign image that holds the savings and retirement accounts of more than twenty million people (including this author). Very few of them were aware that they owned a piece of a factory from the roof of which workers jumped to their deaths in despair. (Vanguard also was the third-largest holder of stock in Pegatron Corporation, Foxconn's biggest rival, and the ninth-largest holder of stock in footwear maker Yue Yuen.) Even funds that claim to be socially responsible have dirty hands; the largest holding of Calvert Investments, "founded on the belief that investment capital, properly stewarded, could improve the world for its less powerful inhabitants," is Apple, Foxconn's partner and biggest customer.[10] And of course, even if you do not own a tiny piece of a giant factory through a retirement fund or a savings account, it is almost certain that you own a product made in one.

The giant factory has left us with a complex legacy and many lessons. It demonstrated in practical, concrete ways the ability of humankind to exert mastery over nature (at least for a while), in the process vastly improving the standard of living for billions of people but also despoiling the earth. It illuminated the deep ties between coercion and freedom, exploitation and material advance. It revealed the beauty to be found not only in the natural world, but in the manmade world, in labor and its products. It demonstrated the deep yearning of working people for control over their lives and a degree of justice, as decade after decade, century after century, they launched struggles against exploiting employers and oppressive states, often against enormous odds. But perhaps, at this moment, the most important lesson of the giant factory is the one that is easiest to forget; that it is possible to reinvent the world. It has been done before, and it can be done again.

Acknowledgments

THIS BOOK WOULD NOT BE POSSIBLE WITHOUT THE work of generations of scholars, journalists, and writers who shared a fascination with big factories and their social and cultural importance. Their publications, cited in my notes, collectively represent a stupendous intellectual achievement, indispensable for understanding the past and present. I owe them all a great debt.

As I finished this study, at the same time that my father, Harold Freeman, reached his one-hundredth birthday, I realized how much his sensibility pervades it. Throughout his life he combined a deep interest in technical matters with a critical political stance and broad familiarity with European and American culture, a kind of Enlightenment out-look once common in the working-class milieu in which he grew up. I vividly remember him taking me along as a child on a work-related visit to a glass factory, watching a worker pull a still glowing-red Coke bottle off the line, stretching and twisting it with tongs for our amusement. In that bit of magic, I suspect, lies the origin of this project.

I have been fortunate to learn about the factory and its implications not only from scholars but from workers and unionists, too. My sum-mer sojourn as an eighteen-year-old in a cosmetics factory opened my eyes to the dense human drama that takes place within the walls of a production site, the combination of boredom, pride, fatigue, and soli-darity; the gossip, storytelling, and arguing; the differing experiences and beliefs rubbing against one another; the skills of work, survival, and maneuver that working women and men have to master. In the years since, in other jobs and in my work with the labor movement, I have been privileged to learn more about what the poet Philip Levine called "What Work Is." I am grateful to the many labor activists and workers who, often without realizing it, enriched my understanding of toil, unionism, politics, and working-class life.

An embryonic version of this study appeared in the journal *New Labor Forum* as one of a series of columns I wrote with Steve Fraser under the heading "In the Rearview Mirror." Steve came up with the idea of a column that looked at historic precedents for current events. His notion of how to link the past and the present sparked the idea for what became *Behemoth*, which I sometimes think of as one of our columns writ very large. The encouragement of my colleagues at the Joseph S. Murphy Institute for Worker Education and Labor Studies at the City University of New York, who heard a lunchtime talk I gave on the history of giant factories, convinced me the subject was worth pursuing further. Bryan Palmer pushed me along when he suggested I submit a version of that talk for publication in *Labour/Le Travail*, the lively, sophisticated journal, of which he was the long-time editor.

A year-long fellowship at the Advanced Research Collaborative at the Graduate Center of the City University of New York made possible research for this book. I am greatly thankful to its director, Donald Robotham, and my fellow fellows for a remarkably stimulating and fruitful year. Additional support for this project was provided by a PSC-CUNY Award, jointly funded by The Professional Staff Congress and The City University of New York.

Mark Levinson from the Service Employees International Union and Cathy Feingold from the AFL-CIO, along with Robert Szewczyk and Dorota Miklos from NSZZ Solidarność, helped arrange my meeting with Solidarity union leaders in Nowa Huta. Stanisław Lebiest, Roman Natkonski, Krysztof Pfister, and their colleagues (with able translation by Piotr Smreczynski) were extraordinarily generous in devoting their time to discussing the history of the plant and their union and giving me a tour of the mill. May Ying Chen and Ruting Chen from the Murphy Institute and Lu Zhang from Temple University made extensive efforts to arrange a visit to Chinese factories. Though in the end the trip proved impossible, I deeply appreciate their attempts and the great deal I learned from them about China.

Many others helped along the way. Early on, Carol Quirke gave me valuable suggestions for reading about industrial photography. Dave Gillespie, John Thayer, and Maayan Brodsky provided research assistance. Josh Brown was extraordinarily generous in helping me with illustrations, sharing his incomparable knowledge of nineteenth-century graphics and personally scanning images to make sure they came out right. My students at Queens College put up with good humor my use of them as guinea pigs for many of the ideas in this book, in a global history course that focused on factories and industrialization. Daniel Esterman accompanied me on a research trip to Lowell and engaged in numerous discussions of this project as it unfolded, providing a sounding board and many good ideas. I also talked about factories on repeated occasions with Edgar Masters, whose long career in the textile industry and efforts to preserve industrial sites have made him a repository of information and insight.

I am especially grateful to colleagues who read drafts of chapters about places I was writing about for the first time: Timothy Alborn (chapter 1), Kate Brown (chapter 5), and Xiaodan Zhang (chapter 7). Their expertise and advice proved invaluable, even when my interpretations differed from theirs. These chapters are much improved as a result of their generosity. Jack Metzgar once again became an unflagging supporter as I worked on this book, reading every chapter in draft, providing detailed comments, and, most importantly, reassuring me that I was on to something when my doubts swelled. No one could ask for a more generous colleague and friend.

Steve Fraser was there at the end, just as he was at the beginning, reading the full manuscript and responding as I have come to take for granted, with comments both detailed and sweeping, raising historical questions and seeing connections I missed and making suggestions for structural changes that greatly strengthened the narrative. His friendship and our collaboration over the years have been enormously important to me. Kim Phillips-Fein put aside other things to read the first chapters of the manuscript as I approached the finish line, helping clarify and deepen them.

Nearly two decades ago, Matt Weiland edited a book I wrote and it was a terrific experience. The opportunity for a reprise has been one of the pleasures of this project. Once again, Matt got right away what I was trying to do, encouraging me to be bolder and more expansive, combining a sense of adventure with some necessary realism. Remy Cawley shepherded me through the publication process with good cheer, good advice, and a remarkable ability to keep track of never-ending details. I want to thank as well William Hudson, for his copyediting; Brian Mulligan, for the beautiful design of the book; and everyone at W. W. Norton for their extraordinary professionalism.

Finally, I want to thank my family for their love and support, especially my partner through the many years, Deborah Ellen Bell, who among many other things read the manuscript for this book and provided her usual good advice, and our wonderful daughters, Julia Freeman Bell and Lena Freeman Bell.

Notes

Introduction

1. Most manufacturing jobs are in factories, but not all. Some are in retail establishments, like bakeries, or even in homes. U.S. Bureau of Labor Statistics, "Employment, Hours and Earnings from the Current Employment Statistics survey (National)," http://data.bls.gov/pdq/SurveyOutputServlet (accessed Sept. 24, 2016).
2. Heather Long, "U.S. Has Lost 5 Million Manufacturing Jobs Since 2000," *CNN Money*, Mar. 29, 2016, http://money.cnn.com/2016/03/29/news/economy/us-manufacturing-jobs/; The World Bank, World Data Bank, "Employment in Industry and World Development Indicators" (based on International Labour Organization data), http://data.worldbank.org/indicator/SL.IND.EMPL.ZS, and http://databank.worldbank.org/data/reports.aspx?source=2&series=SL.IND.EMPL.ZS&country= (accessed Sept. 24, 2016); Central Intelligence Agency, *The World Factbook, 2017* (New York: Skyhorse Publishing, 2016), 179.
3. For life on the eve of the Industrial Revolution, see Fernand Braudel, *The Structures of Everyday Life: Civilization and Capitalism, 15th–18th Century,* vol. 1 (New York: Harper and Row, 1981) (French life expectancy, 90), and E. J. Hobsbawm, *The Age of Revolution, 1789–1848* (New York: New American Library, 1962), 22–43. See also Roderick Floud, Kenneth Wachter, and Annabel Gregory, *Height, Health and History: Nutritional Status in the United Kingdom, 1750–1980* (Cambridge: Cambridge University Press, 1990), 292; Thomas Piketty, *Capital in the Twenty-First Century* (Cambridge, MA: Harvard University Press, 2014), 71–72; and Central Intelligence Agency, *World Factbook, 2017,* 303, 895, 943.

4. Tim Strangleman, "'Smokestack Nostalgia,' 'Ruin Porn' or Working-Class Obituary: The Role and Meaning of Deindustrial Representation," *International Labor and Working-Class History* 84 (Fall 2013), 23–37; Marshall Berman, "Dancing with America: Philip Roth, Writer on the Left," *New Labor Forum* 9 (Fall–Winter 2001), 53–54.

5. "modern, adj. and n." and "modernity, n." OED Online. September 2016. Oxford University Press. http://www.oed.com/view/Entry/120618 (accessed September 17, 2016); Raymond Williams, *Keywords: A Vocabulary of Culture and Society,* rev. ed. (New York: Oxford University Press, 1983), 208–09; Jürgen Habermas, "Modernity: An Unfinished Project," in M. Passerin d'Entrèves and Seyla Benhabib, eds., *Habermas and the Unfinished Project of Modernity: Critical Essays on the Philosophical Discourse of Modernity* (Cambridge, MA: MIT Press, 1997); Peter Gay, *Modernism: The Lure of Heresy, From Baudelaire to Beckett and Beyond* (New York: W. W. Norton, 2008).

6. Size can be measured in different ways. I have defined it by number of employees. As a labor historian, that seems natural, coming from an interest in the lived experience of workers and class relations. There are other useful ways to define scale that would lead to the selection of a different set of factories to study. If we were to look at the size of factory buildings, in the current era the massive aircraft factories of Boeing and Airbus would rise to the fore, huge structures that go on and on but have within them fewer workers than many more compact plants. To understand the ecological impact of large factories, we might define size by the acreage of the sites on which production facilities are located. By that standard, chemical plants and, especially, atomic-fuel and weapons complexes exceed in size most of the factories discussed in this book. My definition of size is somewhat arbitrary, but it serves well the focus of this study on the linkage between the factory and modernity.

7. Gillian Darley, *Factory* (London: Reaktion Books, 2003), and Nina Rappaport, *Vertical Urban Factory* (New York: Actar, 2016) are exceptions, but are heavily architectural in their tilt.

Chapter 1
"LIKE MINERVA FROM THE BRAIN OF JUPITER"

1. Prior to 1721, only a few British industries had centralized production facilities and these, by later standards, were quite small, like the Nottingham framework knitting workshops that employed several dozen workers apiece.

In Central and Western Europe, there were a few large-scale, unmechanized manufacturing operations. Maxine Berg, *The Age of Manufactures: Industry, Innovation and Work in Britain, 1700–1820* (Oxford: Basil Blackwell, 1985), 212; Fernand Braudel, *The Wheels of Commerce: Civilization and Capitalism, 15th–18th Century,* vol. II (New York: Harper & Row, 1982), 329–38. U.S. figure calculated from 1850 census data in U.S. Census Office, *Manufacturers of the United States in 1860* (Washington, D.C., 1865), 730.

2. The Derby silk mill is generally considered the first factory in England, the pioneer in the Industrial Revolution. There were at least a few earlier production facilities that had some if not all the characteristics of modern factories, including the sixteenth-century silk mills in Bologna, which developed some of the machinery and organization that the Lombes later copied. Anthony Calladine, "Lombe's Mill: An Exercise in Reconstruction," *Industrial Archeology Review* XVI, 1 (Autumn 1993), 82, 86.

3. Calladine, "Lombe's Mill," 82, 89; William Henry Chaloner, *People and Industries* (London, Frank Cass and Co., Ltd., 1963), 14–15. An 1891 fire destroyed most of the building, which was reconstructed on a smaller scale. It now houses the Derby Silk Mill museum.

4. S. R. H. Jones, "Technology, Transaction Costs, and the Transition to Factory Production in the British Silk Industry, 1700–1870," *Journal of Economic History* XLVII (1987), 75; Chaloner, *People and Industries,* 9–18; Calladine, "Lombe's Mill," 82, 87–88; R. B. Prosser and Susan Christian, "Lombe, Sir Thomas (1685–1739)," rev. Maxwell Craven, Susan Christian, *Oxford Dictionary of National Biography* (Oxford: Oxford University Press, 2004); online ed., Jan. 2008, http://www.oxforddnb.com/view/article/16956.

5. John Guardivaglio, one of the Italian workers who had come back with John Lombe, helped set up the mill near Manchester. Tram could be made from raw silk imported from Persia, easier to get than the higher-quality Italian or Chinese silk needed for organzine. Calladine, "Lombe's Mill," 87, 96–97; Berg, *Age of Manufactures,* 202–03; Jones, "Technology, Transaction Costs, and the Transition to Factory Production," 77.

6. Daniel Defoe, *A Tour Thro' the Whole Island of Great Britain,* 3rd. ed., vol. III (London: J. Osborn, 1742), 67; Charles Dickens, *Hard Times for These Times* ([1854] London: Oxford University Press, 1955), 7, 1.

7. James Boswell, *The Life of Samuel Johnson,* vol. III (London: J.M. Dent & Sons, 1906), 121.

8. Though India was the most prominent center of cotton textile production, there were others, including Southeast Asia, the Arabian Gulf, and the Ottoman Empire, where artisans turned out imitations of Indian cottons.

Prasannan Parthasarathi, "Cotton Textiles in the Indian Subcontinent, 1200–1800," 17–41, and Giorgio Riello, "The Globalization of Cotton Textiles: Indian Cottons, Europe, and the Atlantic World, 1600–1850," 274, in *The Spinning World: A Global history of Cotton Textiles, 1200–1850*, ed. Riello and Parthasarathi (Oxford: Oxford University Press, 2009), 17–41.

9. Giorgio Riello, *Cotton: The Fabric that Made the Modern World* (Cambridge: Cambridge University Press, 2013), 126; Andrew Ure, *The Philosophy of Manufactures or An Exposition of the Scientific, Moral, and Commercial Economy of the Factory System of Great Britain* (1835; New York: Augustus M. Kelley, 1967), 12.

10. D. T. Jenkins, "Introduction," in D. T. Jenkins, *The Textile Industries* (Volume 8 of the *Industrial Revolutions*, ed. R. A. Church and E. A. Wrigley) (Cambridge, MA: Blackwell, 1994), xvii; Riello, *Cotton*, 127.

11. Riello, *Cotton*, 172–73, 176; Berg, *Age of Manufactures*, 205.

12. Fustians were easier to produce than all-cotton fabric because flax warps were less likely than cotton to break during weaving.

13. Riello, "The Globalization of Cotton Textiles, 337–39; Riello, *Cotton*, 217, 219.

14. In the 1850s, the United States supplied 77 percent of the raw cotton imported by Britain, 90 percent by France, 92 percent by Russia, and 60 percent by the German states. Between 1820 and 1860 the number of slaves in Mississippi and Louisiana, mostly growing cotton, rose from 101,878 to 768,357. R. S. Fitton and A. P. Wadsworth, *The Strutts and the Arkwrights 1758–1830: A Study of the Early Factory System* (Manchester: Manchester University Press, 1958), 347–48; Riello, *Cotton*, 188, 191, 195 (Marx quote), 200–207, 259; Frederick Douglass, "What to the Slave Is the Fourth of July?," in *Frederick Douglass: Selected Speeches and Writings*, ed. Philip S. Foner (Chicago: Lawrence Hill, 1999), 197; Sven Beckert, *Empire of Cotton: A Global History* (New York: Knopf, 2014), 243; Walter Johnson, *River of Dark Dreams: Slavery and Empire in the Cotton Kingdom* (Cambridge, MA: Harvard University Press, 2013), 256.

15. Edward Baines, *History of the Cotton Manufacture in Great Britain* (London: H. Fisher, R. Fisher, and P. Jackson, [1835]), 11; R. L. Hills, "Hargreaves, Arkwright and Crompton, 'Why Three Inventors?'" *Textile History* 10 (1979), 114–15.

16. Baines, *History of the Cotton Manufacture*, 115; Deborah Valenze, *The First Industrial Woman* (New York: Oxford University Press, 1995), 78; David S. Landes, *The Unbound Prometheus: Technological Change and Industrial Development in Western Europe from 1750 to the Present* (Cambridge: Cam-

bridge University Press, 1969), 57. European commentators and historians long claimed that Indian wages were far below British ones, leading to lower prices for cotton products, but recently some historians have challenged this view. For a restatement of the orthodox position, see Beckert, *Empire of Cotton*, 64; for a reassessment suggesting near parity of wages, see Prasannan Parthasarathi, *Why Europe Grew Rich and Asia Did Not: Global Economic Divergence, 1600–1850* (Cambridge: Cambridge University Press, 2011), 35–46.

17. Jenkins, "Introduction," x; Franklin F. Mendels, "Proto-Industrialization: The First Phase of the Industrialization Process," *Journal of Economic History* XXXII (1972), 241–61; S. D. Chapman, "Financial Restraints on the Growth of Firms in the Cotton Industry, 1790–1850," *Textile History* 5 (1974), 50–69; Berg, *Age of Manufactures*, 182.

18. Hills, "Hargreaves, Arkwright and Crompton," 118–23; Berg, *Age of Manufactures*, 236; Fitton and Wadsworth, *The Strutts and the Arkwrights*, 61–68, 76–78, 94–97; Adam Menuge, "The Cotton Mills of the Derbyshire Derwent and Its Tributaries," *Industrial Archeology Review* XVI (1) (Autumn 1993), 38.

19. Berg, *Age of Manufactures*, 236, 239, 244, 248, 258; George Unwin, *Samuel Oldknow and the Arkwrights* (Manchester: Manchester University Press, 1924), 30–32, 71, 124–25; Landes, *Unbound Prometheus*, 85; E. P. Thompson, *The Making of the English Working Class* ([1963] London: Pelican Books, 1968), 327, 335; Jones, "Technology, Transaction Costs, and the Transition to Factory Production," 89–90.

20. Chaloner, *People and Industries*, 14–15; Fitton and Wadsworth, *The Strutts and the Arkwrights*, 98–99, 192–95, 224–25.

21. Small four-spindle, hand-powered spinning frames, built from Arkwright's plans for a demonstration model, can be seen at the museums in Cromford and Belper. Hills, "Hargreaves, Arkwright and Crompton," 121; Berg, *Age of Manufactures*, 236, 239, 242, 246; Menuge, "The Cotton Mills of the Derbyshire Derwent," 56 (Arkwright quote).

22. John S. Cohen, "Managers and Machinery: An Analysis of the Rise of Factory Production," *Australian Economic Papers* 20 (1981), 27–28; Berg, *Age of Manufactures*, 19, 24, 40–42.

23. Jenkins, "Introduction," xv.

24. Berg, *Age of Manufactures*, 40–41, 231–32, 282–83; Pat Hudson, *The Genesis of Industrial Capital: A Study of the West Riding Wool Textile Industry c. 1750–1850* (Cambridge: Cambridge University Press, 1986), 137; Jones, "Technology, Transaction Costs, and the Transition to Factory Production,"

89–90; Roger Lloyd-Jones and A. A. Le Roux, "The Size of Firms in the Cotton Industry: Manchester 1815–1840," *The Economic History Review*, new series, vol. 33, no. 1 (Feb. 1980), 77.

25. V. A. C. Gatrell, "Labour, Power, and the Size of Firms in Lancashire Cotton in the Second Quarter of the Nineteenth Century," *Economic History Review*, new series, vol. 30, no. 1 (Feb. 1977), 96, 98, 112; Jenkins, "Introduction," xv.

26. Berg, *Age of Manufactures*, 23–24; Thompson, *Making of the English Working Class*, 208–11; Robert Gray, *The Factory Question and Industrial England, 1830–1860* (Cambridge: Cambridge University Press, 1996), 3–4.

27. Charles Babbage, *On the Economy of Machinery and Manufacturers*, 4th ed. (London: Charles Knight, 1835), 211–23.

28. Gatrell, "Labour, Power, and the Size of Firms," 96–97, 108; Alfred Marshall, *Principles of Economics* (1890; London: Macmillan and Co., Ltd., 1920), 8th ed., IV.XI.7, http://www.econlib.org/library/Marshall/marP25.html#Bk. IV,Ch.XI.

29. Baines, *History of the Cotton Manufacture*, 184–85.

30. Landes, *Unbound Prometheus*, 41; Jones, "Technology, Transaction Costs, and the Transition to Factory Production," 71–74; Jenkins, "Introduction," xiii; Berg, *Age of Manufactures*, 23–24, 190, 246; Hudson, *Genesis of Industrial Capital*, 70–71. Marx discussed the issue of economies of scale and the rise of the factory system at great length in Karl Marx, *Capital: A Critique of Political Economy*, vol. 1 ([1867] New York: International Publishers, 1967), chap. 13 and 14 ("Cooperation" and "Division of Labour and Manufacture").

31. Jenkins, "Introduction," x–xii; Berg, *Age of Manufactures*, 24; Hudson, *Genesis of Industrial Capital*, 81, 260; Thompson, *Making of the English Working Class*, 299, 302.

32. Gatrell, "Labour, Power, and the Size of Firms," 96–97, 107.

33. On British forms of wealth, see Thomas Piketty, *Capital in the Twenty-First Century* (Cambridge, MA: Harvard University Press, 2014), 113–20, 129–31. Willersley Castle now is a Christian Guild hotel. Fitton and Wadsworth, *The Strutts and the Arkwrights*, 91, 94–98, 102, 169, 246; R. S. Fitton, *The Arkwrights: Spinners of Fortune* ([1989] Matlock, Eng.: Derwent Valley Mills Educational Trust, 2012), 224–96; Frances Trollope, *The Life and Adventures of Michael Armstrong the Factory Boy* ([1840] London: Frank Cass and Company Limited, 1968), quote on 76.

34. Local church towers, however, did rival the mills in height. Mark Girouard, *Cities & People: A Social and Architectural History* (New Haven, CT: Yale University Press, 1985), 211–18; Thomas A. Markus, *Buildings and Power:*

Freedom and Control in the Origin of Modern Building Types (London: Routledge, 1993), 263.

35. Fitton, *The Arkwrights*, 30, 50, 81.

36. Fitton, *The Arkwrights*, 30, 81; Thomas A. Markus, "Factories, to 1850," *The Oxford Companion to Architecture*, vol. 1, ed. Patrick Goode (New York: Oxford University Press, 2009), 304–05; Fitton and Wadsworth, *The Strutts and the Arkwrights*, 200–207, 211–12; Malcolm Dick, "Charles Bage, the Flax Industry and Shrewsbury's Iron-Framed Mills," accessed Mar. 29, 2017, http://www.revolutionaryplayers.org.uk/charles-bage-the-flax-industry-and-shrewsburys-iron-framed-mills/; Markus, *Buildings and Power*, 266–67, 270–71, 281–82; Menuge, "The Cotton Mills of the Derbyshire Derwent," 52–56.

37. A. J. Taylor, "Concentration and Specialization in the Lancashire Cotton Industry, 1825–1850," *Economic History Review*, 2nd series, I (1949), 119–20; Markus, *Buildings and Power*, 275. Not all power looms were situated in sheds; some manufacturers built multistory weaving mills. See Colum Giles, "Housing the Loom, 1790–1850: A Study of Industrial Building and Mechanization in a Transitional Period," *Industrial Archeology Review* XVI (1) (Autumn 1993), 30–33. On the spread of the sawtooth roof, first called the "weave shed roof," to the United States, see Betsy Hunter Bradley, *The Works: The Industrial Architecture of the United States* (New York: Oxford University Press, 1999), 192–93.

38. The first Cromford mills, though near the Derwent, were powered by a sough draining a lead mine and a brook, not the river itself. Fitton, *The Arkwrights*, 28–29.

39. Steam power was first used in a cotton mill in 1789, but water remained the most common power source for several decades. An 1870 industrial census found that cotton mills used more power from steam engines than any other industry. Fitton and Wadsworth, *The Strutts and the Arkwrights*, 103; Unwin, *Samuel Oldknow*, 119; Markus, *Buildings and Power*, 265–66; Parthasarathi, *Why Europe Grew Rich and Asia Did Not*, 155; Dickens, *Hard Times*, 22, 69; W. Cooke Taylor, *Notes of a Tour in the Manufacturing Districts of Lancashire*, 2nd ed. (London: Duncan and Malcolm, 1842), 1–2.

40. In the first report of the Factory Commission, Edwin Chadwick described an elevator as "an ascending and descending room, moved by steam." Ure, *The Philosophy of Manufactures*, 32–33, 44–54 ("upright tunnels" on 45); Markus, *Buildings and Power*, 275, 280–81; Gray, *The Factory Question*, 92–93.

41. The Round Mill, built between 1803 and 1813, remained standing until 1959, when in the course of its demolition four workers were killed.

Fitton and Wadsworth, *The Strutts and the Arkwrights*, 221; Markus, *Buildings and Power*, 125; Humphrey Jennings, *Pandemonium, 1660– 1886: The Coming of the Machine as Seen by Contemporary Observers*, ed. Mary-Lou Jennings and Charles Madge (New York: Free Press, 1985), 98; Belper Derbyshire, Historical & Genealogical Records, "Belper & the Strutts: The Mills," July 20, 2011, http://www.belper -research.com/strutts_mills/mills.html.

42. The housing Arkwright built in Cromford is still occupied. The row houses had lofts for weavers, who bought yarn from Arkwright and whose wives and children worked in his mill. Fitton, *The Arkwrights*, 29, 187; Arkwright Society presentation at Cromford Mills, May 15, 2015; Fitton and Wadsworth, *The Strutts and the Arkwrights*, 97, 102–04, 246; Chris Aspin, *The First Industrial Society; Lancashire, 1750–1850* (Preston, UK: Carnegie Publishing, 1995), 184; Unwin, *Samuel Oldknow and the Arkwrights*, 95.

43. Fitton and Wadsworth, *The Strutts and the Arkwrights*, 246, 252; Unwin, *Samuel Oldknow and the Arkwrights*, 191; Fredrich Engels, *The Condition of the Working Class in England*, trans. W. O. Henderson and W. H. Chaloner (Stanford, CA: Stanford University Press, 1958), 205.

44. Fitton and Wadsworth, *The Strutts and the Arkwrights*, 240–44; Unwin, *Samuel Oldknow and the Arkwrights*, 178.

45. Ure, *The Philosophy of Manufactures*, 150, 283–84, 312; Fitton, *The Arkwrights*, 146, 151; John Brown, *A Memoir of Robert Blincoe, An Orphan Boy* (1832), reprinted in James R. Simmons, Jr., ed., *Factory Lives: Four Nineteenth-Century Working-Class Autobiographies* (Peterborough, ON: Broadview Editions, 2007), 169; Cohen, "Managers and Machinery," 25; Engels, *The Condition of the Working Class in England*, 174, 199; Marx, *Capital*, vol. 1, 422. The classic study of the change from task-oriented to time-oriented work is E. P. Thompson, "Time, Work-Discipline, and Industrial Capitalism," *Past and Present* 38 (Dec. 1967), pp. 56–97.

46. Landes, *Unbound Prometheus*, 43; Ellen Johnston, *Autobiography* (1869), reprinted in Simmons, Jr., ed., *Factory Lives*, 308; Aspin, *First Industrial Society*, 92; "knocker, n." OED Online. September 2014. Oxford University Press. http://www.oed.com/view/Entry/104097; "knock, v." OED Online. September 2014. Oxford University Press. http://www.oed.com/view/Entry/104090.

47. Fitton and Wadsworth, *The Strutts and the Arkwrights*, 97; Gray, *The Factory Question*, 136; Giorgio Riello and Patrick K. O'Brien, "The Future Is Another Country: Offshore Views of the British Industrial Revolution," *Journal of Historical Sociology* 22 (1) (March 2009), 4–5.

48. Taylor, *Notes of a Tour*, 4.

49. Robert Southey, *Journal of a Tour in Scotland in 1819*, quoted in Jennings, *Pandemonium*, 156; Steven Marcus, *Engels, Manchester, and the Working Class* (New York: Random House, 1974), 34–40, 60–61; Riello and O'Brien, "The Future Is Another Country," 6; Benjamin Disraeli, *Sybil, or the Two Nations* (London: Henry Colburn, 1845), 195; Karl Marx, *The Eighteenth Brumaire of Louis Bonaparte* (1852; New York: International Publishers, 1963), 15.

50. Trollope, *The Life and Adventures of Michael Armstrong*, 236–37; Flora Tristan, *Promenades dans Londres* (Paris, 1840), quoted in Riello and O'Brien, "The Future Is Another Country," 5.

51. Dickens, *Hard Times*, 69; Aspin, *First Industrial Society*, 4, 239–41.

52. It was a measure of how quickly the system was spreading that Taylor used the metaphor of machinery to describe society, a usage unusual before the eighteenth century. Gray, *The Factory Question*, 23–24; Thompson, *Making of the English Working Class*, 209; Taylor, *Notes of a Tour*, 4–5; "machinery, n." OED Online. September 2016. Oxford University Press, http://www. oed.com/view/Entry/111856.

53. Thompson, *Making of the English Working Class*, 341; Ure, *The Philosophy of Manufactures*, 20–22, 474.

54. Fitton and Wadsworth, *The Strutts and the Arkwrights*, 226; Katrina Honeyman, "The Poor Law, the Parish Apprentice, and the Textile Industries in the North of England, 1780–1830," *Northern History* 44 (2) (Sept. 2007), 127.

55. Brown, *Memoir of Robert Blincoe*, 115–18, 132, 173; William Dodd, *A Narrative of the Experience and Sufferings of William Dodd, A Factory Cripple, Written by Himself* (1841), reprinted in Simmons, Jr., ed., *Factory Lives*, 191, 193–95; Fitton and Wadsworth, *The Strutts and the Arkwrights*, 98–99, 103, 226; Fitton, *The Arkwrights*, 152, 160–61; Honeyman, "The Poor Law," 123–25; Ure, *The Philosophy of Manufactures*, 171, 179–80, 299, 301; Jennings, *Pandemonium*, 214–15.

56. Some mills withheld part of the wages of workers on contract until the end of each quarter as further insurance against their departure. Fitton and Wadsworth, *The Strutts and the Arkwrights*, 104–06, 226, 233; Aspin, *First Industrial Society*, 53, 104.

57. Parthasarathi, *Why Europe Grew Rich and Asia Did Not*, 3–4, 53–54. See, for example, Thomas E. Woods, Jr., "A Myth Shattered: Mises, Hayek, and the Industrial Revolution," Nov. 1, 2001, Foundation for Economic Education, https://fee.org/articles/a-myth-shattered-mises-hayek-and-the-industrial-revolution/; "Wake Up America," Freedom: A History of US (PBS), accessed Dec. 8, 2016, http://www.pbs.org/wnet/historyofus/web04/.

58. Livesey quoted in Aspin, *First Industrial Society*, 86. See also, Brown, *Memoir of Robert Blincoe*, 91, 109, 138–39.

59. Trollope, *The Life and Adventures of Michael Armstrong*, quote on 186.

60. The equation of British factory workers with West Indian slaves was used not only by critics of the factory system but also by defenders of slavery, who argued that slaves were actually better off than mill workers. Thompson, *Making of the English Working Class*, 220; Engels, *Condition of the Working Class in England*, 202, 204, 207–08; Disraeli, *Sybil*, 198; Catherine Gallagher, *The Industrial Reformation of English Fiction: Social Discourse and Narrative Form, 1832–1867* (Chicago: University of Chicago Press, 1985), 1–2.

61. Southey, *Journal of a Tour in Scotland in 1819*, quoted in Jennings, *Pandemonium*, 157–58; Robert Southey, *Espiella's Letters*, quoted in Aspin, *First Industrial Society*, 53.

62. Gallagher, *Industrial Reformation of English Fiction*, 6–21 (quotes on 7 and 10).

63. Jennings, *Pandemonium*, 230; Taylor, *Notes of a Tour*, 1–2, 30.

64. Marcus, *Engels, Manchester, and the Working Class*, 45–46.

65. Jennings, *Pandemonium*, 231.

66. Johnson, *River of Dark Dreams*, 154–57, 180–83; Paul L. Younger, "Environmental Impacts of Coal Mining and Associated Wastes: A Geochemical Perspective," *Geological Society, London, Special Publications* 236 (2004), 169–209.

67. William Blake, *Collected Poems*, ed. W. B. Yeats ([1905] London: Routledge, 2002), 211–12. Blake's original manuscript, with the punctuation used here, can be seen at http://en.wikipedia.org/wiki/And_did_those_feet_in_ancient_time#mediaviewer/File:Milton_preface.jpg (accessed Dec. 6, 2016). Steven E. Jones, *Against Technology: From the Luddites to Neo-Luddism* (New York: Routledge, 2006), 81–96.

68. By 1881, the Lancashire population had doubled again, to 630,323. GB Historical GIS / University of Portsmouth, Lancashire through time | Population Statistics | Total Population, *A Vision of Britain through Time* (accessed Oct. 5, 2016), http://www.visionofbritain.org.uk/unit/10097848/cube/TOT_POP. Engels, *The Condition of the Working Class in England*, 16; Tristram Hunt, *Marx's General: The Revolutionary Life of Friedrich Engels* (New York: Metropolitan Books, 2009), 78–79.

69. Landes, *Unbound Prometheus*, 116–17.

70. Taylor, *Notes of a Tour*, 6–7. For a different view, stressing the infection of

both mill owners and workers by greed, see Robert Owen, *Observations on the Effect of the Manufacturing System*, 2nd ed. (London: Longman, Hart, Rees, and Orml, 1817), 5–9.

71. Engels wrote this not long after leaving his first stint at his family's cotton mill in Manchester, a job he himself abhorred and was to return to for another two decades. Engels, *Condition of the Working Class in England*, 9–12, 153, 174, 199–202.

72. *The Condition of the Working Class in England* was an enormously influential book, both in the development of Marxism and in perceptions of Manchester and the Industrial Revolution. However, it had no immediate impact in the English-speaking world, since it did not appear in English until 1886, more than forty years after its publication in German, when an American edition came out. It was not published in England until 1892. Engels, *Condition of the Working Class in England*, 134–38; Thompson, *Making of the English Working Class*, 209; Hunt, *Marx's General*, 81, 100, 111–12, 312.

73. For the history of debate over factory legislation, see Gray, *The Factory Question*.

74. Marx, *Capital*, vol. 1, 418; Ure, *The Philosophy of Manufactures*, 17–18, 171, 179–80, 290, 299–301.

75. Taylor, *Notes of a Tour*, 3–4, 46, 237–38, 330.

76. Thomas Carlyle, *Chartism*, quoted in Jennings, *Pandemonium*, 35. Marx and Engels shared the belief that the rise of the factory system represented progress for mankind, in their eyes laying the basis for a new, more democratic, egalitarian, and productive social system. See, for example, Hunt, *Marx's General*, 323–24.

77. Gray, *The Factory Question*, 100–101, 103–04; Ure, *The Philosophy of Manufactures*, 295.

78. Taylor, *Notes of a Tour*, 80–82, 223–24; Ure, *The Philosophy of Manufactures*, 334–38; Engels, *The Condition of the Working Class in England*, 27, 156, 278.

79. Gray, *The Factory Question*; Valenze, *The First Industrial Woman*, 5.

80. B. L. Hutchins and A. Harrison, *A History of Factory Legislation* (London: P.S. King & Son, 1911).

81. Gray, *The Factory Question*, 23–24, 59–60, 72, 88 (quote from Factory Commission First Report), 130; Michael Merrill, "How Capitalism Got Its Name," *Dissent* (Fall 2014), 87–92.

82. Engels, *The Condition of the Working Class in England*, 195.

83. Marx devoted Chapter X of the first volume of *Capital* to "The Working-Day," capital's "vampire thirst for the living blood of labour," including

a detailed discussion of the Factory Acts. Marx, *Capital*, vol. 1, 231–302 ("struggle" on 235; "vampire" on 256). Engels analyzed the Factory Acts in *The Condition of the Working Class in England*, 191–99.

84. Marx, *Capital*, vol. 1, 219; Hunt, *Marx's General*, 1, 7, 179, 198, 234. As Hunt repeatedly points out, Engels' years as a cotton mill manager supplied Marx not only with detailed information about how the business worked but with the financial support he needed to write *Capital*.

85. Janice Carlisle, "Introduction," in Simmons, Jr., ed., *Factory Lives*, 27–28. See also David Vincent, *Bread, Knowledge, and Freedom: A Study of Nineteenth-Century Working-Class Autobiography* (London: Europa Publications, 1981), and Kevin Binfield, ed., *Writings of the Luddites* (Baltimore, MD, and London: Johns Hopkins Press, 2004) for how limited the sources are for working-class views of the factory system.

86. In *Against Technology*, Steven E. Jones traces the changing understanding of Luddism in British and American culture up through the twentieth century.

87. Berg, *Age of Manufactures*, 262; E. J. Hobsbawm, "The Machine Breakers," in *Labouring Men: Studies in the History of Labour* ([1964] Garden City, NY: Anchor Books, 1967), 7–26; Fitton, *The Arkwrights*, 51, 53–55.

88. There is an extensive literature of Luddism. Particularly useful were Hobsbawm, "The Machine Breakers"; Thompson, *Making of the English Working Class*, chap. 14 ("An Army of Redressers"); and Kevin Binfield, ed., *Writings of the Luddites* (quoted letter on 74).

89. Thompson, *Making of the English Working Class*, 570–91, 608–18.

90. Maxine Berg, *The Age of Manufactures*, 42, 259; Aspin, *First Industrial Society*, 67; Thompson, *Making of the English Working Class*, 211, 297–346, 616–21; Marx, *Capital*, vol. I, 431–32.

91. Jones, *Against Technology*, 9, 47; Hobsbawm, "The Machine Breakers," 9–16.

92. Thompson, however, questioned Engels's depiction of cotton workers making up the nucleus of the emerging labor movement. Aspin, *First Industrial Society*, 55; Engels, *Condition of the Working Class in England*, 24, 137, 237; Thompson, *Making of the English Working Class*, 211, 213.

93. Not only were workers unable to vote but also the districts in which mills were located were vastly underrepresented in Parliament as a result of the way seats were apportioned. Aspin, *First Industrial Society*, 56–57, 153–54; Henry Pelling, *A History of British Trade Unionism* (Hammondsworth, UK: Penguin Books, 1963), 18–19.

94. Pelling, *History of British Trade Unionism*, 24–29; Beckert, *Empire of Cotton*, 196.

95. Hobsbawm summarizes the major outbreaks of unrest in Britain between 1800 and 1850 in *Labouring Men*, 155. See also Ure, *The Philosophy of Manufactures*, 287, 366–67; Pelling, *History of British Trade Unionism*, 29–33, 36–37, 43–44, 46–49; and Thompson, *Making of the English Working Class*, 308, 706–08, 734–68.

96. Landes, *Unbound Prometheus*, 48–50, 62, 71. Walt Rostow made a similar claim in W. W. Rostow, *The Stages of Economic Growth: A Non-Communist Manifesto* (Cambridge: Cambridge University Press, 1960), 33–34, 54.

97. See, for example, Ludwig Von Mises, *Human Action: A Treatise on Economics* (Auburn, AL: Ludwig Von Mises Institute, 1998), 613–19. Von Mises writes of early factories, "The factory owners did not have the power to compel anybody to take a factory job," ignoring the fact that the state performed that function for them. On hanging Luddites, see Thompson, *Making of the English Working Class*, 627–28, and Lord Byron's eloquent speech in the House of Lords against making machine breaking a capital crime, http://www.luddites200.org.uk/LordByronspeech.html (accessed Oct. 7, 2016).

98. Patrick Joyce, *Work, Society and Politics: The Culture of the Factory in Later Victorian England* (New Brunswick, NJ: Rutgers University Press, 1980), 55; Leo Marx, *The Machine in the Garden: Technology and the Pastoral Ideal in America* (New York: Oxford University Press, 1964), 194; Aspin, *First Industrial Society*, 15–17, 23–30; *Mechanics'* magazine, Sept. 25, 1830, reprinted in Jennings, *Pandemonium*, 176–79; J. C. Jeaffreson and William Pole, *The Life of Robert Stephenson, F.R.S.*, vol. 1 (London: Longmans, Green, Reader, and Dyer, 1866), 141; Tony Judt, "The Glory of the Rails" and "Bring Back the Rails!," *The New York Review of Books,* vol. 57, no. 20 (Dec. 23, 2010), and vol. 58, no. 1 (Jan. 13, 2011).

99. Timothy L. Alborn, *Conceiving Companies; Joint-Stock Politics in Victorian England* (London and New York: Routledge, 1998), 182–83; Jennings, *Pandemonium*, 311–12; Landes, *Unbound Prometheus*, 121.

100. G. W. Hilton, "The Truck Act of 1831," *The Economic History Review*, new series, vol. 10, no. 3 (1958): 470–79; Hutchins and Harrison, *History of Factory Legislation*, 43–70; Hunt, *Marx's General*, 184–86.

101. Gray, *Factory Question*, 140, 163; Aspin, *First Industrial Society*, 185. On paternalism, see Joyce, *Work, Society and Politics*, esp. 135–53, 168–71, 185.

102. Brontë, *Shirley*, 487–88; Pelling, *History of British Trade Unionism*, 43–49; Carlisle, "Introduction," in Simmons, Jr., ed., *Factory Lives*, 63–65.

103. Engels, "Principles of Communism," quoted in Hunt, *Marx's General*, 144.

Chapter 2
"THE LIVING LIGHT"

1. Charles Dickens, *American Notes for General Circulation* (London: Chapman and Hall, 1842), 152–64 (quote on 164); [John Dix], *Local Loiterings and Visits in the Vicinity of Boston* (Boston: Redding & Co., 1845), 44; Michael Chevalier, *Society, Manner and Politics in the United States: Being a Series of Letters on North America* (Boston: Weeks, Jordan and Company, 1839), 128–44 (quotes on 136, 142, 143); Anthony Trollope, *North America* ([1862] New York: Knopf, 1951), 247–55 (quote on 250).

2. Marvin Fisher, *Workshops in the Wilderness; The European Response to American Industrialization, 1830–1860* (New York: Oxford University Press, 1967), 32–43, 92–95, 105–08; Dix, *Local Loiterings*, 48–49, 75, 79; Chevalier, *Society, Manner and Politics in the United States*, 133, 137.

3. Caroline F. Ware, *The Early New England Cotton Manufacture: A Study in Industrial Beginnings* ([1931] New York: Russell & Russell, 1966), 17–18, 30. Three of the most important histories of the New England textile industry were written by women: Ware's *Early New England Cotton Manufacture*; Vera Shlakman, *Economic History of a Factory Town; A Study of Chicopee, Massachusetts* (Northampton, MA: Department of History of Smith College, 1936); and Hannah Josephson, *The Golden Threads; New England's Mill Girls and Magnates* (New York: Duell, Sloan and Pearce, 1949). At the time, economic history (and academic scholarship more generally) was almost exclusively a male enterprise. Perhaps they were drawn to the subject by the large number of female textile workers. In an appreciation of their contributions, Herbert Gutman and Donald Bell wrote that the three "extended the boundaries of American working-class history beyond those fixed by John R. Commons and others described as this subject's founding fathers. Their books . . . offered new ways to *think* about working-class history. . . . Their perspectives differed, but all asked new questions about the early history of New England capitalism and wage labor." Long before the current vogue in the history of capitalism, these extraordinary scholars were writing just that. Herbert G. Gutman and Donald H. Bell, eds., *The New England Working Class and the New Labor History* (Urbana: University of Illinois Press, 1987), xii.

4. George Rogers Taylor, "Introduction," in Nathan Appleton and Samuel Batchelder, *The Early Development of the American Cotton Textile Industry* ([1858 and 1863] New York: Harper & Row, 1969), xiv.

5. George S. White, *Memoir of Samuel Slater: The Father of American Manu-*

factures: Connected with a History of the Rise and Progress of the Cotton Manufacture in England and America (Philadelphia: Printed at No. 46, Carpenter Street, 1836), 33–42; Sven Beckert, *Empire of Cotton: A Global History* (New York: Knopf, 2014), 152–54; Ware, *Early New England Cotton Manufacture*, 19–23; Betsy W. Bahr, "New England Mill Engineering: Rationalization and Reform in Textile Mill Design, 1790–1920," Ph.D. dissertation, University of Delaware, 1987, 13–16.

6. Ware, *Early New England Cotton Manufacture*, 26–27, 29–30, 60, 82, 227.
7. Following the English example, Slater and other southern New England mill owners set up Sunday schools for their child workers. Ware, *Early New England Cotton Manufacture*, 22–23, 28, 30–32, 245–47, 284–85; Samuel Batchelder, *Introduction and Early Progress of the Cotton Manufacture in the United States* (Boston: Little, Brown and Company, 1863), in Appleton and Batchelder, *Early Development of the American Cotton Textile Industry*, 46, 74.
8. Ware, *Early New England Cotton Manufacture*, 17, 28, 50–55.
9. Nathan Appleton, *Introduction of the Power Loom, and Origin of Lowell* (Lowell, MA: Proprietors of the Locks and Canals on Merrimack River, 1858), in Appleton and Batchelder, *Early Development of the American Cotton Textile Industry*, 7; Robert Brook Zevin, "The Growth of Cotton Textile Production After 1815," in Robert W. Fogel and Stanley L. Engerman, eds., *The Reinterpretation of American Economic History* (New York: Harper & Row, 1971), 139; Taylor, "Introduction," in Appleton and Batchelder, *Early Development of the American Cotton Textile Industry*, 9. Lowell also was in contact with machinists in Rhode Island who could build spinning equipment. See, for example, Wm. Blackburns to Francis Cabot Lowell, June 2, 1814, Loose Manuscripts, box 6, Old B7 F7.19, Francis Cabot Lowell (1775–1817) Papers, Massachusetts Historical Society, Boston, Massachusetts.
10. Director's Records, Volume 1, MSS:442, 1–2, Boston Manufacturing Company Records, Baker Library Historical Collections, Harvard Business School, Allston, Massachusetts; Robert F. Dalzell, Jr., *Enterprising Elite: The Boston Associates and the World They Made* (Cambridge, MA: Harvard University Press, 1987), 8–10, 26; Ware, *Early New England Cotton Manufacture*, 63, 138, 147–48.
11. Carding was done on the first floor, spinning on the second, and weaving on the third and fourth. In 1820, after it built a second mill, Boston Manufacturing employed about 230 to 265 workers, of whom roughly 85 percent were women and only 5 percent "boys." Appleton, *Introduction of the Power Loom*, 1; Richard M. Candee, "Architecture and Corporate Planning in the Early Waltham System," in Robert Weible, *Essays from the Lowell Conference on*

Industrial History 1982 and 1983: The Arts and Industrialism, The Industrial City (North Andover, MA: Museum of American Textile History, 1985), 19, 24, 26; U.S. Department of Interior, National Park Service, "National Register of Historical Places Inventory—Nomination Form," Boston Manufacturing Company (accessed Jan. 16. 2015), http://pdfhost.focus.nps.gov/docs/NHLS/Text/77001412.pdf; Ware, *Early New England Cotton Manufacture*, 64.

12. Peter Temin, "Product Quality and Vertical Integration in the Early Cotton Textile Industry," *Journal of Economic History* XVIII (1988), 893, 897; Appleton, *Introduction of the Power Loom*, 9–12; Beckert, *Empire of Cotton*, 147; Ware, *Early New England Cotton Manufacture*, 65, 70–72; Zevin, "The Growth of Cotton Textile Production," 126–27.

13. The original two Waltham mills are still standing, but in altered form, having had their pitched roofs replaced by flat ones and the space between them filled in by subsequent construction. Ware, *Early New England Cotton Manufacture*, 66; Candee, "Architecture and Corporate Planning in the Early Waltham System," 24–25; "National Register of Historical Places Inventory—Nomination Form," Boston Manufacturing Company.

14. While the first mill established the basic framework for production, the second mill established the physical template for future mills. Candee, "Architecture and Corporate Planning in the Early Waltham System," 29, 34; Appleton, *Introduction of the Power Loom*, 14.

15. Batchelder, *Introduction and Early Progress of the Cotton Manufacture*, 81; Dalzell, Jr., *Enterprising Elite*, 30–31, 50; Appleton, *Introduction of the Power Loom*, 9; Ware, *Early New England Cotton Manufacture*, 83; Laurence Gross, *The Course of Industrial Decline: The Boott Cotton Mills of Lowell, Mass., 1835–1955* (Baltimore: Johns Hopkins University Press, 1993), 12; Thomas Dublin, *Women at Work: The Transformation of Work and Community in Lowell, Massachusetts, 1826–1860* (New York: Columbia University Press, 1979), 59; Betsy Hunter Bradley, *The Works: The Industrial Architecture of the United States* (New York: Oxford University Press, 1999), 93.

16. Ware, *Early New England Cotton Manufacture*, 63, 139, 145, 184; Gross, *Course of Industrial Decline*, 6–7, 229.

17. Recent accounts that stress the global nature of the cotton industry include Prasannan Parthasarathi and Giorgio Riello, eds., *The Spinning World: A Global History of Cotton Textiles, 1200–1850* (Oxford: Oxford University Press, 2009); Riello, *Cotton: The Fabric that Made the Modern World* (Cambridge: Cambridge University Press, 2013); and Beckert, *Empire of Cotton*. On U.S. cotton exports, see Ware, *Early New England Cotton Manufacture*, 189–91.

18. Gross, *Course of Industrial Decline*, 4–5; Appleton, *Introduction of the Power Loom*, 23–24; Minutes: Directors, 1822–1843, shelf number 1, Merrimack Manufacturing Company Records, Baker Library, HBS, 5, 15; Shlakman, *Economic History of a Factory Town*, 36.

19. Mills of this design still can be seen across large parts of New England, many now converted to condominiums, office space, warehouses, artists' studios, museums, or cultural centers or sitting abandoned.

20. Merrimack purchased from Boston Manufacturing the right to use machinery it had designed and patented. All the space in the five mill buildings was not initially filled with equipment, but the company soon purchased additional machinery. Minutes: Directors, 1822–1843, shelf number 1, Merrimack Manufacturing Company Records, 5, 51–54; Bradley, *The Works*, 93, 113–14, 125–28, 133–35, 139; Bahr, "New England Mill Engineering," 13, 21, 27, 40–41, 44–45; Gross, *Course of Industrial Decline*, 7. For an example of the concern about fire, see the 1829 report by a committee of the Merrimack Board of Directors about measures "to render the mills at Lowell more secure from fire," in Minutes: Directors, 1822–1843, shelf number 1, Merrimack Manufacturing Company Records, 61, 63–65.

21. Dalzell, Jr., *Enterprising Elite*, 47–50, 47–50; Thomas Dublin, *Farm to Factory: Women's Letters, 1830–1860* (New York: Columbia University Press, 1981), 5–8; Appleton, *Introduction of the Power Loom*, 28–29; Samuel Batchelder to Nathan Appleton, Sept. 25, 1824, and William Appleton to Samuel Batchelder, Oct. 8, 1824, in Minute Books, v.a – Directors, 1824–1857; Proprietors, 1824–64, Hamilton Manufacturing Company Records, Baker Library; F-1 Records 1828–1858, 26–27, Bigelow Stanford Carpet Co. collection, Lowell Manufacturing Company records, Baker Library; Shlakman, *Economic History of a Factory Town*, 38, 42. A list of the various Lowell textile firms, their officers, and principal stockholders appears in Shlakman, 39–42.

22. Candee, "Architecture and Corporate Planning in the Early Waltham System," 25–30.

23. Minutes: Directors, 1822–1843, Merrimack Manufacturing Company Records, 23, 25–26, 81; Candee, "Architecture and Corporate Planning in the Early Waltham System," 38–39; Appleton, *Introduction of the Power Loom*, 24; Lowell Manufacturing Company Records, 1828–1858, 66–68; Dublin, *Farm to Factory*, 5–8; U.S. Bureau of the Census, "Population of the 100 Largest Urban Places: 1840," June 15, 1998, https://www.census.gov/population/www/documentation/twps0027/tab07.txt.

24. Shlakman, *Economic History of a Factory Town*, 25–26, 36–37, 39–42.

25. Local workers were hired for construction work associated with the mills.

But even for that, some outside workers—like Irish canal diggers—were brought in. Dalzell, Jr., *Enterprising Elite*, x, xi, 56, Shlakman, *Economic History of a Factory Town*, 24–25, 49, 64–65; Tamara K. Hareven and Randolph Lanenbach, *Amoskeag: Life and Work in an American Factory City* (New York: Pantheon Books, 1978), 16.

26. The Lowell Machine Shop, separated from Canals and Locks in 1845, employed 550 workers, making it a giant among machine shops, turning out not only textile equipment but also planning machines, steam boilers, mill shafting, and even locomotives. Alfred D. Chandler, Jr., *The Visible Hand: The Managerial Revolution in American Business* (Cambridge, MA: Harvard University Press, 1977), 60; United States Census Office, *Manufacturers of the United States in 1860* (Washington, DC: 1865), 729; "Statistics of Lowell Manufactures. January, 1857. Compiled from authentic sources." [Lowell, 1857], Library of Congress (accessed Jan. 28, 2015), http://memory.loc.gov/cgi-bin/query/h?ammem/rbpebib:@field(NUMBER+@band[rbpe+0620280a]); David R. Meyer, *Networked Machinists: High-Technology Industries in Antebellum America* (Baltimore: Johns Hopkins University Press, 2006), 205.

27. Some companies did introduce new types of spinning machines, which could operate at higher speeds. Dublin, *Farm to Factory*, 5–8; Dalzell, Jr., *Enterprising Elite*, 55, 69–71; Daniel Nelson, *Managers and Workers: The Origins of the New Factory System in the United States, 1880–1920* (Madison: University of Wisconsin Press, 1975), 6; Gross, *Course of Industrial Decline*, 37, 42; Candee, "Architecture and Corporate Planning in the Early Waltham System," 34, 38.

28. The companies eventually replaced their original waterwheels with more efficient water turbines. Even after the Civil War, they only gradually installed steam engines. As late as the 1890s, the Boott mills were getting half their power from water. Gross, *Course of Industrial Decline*, 19–20, 42–43; Ware, *Early New England Cotton Manufacture*, 144–45. For a comparison of the cost of steam and water power, see "Difference between the cost of power to be used at Dover the next 15 years and a full supply of water," Box 6, Vol. III–IX, Nov. 1847, Amos Lawrence Papers, Massachusetts Historical Society.

29. Shlakman, *Economic History of a Factory Town*, 37; Ware, *Early New England Cotton Manufacture*, 86–87; "Statistics of Lowell Manufactures. January, 1857."

30. Gross, *Course of Industrial Decline*, 30–31, 37, 42, 50–53; David A. Zonderman, *Aspirations and Anxieties: New England Workers & The Mechanized Factory System, 1815–1850* (New York: Oxford University Press, 1992), 69–70; Chandler, *The Visible Hand*, 68–71.

31. Hareven and Lanenbach, *Amoskeag*, 9–10, 13–16.
32. Even after it ballooned in size, Amoskeag continued to be run by a single treasurer, working out of Boston. Hareven and Lanenbach, *Amoskeag*, 16. The best overview of the development of American management remains Chandler, *The Visible Hand*.
33. *New-York Daily Tribune*, Jan. 17, 1844. For the experience of young women in British cotton mills, see Deborah Valenze, *The First Industrial Woman* (New York: Oxford University Press, 1995), 103–11.
34. Dalzell, Jr., *Enterprising Elite*, 31–34; Robert S. Starobin, *Industrial Slavery in the Old South* (New York: Oxford University Press, 1970), 13; Ware, *Early New England Cotton Manufacture*, 12–13, 198–99, 203. Slave labor also was used in the cotton industry in Egypt, where the first mechanized equipment was introduced at the same time as the Waltham mills were being built. Beckert, *Empire of Cotton*, 166–68.
35. Dublin, *Women at Work*, 5, 31–34, 141; Zonderman, *Aspirations and Anxieties*, 131, 270–71, 276; Dublin, *Farm to Factory*, 13–14; Ware, *Early New England Cotton Manufacture*, 217–18.
36. Dublin, *Women at Work*, 26, 31, 64–65; Zonderman, *Aspirations and Anxieties,* 130, 138–40. When in 1826 Merrimack Manufacturing was planning to print calicoes, it sent its treasurer, Kirk Boott, to England "for the purpose of procuring a first rate Engraver, or such as he can get," as well as to gather information "which he may think will be useful in manufacturing, printing or machine building." Minutes: Directors, 1822–1843, shelf number 1, Merrimack Manufacturing Company Records, 32–33. 1857 percentage calculated from "Statistics of Lowell Manufactures. January, 1857."
37. Ware, *Early New England Cotton Manufacture*, 212–15, 220–21; Thomas Dublin, *Transforming Women's Work: New England Lives in the Industrial Revolution* (Ithaca, NY: Cornell University Press, 1994), 82–83, 89; Shlakman, *Economic History of a Factory Town*, 49; Zonderman, *Aspirations and Anxieties*, 163–64, 166–68.
38. Ware, *Early New England Cotton Manufacture*, 224–25; Zonderman, *Aspirations and Anxieties*, 256–57. Thomas Dublin found among a sample of workers at the Hamilton mill in Lowell that those who never married worked on average 3.9 years; those who did, 2.4. Dublin, *Farm to Factory*, 110.
39. *Burlington Free Press*, Dec. 5, 1845; Ware, *Early New England Cotton Manufacture*, 200, 263; "Regulations to Be Observed by All Persons Employed in the Factories of the Middlesex Company" (1846); "General Regulations, to Be Observed by All Persons Employed by the Lawrence Manufacturing

Company, In Lowell" (1833); "Regulations to Be Observed by All Persons Employed by the Lawrence Manufacturing Company" (1838); and "Regulations for the Boarding Houses of the Middlesex Company" (n.d.), all in Osborne Library, American Textile History Museum, Lowell, Massachusetts; Zonderman, *Aspirations and Anxieties,* 150, 152, 157–60; Dublin, *Women at Work,* 78–79.

40. Zonderman, *Aspirations and Anxieties,* 66–67, 90; Shlakman, *Economic History of a Factory Town,* 59; Friedrich Engels, *The Condition of the Working Class in England,* trans. W. O. Henderson and W. H. Chaloner (Stanford, CA: Stanford University Press, 1958), 30–87.

41. Augusta Harvey Worthen, *The History of Sutton, New Hampshire: Consisting of the Historical Collections of Erastus Wadleigh, Esq., and A. H. Worthen,* 2 parts (Concord, NH: Republican Press Association, 1890), 192, quoted in Dublin, *Women at Work,* 55; population of Sutton from New Hampshire Office of Energy and Planning, State Data Center (accessed Feb. 6, 2015), https://www.nh.gov/oep/data-center/documents/1830-1920-historic.pdf; Harriet H. Robinson, *Loom and Spindle, or Life Among the Early Mill Girls* (New York: Thomas Y. Crowell, 1898), 69–70; Dublin, *Transforming Women's Work,* 111–18.

42. Zonderman, *Aspirations and Anxieties,* 8. Dublin's *Farm to Factory* presents an excellent selection of letters from female mill workers.

43. *The Lowell Offering and Magazine,* May 1843, 191; Dublin, *Farm to Factory,* 69, 73; Zonderman, *Aspirations and Anxieties,* 22–27, 30, 38–40.

44. *The Lowell Offering and Magazine,* January 1843, 96; Zonderman, *Aspirations and Anxieties,* 42–43, 78–79, 82–83, 113–14.

45. According to Harriet Robinson, in 1843 there were "fourteen regularly organized religious societies" in Lowell. Robinson, *Loom and Spindle,* 78; Zonderman, *Aspirations and Anxieties,* 97; Dublin, *Farm to Factory,* 80–81; Ware, *Early New England Cotton Manufacture,* 256–59.

46. Ware, *Early New England Cotton Manufacture,* 38, 85–86, 110, 112; Shlakman, *Economic History of a Factory Town,* 98–101, 103–07; Dublin, *Women at Work,* 136–37. As Dublin points out, the expiration of patents taken out by the Waltham-Lowell group and advances in equipment design elsewhere made it easier for new companies to compete. On the relative cost of raw cotton and labor, see, for example, "Boston Manufacturing Company Memo of Cloth Made and Cost of Same . . . 25th August 1827 to 30th August 1828" and "Appleton Co. Mem. of Cloth Made to May 30, 1829," both in box 1, folder 16, vol. 42, Patrick Tracy Jackson Papers, Massachusetts Historical Society.

47. Minutes: Directors, 1822–1843, Merrimack Manufacturing Company Records, 142; Shlakman, *Economic History of a Factory Town*, 98–99; Dublin, *Women at Work*, 89–90, 98, 109–11, 137.

48. Dublin, *Women at Work*, 90–102.

49. Dublin, *Women at Work*, 93–96; Robinson, *Loom and Spindle*, 84. One version of the original song began: "What a pity that such a pretty girl as I, Should be sent to a nunnery to pine away and die!" with the chorus: "So I won't be a nun, I cannot be a nun! I'm so fond of pleasure that I cannot be a nun." https://thesession.org/tunes/3822 (accessed Feb. 7, 2015). For the growth of the labor movement before the Civil War, the most comprehensive account remains John R. Commons et al., *History of Labor in the United States*, vol. I ([1918] New York: Augustus M. Kelley, 1966).

50. There were a few later strikes in other mill towns and a small strike by immigrant workers in Lowell in 1859. Zonderman, *Aspirations and Anxieties*, 235, 241; *New-York Daily Tribune*, May 14, 1846; Fisher, *Workshops in the Wilderness*, 146–47; Dublin, *Women at Work*, 203–05.

51. Massachusetts restricted children under twelve and Connecticut children under fourteen to ten hours work a day. New Hampshire established ten hours as a day's work for everyone, but allowed contracts calling for longer working hours, rendering the law all but meaningless. Zonderman, *Aspirations and Anxieties*, 242–49; Dublin, *Women at Work*, 108–22.

52. Much later on, left-leaning historians perhaps made too much of the walkouts and agitation. For extended discussions of the protests which emphasize their importance, see, for example, Zonderman, *Aspirations and Anxieties,* and Dublin, *Women at Work.* By contrast, Ware, earlier, was somewhat dismissive of the turnouts, which she wrote "were really less strikes than demonstrations, unorganized outbursts led by a few inflammatory spirits who had little idea what they were to achieve but who raised the girls to a state of great excitement" and noted that "Public sentiment did not generally support 'striking females.'" Ware, *Early New England Cotton Manufacture*, 275, 277.

53. David Crockett, *An Account of Col. Crockett's Tour to the North and Down East* (Philadelphia: E. L. Carey and A. Hart, 1835), 91–99; John F. Kasson, *Civilizing the Machine: Technology and Republican Values in America, 1776–1900* (New York: Penguin, 1977), 81; Zonderman, *Aspirations and Anxieties*, 208.

54. For extended discussions of this evolution, see Leo Marx, *The Machine in the Garden: Technology and the Pastoral Ideal in America* (New York: Oxford University Press, 1964), and Kasson, *Civilizing the Machine*, esp. chap. 1 and 2. See also, Lawrence A. Peskin, "How the Republicans Learned to Love

Manufacturing: The First Parties and the 'New Economy,'" *Journal of the Early Republic* 22 (2) (Summer, 2002), 235–62, and Jonathan A. Glickstein, *Concepts of Free Labor in Antebellum America* (New Haven, CT: Yale University Press, 1991), esp. 233–35.

55. John G. Whittier, "The Factory Girls of Lowell," in *Voices of the True-Hearted* (Philadelphia: J. Miller M'Kim, 1846), 40–41.

56. Seth Luther, *An Address to the Working Men of New England on the State of Education and on the Condition of the Producing Classes in Europe and America*, 2nd ed. (New York: George H. Evans, 1833), 19.

57. Fisher, *Workshops in the Wilderness*, 165; Emerson quoted in Kasson, *Civilizing the Machine*, 124–25. Earlier, Emerson had hailed manufacturing for freeing New England from the need to farm under uncongenial conditions: "Where they have sun, let them plant; we who have it not, will drive our pens and water-wheels." Ralph Waldo Emerson, Edward Waldo Emerson, and Waldo Emerson Forbes, *Journals of Ralph Waldo Emerson with Annotations*, vol. IV (Boston: Houghton Mifflin, 1910), 209.

58. Zonderman, *Aspirations and Anxieties*, 115–18.

59. As late as 1853, there were over 1,800 children under fifteen working in Rhode Island manufacturing establishments, including 621 between ages nine and twelve and 59 under the age of nine. Luther, *An Address to the Working Men of New England*, 10, 21–22, 30; Ware, *Early New England Cotton Manufacture*, 210; Jonathan Prude, *The Coming of Industrial Order: Town and Factory Life in Rural Massachusetts, 1810–1860* (Cambridge: Cambridge University Press, 1983), 86, 213.

60. Trollope, *North America*, 253; John Robert Godley, *Letters from America*, vol. 1 (London: John Murray, 1844), 7–11; Edward Bellamy, "How I Wrote 'Looking Backwards,'" in *Edward Bellamy Speaks Again* (Chicago: Peerage Press, 1937), 218, quoted in Kasson, *Civilizing the Machine*, 192. Relative industry size from Beckert, *Empire of Cotton*, 180.

61. Herman Melville, "The Paradise of Bachelors and the Tartarus of Maids," *Harper's* magazine, Apr. 1855, 670–78; Scott Heron, "Harper's Magazine as Matchmaker: Charles Dickens and Herman Melville," *Browsings: The Harper's Blog*, Jan. 13, 2008, http://harpers.org/blog/2008/01/harpers-magazine-dickens-and-melvilles-paradise-of-bachelors/.

62. Kasson, *Civilizing the Machine*, 90–93.

63. Luther, *An Address to the Working Men of New England*, 29.

64. Fisher, *Workshops in the Wilderness*, 115–16, 119, 130–35, 139–41, 146.

65. Dublin, *Women at Work*, 139–40.

66. U.S. Bureau of the Census, *Historical Statistics of the United States: Colo-*

nial Times to 1970, Bicentennial Edition Part 1 (Washington, D.C.: U.S. Government Printing Office, 1975), 106; Ware, *Early New England Cotton Manufacture*, 227–232; Dublin, *Women at Work*, 138–39.

67. Dublin, *Women at Work*, 134, 140–44, 155, 198; Gross, *Course of Industrial Decline*, 83.

68. Dublin, *Farm to Factory*, 187; Gross, *Course of Industrial Decline*, 37, 42–43, 79; Nelson, *Managers and Workers*, 6; Ardis Cameron, *Radicals of the Worst Sort: Laboring Women in Lawrence, Massachusetts, 1860–1912* (Urbana: University of Illinois Press, 1993), xiv–xv, 28, 75; Hareven and Lanenbach, *Amoskeag*, 10.

69. Dublin, *Farm to Factory*, 187; Gross, *Course of Industrial Decline*, 80, 142; Hareven and Lanenbach, *Amoskeag*, 18–19, 202–03; Cameron, *Radicals of the Worst Sort*, 29–30, 75, 82–83, 97 (quote).

70. Accounts of the death toll from the Pemberton collapse vary considerably, from 83 to 145. Clarisse A. Poirier, "Pemberton Mills 1852–1938: A Case Study of the Industrial and Labor History of Lawrence, Massachusetts," Ph.D. dissertation, Boston University, 1978, 81–84, 191–93; *Polynesian* [Honolulu], Mar. 3, 1860; *New York Times*, Jan. 12, 1860, and Feb. 4, 1860; *The Daily Dispatch* [Richmond, Virginia], Jan. 16, 1860; *The Daily Exchange* [Baltimore], Jan. 12, 1860; *New-York Daily Tribune*, Jan. 16, 1860; Alvin F. Oickle, *Disaster in Lawrence: The Fall of the Pemberton Mill* (Charleston, SC: History Press, 2008); Bahr, "New England Mill Engineering, 68–71; Cameron, *Radicals of the Worst Sort*, 18–19.

71. By the time Hine visited Amoskeag, children under age sixteen actually made up only a small part of the New England textile workforce: 2.0 percent in New Hampshire, 5.7 percent in Massachusetts, and 6.0 percent in Rhode Island, compared to 10.4 percent nationally and 20.3 percent in Mississippi. Hareven and Lanenbach, *Amoskeag*, 33; Arden J. Lea, "Cotton Textiles and the Federal Child Labor Act of 1916," *Labor History* 16 (4) (Fall 1975), 492.

72. Gross, *Course of Industrial Decline*, 88–90; Cameron, *Radicals of the Worst Sort*, 7, 47–62, 77.

73. In the racialist language of the day, which many socialists shared, Berger went on to say, "White men and women of any nationality will endure a certain degree of slavery, but no more. The limit of endurance seems to have been reached in Lawrence." House Committee on Rules, *The Strike at Lawrence, Hearings before the Committee on Rules of the House of Representatives on House Resolutions 409 and 433, March 2–7, 1912* (Washington, D.C.: Government Printing Office, 1912), 10–11. There is a large literature on the 1912 strike. An excellent account can be found in Melvyn Dubofsky, *We Shall Be*

All: A History of the Industrial Workers of the World (Chicago: Quadrangle Books, 1969).

74. Hareven and Lanenbach, *Amoskeag*, 11, 336; Gross, *Course of Industrial Decline*, 165, 190–95, 225–29; Mary H. Blewett, *The Last Generation: Work and Life in the Textile Mills of Lowell, Massachusetts, 1910–1960* (Amherst: University of Massachusetts Press, 1990).

75. British population does not include Ireland. Chandler, Jr., *Scale and Scope*, 4; B. R. Mitchell, *International Historical Statistics: Europe, 1750–1993* (London: Macmillan Reference, 1998), 4, 8; U.S. Bureau of the Census, *Historical Statistics of the United States: Colonial Times to 1970*, 8.

Chapter 3
"THE PROGRESS OF CIVILIZATION"

1. Joshua Freeman et al., *Who Built America? Working People and the Nation's Economy, Politics, Culture, and Society*, vol. 2 (New York: Pantheon Books, 1992), xii–xx; *Frank Leslie's Illustrated Newspaper*, May 20, 1876; J. S. Ingram, *The Centennial Exposition, Described and Illustrated* (Philadelphia: Hubbard Bros., 1876); Linda P. Gross and Theresa R. Snyder, *Philadelphia's 1876 Centennial Exhibition* (Charleston, SC: Arcadia Publishing, 2005); John E. Findling, ed., *Historical Dictionary of World's Fairs and Expositions, 1851–1988* (New York: Greenwood Press, 1990), 57–59; Robert W. Rydell, *All the World's a Fair: Visions of Empire at American International Expositions, 1876–1916* (Chicago: University of Chicago Press, 1987), 9–37; Centennial Photographic Co., "[Saco] Water Power Co.—Cotton Machinery," Centennial Exhibition Digital Collection Philadelphia 1876, Free Library of Philadelphia, CEDC No. c032106 (accessed Mar. 20, 2015), http://libwww.library.phila.gov/CenCol/Details.cfm?ItemNo=c032106. See also Bruni Giberti, *Designing the Centennial: A History of the 1876 International Exhibition in Philadelphia* (Lexington: University of Kentucky Press, 2002).

2. On the national divides at the time of the exhibition, see Freeman et al., *Who Built America?* vol. 2, xx–xxiv.

3. When Whitman visited the Centennial Exhibition, he reportedly sat for a half hour in silence before the Corliss engine. Leo Marx, *The Machine in the Garden: Technology and the Pastoral Ideal in America* (New York: Oxford University Press, 1964), 150–58, 163–64; Andrea Sutcliffe, *Steam:*

The Untold Story of America's First Great Invention (New York: Palgrave, 2004); Walter Johnson, *River of Dark Dreams: Slavery and Empire in the Cotton Kingdom* (Cambridge, MA: Harvard University Press, 2013), 73–96; Edmund Flagg, *The Far West: or, A Tour Beyond the Mountains*, vol. 1 (New York: Harper & Brothers, 1838), 17–18; John F. Kasson, *Civilizing the Machine: Technology and Republican Values in America, 1776–1900* (New York: Penguin, 1977), 141; Robert W. Rydell, *All the World's a Fair*, 15–16.

4. Walt Whitman, *Two Rivulets: Including Democratic Vistas, Centennial Songs, and Passage to India* (Camden, NJ: [Walt Whitman], 1876), 25–26; Marx, *Machine in the Garden*, 27. There is a very large literature on the railroad and modernity. See, for example, Wolfgang Schivelbusch, *The Railway Journey: The Industrialization of Time and Space in the 19th Century* (Berkeley: University of California Press, 1986).

5. Giberti, *Designing the Centennial*, 2–3; "Manufactures of Massachusetts," *The North American Review* 50 (106) (Jan. 1840), 223–31.

6. The Crystal Palace burned down in 1936. Jeffrey A. Auerbach, *The Great Exhibition of 1851: A Nation on Display* (New Haven, CT: Yale University Press, 1999); Benjamin quoted in Robert W. Rydell, *Worlds of Fairs: The Century-of-Progress Expositions* (Chicago: University of Chicago Press, 1993), 15.

7. Many exhibits for the New York fair were not ready when it opened, damping down attendance. Unlike the profitable original, it ended in bankruptcy. Charles Hirschfeld, "America on Exhibition: The New York Crystal Palace," *American Quarterly* 9 (2, pt. 1) (Summer 1957), 101–16.

8. Pauline de Tholozany, "The Expositions Universelles in Nineteenth Century Paris," Brown University Center for Digital Scholarship, http://library.brown.edu/cds/paris/worldfairs.html (accessed Mar. 27, 2015). For a list of nineteenth- and twentieth-century international expositions and fairs, see Findling, ed., *Historical Dictionary of World's Fairs and Expositions*, 376–81.

9. *Report of the Board of Commissioners Representing the State of New York at the Cotton States and International Exposition held at Atlanta, Georgia, 1895* (Albany, NY: Wynkoop Hallenbeck Crawford Co, 1896), quote on page 205; C. Vann Woodward, *Origins of the New South, 1877–1913: A History of the South* (Baton Rouge: Louisiana State University, 1951), 123–24.

10. Jill Jonnes, *Eiffel's Tower: The Thrilling Story Behind Paris's Beloved Monument and the World's Fair Where Buffalo Bill Beguiled Paris, the Artists Quarreled, and Thomas Edison Became a Count* (New York: Viking, 2009); "Origins and Construction of the Eiffel Tower,"

http://www.toureiffel.paris/en/everything-about-the-tower/themed-files/69.html, and "All You need to Know About the Eiffel Tower," http://www.toureiffel.paris/images/PDF/about_the_Eiffel_Tower.pdf (both accessed Oct. 21, 2016); Roland Barthes, *The Eiffel Tower and Other Mythologies* ([1979] Berkeley: University of California Press, 1997), 8–14.

11. Letter published in *Le Temps*, Feb. 14, 1887, reprinted in "All You Need to Know About the Eiffel Tower."

12. "Représentation de la tour Eiffel dans l'art," http://fr.wikipedia.org/wiki/Repr%C3%A9sentation_de_la_tour_Eiffel_dans_l%27art; and Michaela Haffner, "Diego Rivera, *The Eiffel Tower, 1914*," the Davis Museum at Wellesley College, https://www.wellesley.edu/davismuseum/artwork/node/37002 (both accessed Apr. 1, 2015). For a different reading of the iconography of the Eiffel Tower, with less emphasis on its importance as a symbol of industrialism and the mechanical age, see Gabriel Insausti, "The Making of the Eiffel Tower as a Modern Icon," in *Writing and Seeing: Essays on Word and Image*, ed. Rui Carvalho Homem and Maria de Fátima Lambert (Amsterdam: Editions Rodopi, 2006).

13. Guillaume Apollinaire, "Zone," translated by Donald Revell, http://www.poets.org/poetsorg/poem/zone. For an alternative, more literal translation by Charlotte Mandell, see http://www.charlottemandell.com/Apollinaire.php (accessed Apr. 2, 2015).

14. Blaise Cendrars, "Elastic Poem 2: Tower," trans. by Tony Baker, *GutCult* 2 (1) (Winter 2004), http://gutcult.com/Site/litjourn3/html/cendrars1.html.

15. The great nineteenth-century expositions were not only about industry and consumer goods. They also celebrated national identity and greatness as manifested in the arts and empire. And empire was tightly linked to ideas of racial hierarchy, a theme that bluntly recurred in fair after fair. Technological and racial advance were inextricably linked. See Auerbach, *Great Exhibition of 1851*, 159–89; Joseph Harris, *The Tallest Tower: Eiffel and the Belle Epoque* (Bloomington, IN: Unlimited Publishing, 2004), 88–89, 107–08; Rydell, *All the World's a Fair*, 21–22; Rydell, *Worlds of Fairs*, 19–22; Findling, ed., *Historical Dictionary of World's Fairs and Expositions*, 79, 181, 183.

16. Guy de Maupassant, *La Vie Errane, Allouma, Toine, and Other Stories* (London: Classic Publishing Company, 1911), 1–4.

17. Auerbach, *Great Exhibition of 1851*, 128–58; Freeman et al., *Who Built America?* vol. 2, xxiii.

18. Auerbach, *Great Exhibition of 1851*, 132, 156; Friedrich Engels to Laura Lafarge, June 11, 1889, www.marxists.org/archive/marx/works/1889/letters/

89_06_11.htm (accessed Apr. 4, 2017); Tristram Hunt, *Marx's General: The Revolutionary Life of Friedrich Engels* (New York: Metropolitan Books, 2009), 335–36.

19. *The Making, Shaping and Treating of Steel*, published by U.S. Steel in ten editions between 1919 and 1985, provides encyclopedic information on iron- and steelmaking, including their history. For a history and analysis of this remarkable volume, see Carol Siri Johnson, "The Steel Bible: A Case Study of 20th Century Technical Communication," *Journal of Technical Writing and Communication* 37 (3) (2007), 281–303. See also Peter Temin, *Iron and Steel in Nineteenth-Century America: An Economic Inquiry* (Cambridge, MA: MIT Press, 1964), 13–17, 83–85.

20. Eric Hobsbawm, *The Age of Capital 1848–1875* (New York: Charles Scribner's Sons, 1975), 39, 54–55; Temin, *Iron and Steel*, 3–5, 14–15, 21. For the difficulties in producing rails, see John Fritz, *The Autobiography of John Fritz* (New York: John Wiley & Sons, 1912), 92–101, 111–15, 121–23, 149. Overman quoted in Paul Krause, *The Battle for Homestead, 1880–1892: Politics, Culture, and Steel* (Pittsburgh, PA: University of Pittsburgh Press, 1992), 47.

21. In addition to iron ore and fuel (charcoal, coke, or sometimes anthracite coal), limestone was put into blast furnaces to help form slag out of impurities. Temin, *Iron and Steel*, 58–62, 96–98, 157–63; U.S. Steel, *The Making, Shaping and Treating of Steel* (Pittsburgh, PA: U.S. Steel, 1957), 221–25.

22. Krause, *Battle for Homestead*, 48–49; David Montgomery, *Workers' Control in America* (Cambridge: Cambridge University Press, 1979), 11–12. For a firsthand account of puddling, see James J. Davis, *The Iron Puddler; My Life in the Rolling Mills and What Came of It* (New York: Grosset & Dunlop, 1922).

23. Temin, *Iron and Steel*, 66–67, 85, 105–06, 109–13; Fritz, *Autobiography of John Fritz*, 91–135; Marvin Fisher, *Workshops in the Wilderness: The European Response to American Industrialization, 1830–1860* (New York: Oxford University Press, 1967), 162–63.

24. Krause, *Battle for Homestead*, 52–65; Temin, *Iron and Steel*, 125–27, 130, 153; David Brody, *Steelworkers in America: The Nonunion Era* (1960; New York: Harper & Row, 1969), 8.

25. Some companies also integrated backward, buying or leasing ore mines and making their own coke. Temin, *Iron and Steel*, 153–69, 190–91; Brody, *Steelworkers in America*, 10–12; William Serrin, *Homestead: The Glory and Tragedy of an American Steel Town* (New York: Random House, 1992), 56–59.

26. Hobsbawm, *Age of Capital*, 213; Harold James, *Krupp: A History of the Legendary German Firm* (Princeton, NJ: Princeton University Press, 2012), 47,

53; Gross and Snyder, *Philadelphia's 1876 Centennial Exhibition*, 83; Schneider Electric, *170 Years of History* (Rueil-Malmaison, France: Schneider Electric, 2005), 3–5, 20–22 (http://www.schneider-electric.com/documents/presentation/en/local/2006/12/se_history_brands_march2005.pdf).

27. Daniel Nelson, *Managers and Workers: Origins of the New Factory System in the United States, 1880–1920* (Madison: University of Wisconsin Press, 1975), 6–7; David Nasaw, *Andrew Carnegie* (New York: Penguin Press, 2006), 405; U.S. Census Office, *Twelfth Census of the United States—1900; Census Reports,* vol. VII—*Manufactures,* part I (Washington, D.C.: U.S. Census Office, 1902), 583, 585, 597.

28. U.S. Steel, *Making, Shaping and Treating of Steel*; Carnegie quoted in Brody, *Steelworkers in America*, 21.

29. Michael W. Santos, "Brother against Brother: The Amalgamated and Sons of Vulcan at the A. M. Byers Company, 1907–1913," *The Pennsylvania Magazine of History and Biography* 111 (2) (Apr. 1987), 199–201; Davis, *Iron Puddler*, 85; John Fitch, *The Steel Workers* (New York: Charities Publication Committee, 1910), 36, 40–44, 48, 52. William Attaway's novel, *Blood on the Forge* ([1941] New York: New York Review of Books, 2005), set in Pittsburgh at the end of World War I, gives a good sense of the rhythms of steelwork, with its alternate periods of exhausting labor and waiting for the next burst of activity.

30. Harry B. Latton, "Steel Wonders," *The Pittsburgh Times,* June 1, 1892, reprinted in David P. Demarest, Jr., ed., *"The River Ran Red": Homestead 1892* (Pittsburgh, PA: University of Pittsburgh Press, 1992), 13–15; Fritz, *Autobiography of John Fritz*, 203; Brody, *Steelworkers in America*, 9; Mark Reutter, *Sparrows Point; Making Steel—The Rise and Ruin of American Industrial Might* (New York: Summit Books, 1988), 18.

31. Fitch, *The Steel Workers*, 3; Nathaniel Hawthorne, *Passages from the English Note-Books of Nathaniel Hawthorne,* vol. 1 (Boston: James R. Osgood and Company, 1872), 370–72. I was pointed to Hawthorne's statement by John F. Kasson, who quotes part of it in *Civilizing the Machine*, 142.

32. Marx, *The Machine in the Garden*, 192, 200, 270–71; Joseph Stella, "In the Glare of the Converter," "In the Light of a Five-Ton Ingot," "At the Base of the Blast Furnace," and "Italian Steelworker" (accessed Apr. 28, 2015), http://www.clpgh.org/exhibit/stell1.html; W. J. Gordon, *Foundry, Forge and Factory with a Chapter on the Centenary of the Rotary Press* (London: Religious Tract Society, 1890), 15; John Commons et al., *History of Labour in the United States,* vol. II ([1918] New York: Augustus M. Kelley, 1966), 80.

33. Hawthorne, *Passages from the English Note-Books*, 371; Thomas G. Andrews,

Killing for Coal: America's Deadliest Labor War (Cambridge, MA: Harvard University Press, 2008), 62; Joseph Stella, "Discovery of America: Autobiographical Notes," quoted in Maurine W. Greenwald, "Visualizing Pittsburgh in the 1900s: Art and Photography in the Service of Social Reform," in Greenwald and Margo Anderson, eds., *Pittsburgh Surveyed: Social Science and Social Reform in the Early Twentieth Century* (Pittsburgh, PA: University of Pittsburgh Press, 1996), 136; Lincoln Steffens, *The Autobiography of Lincoln Steffens* (New York: Harcourt, Brace and Company, 1931), 401.

34. Nasaw, *Carnegie*, 164; Mary Heaton Vorse, *Men and Steel* (New York: Boni and Liveright, 1920), 12; Sharon Zukin, *Landscapes of Power: From Detroit to Disney World* (Berkeley: University of California Press, 1991), 60; Gunther quoted in Reutter, *Sparrows Point*, 9.

35. For a vivid account of the tumultuous struggles of the Gilded Age, see Steve Fraser, *The Age of Acquiescence; The Life and Death of American Resistance to Organized Wealth and Power* (New York: Little, Brown and Company, 2015), chap. 4–6, especially chap. 5 on industrial strife.

36. The Amalgamated Association of Iron and Steel Workers was created by an 1876 merger of the Sons of Vulcan with two unions of rolling mill workers. Brody, *Steelworkers in America*, 50–53; Preamble to the Constitution of the Amalgamated Association of Iron and Steel Workers, reprinted in Demarest, Jr., ed., *"The River Ran Red,"* 17; David Montgomery, *The Fall of the House of Labor: The Workplace, the State, and American Labor Activism, 1865–1925* (Cambridge: Cambridge University Press, 1987), 9–22.

37. Some companies continued to just make iron goods, without the intensely competitive ethos of the dominant steel producers. Montgomery, *Fall of the House of Labor*, 22–36; Brody, *Steelworkers in America*, 1–10, 23–28, 31–32.

38. Krause, *Battle for Homestead*, 177–92; Nasaw, *Carnegie*, 314–26.

39. Nasaw, *Carnegie*, 363–72. See also Krause, *Battle for Homestead*, 240–51.

40. Joshua B. Freeman, "Andrew and Me," *The Nation*, Nov. 16, 1992; Nasaw, *Carnegie*, 406.

41. Frick had made a fortune producing coke before joining forces with Carnegie. Most of the charges against workers were dropped after acquittals in the first trials. *The Local News*, July 2, 1892, *New York Herald*, July 7, 1892, *Pittsburgh Commercial* Gazette, July 25, 1892, and Robert S. Barker, "The Law Takes Sides," all in Demarest, Jr., ed., *"The River Ran Red,"* a wonderful compilation of essays, contemporary accounts, photographs, and drawings about the 1892 battle; Freeman, "Andrew and Me"; Krause, *Battle for Homestead*; Nasaw, *Carnegie*, 405–27.

42. Russell W. Gibbons, "Dateline Homestead," and Randolph Harris, "Photog-

raphers at Homestead in 1892," in Demarest, Jr., ed., *The River Ran Red,*" 158–61.

43. Nasaw, *Carnegie*, 469; Anne E. Mosher, *Capital's Utopia: Vandergrift, Pennsylvania, 1855–1916* (Baltimore, MD: Johns Hopkins University Press, 2004), 66–67; Montgomery, *Fall of the House of Labor*, 41; Brody, *Steelworkers in America*, 56–58, 60–75.

44. Hamlin Garland, "Homestead and Its Perilous Trades; Impressions of a Visit," *McClure's Magazine* 3 (1) (June 1894), in Demarest, Jr., ed., *"The River Ran Red,"* 204–05; Dreiser in Nasaw, *Carnegie*, 470; Fitch, *The Steel Workers*, 214–29; Serrin, *Homestead*, 175–76.

45. Floyd Dell, "Pittsburgh or Petrograd?" *The Liberator* 2 (11) (Dec. 1919), 7–8.

46. Bethlehem Steel later purchased the Sparrows Point mill, which during the 1950s was the largest steel complex in the world. Mosher, *Capital's Utopia*, 73–74; Reutter, *Sparrows Point*, 10, 55–71.

47. Mosher, *Capital's Utopia*, 73–127.

48. Brody, *Steelworkers in America*, 87–89; Mosher, *Capital's Utopia*, 74, 102; Reutter, *Sparrows Point*, 50.

49. For many years after its formation, U.S. Steel functioned essentially as a holding company, with its many subsidiaries operating independently. Alfred D. Chandler, Jr., *The Visible Hand: The Managerial Revolution in American Business* (Cambridge, MA: Harvard University Press, 1977), 359–62; Nasaw, *Carnegie*, 582–88.

50. To prevent workers from sieging or seizing the mill, U.S. Steel redirected a river on the site into a concrete channelway, a moat separating the plant from the town. James B. Lane, *"City of the Century": A History of Gary, Indiana* (Bloomington: Indiana University Press, 1978), 27–37; Brody, *Steelworkers in America*, 158; Mosher, *Capital's Utopia*, 177; S. Paul O'Hara, *Gary, the Most American of All American Cities* (Bloomington: Indiana University Press, 2011), 19–20, 38–53.

51. Richard Edwards, *Contested Terrain: The Transformation of the Workplace in the Twentieth Century* (New York: Basic Books, 1979), 25.

52. In Taylor's account, all the iron loaders eventually achieved the high rate, but independent evidence indicates that only one worker was able to carry anything like forty-seven tons of pig iron a day over an extended period. Daniel Nelson, *Frederick W. Taylor and the Rise of Scientific Management* (Madison: University of Wisconsin Press, 1980); Montgomery, *The Fall of the House of Labor*, esp. chap. 6; Harry Braverman, *Labor and Monopoly Capital: The Degradation of Work in the Twentieth Century* (New York: Monthly Review Press, 1974), 85–123. See also

Charles D. Wrege and Ronald G. Greenwood, *Frederick W. Taylor, the Father of Scientific Management: Myth and Reality* (Homewood, IL: Business One Irwin, 1991).

53. Brody, *Steelworkers in America*, 31–40, 170–73; U. S. Steel, *Making, Shaping and Treating of Steel*, 314; Fitch, *Steel Workers*, 43, 60, 166–81.

54. Fitch, *The Steel Workers*, 57–64.

55. Steel mills in Maryland also hired a substantial number of black workers. Homestead was something of an exception in the strong solidarity between the Eastern European laborers and the English-speaking skilled workers, before and during the 1892 clash. Brody, *Steelworkers in America*, 96–111, 135–37; Henry M. McKiven, *Iron and Steel: Class, Race, and Community in Birmingham, Alabama, 1875–1920* (Chapel Hill: University of North Carolina Press, 1995), 41; Paul Kraus, "East-Europeans in Homestead," in Demarest, Jr., ed., *"The River Ran Red,"* 63–65. For an evocative portrait of Slovak steelworkers in Braddock, Pennsylvania, see Thomas Bell's novel *Out of This Furnace* ([1941] Pittsburgh, PA: University of Pittsburgh Press, 1976).

56. Strictly speaking, these were not steelworkers; they worked in a factory that built steel railway cars. Brody, *Steelworkers in America*, 125, 145–70; Philip S. Foner, *History of the Labor Movement in the United States,* vol. IV: *The Industrial Workers of the World, 1905–1917* (New York: International Publishers, 1965), 281–305.

57. "Labor," in Eric Foner and John A. Garrity, eds., *The Reader's Companion to American History* (Boston: Houghton Mifflin, 1991), 632; Steven Fraser, *Labor Will Rule: Sidney Hillman and the Rise of American Labor* (New York: Free Press, 1991), 146–47.

58. Andrew Carnegie, "Wealth," *The North American Review* 148 (391) (1889): 654.

59. Montgomery, *Fall of the House of Labor*, 88; Whiting Williams, *What's on the Worker's Mind, By One Who Put on Overalls to Find Out* (New York: Charles Scribner's Sons, 1920); "WILLIAMS, WHITING," in *The Encyclopedia of Cleveland History* (accessed May 5, 2015), http://ech.case.edu/cgi/article.pl?id=WW1; Nasaw, *Carnegie*, 386. There is a vast literature on Progressive Era reform. A good place to start is Michael McGeer, *Fierce Discontent: The Rise and Fall of the Progressive Movement in America, 1870–1920* (New York: Oxford University Press, 2005).

60. The Pittsburgh Survey examined the whole region and its economy, but steel dominated the study and was the main subject of several volumes. Greenwald and Anderson, eds., *Pittsburgh Surveyed*.

61. In 1920, the Supreme Court dismissed the antitrust case against U.S. Steel.

Brody, *Steelworkers in America*, 147, 154, 161–71; Fitch, *The Steel Workers*, 178–79.

62. Melvyn Dubofsky, *The State and Labor in Modern America* (Chapel Hill: University of North Carolina Press, 1994), 61–76; union data calculated from U.S. Bureau of the Census, *Historical Statistics of the United States, Colonial Times to 1970, Bicentennial Edition,* part 1 (Washington, D.C.: U.S. Government Printing Office, 1975), 126, 177; Fraser, *Labor Will Rule*, 121–40, 144 (quote).

63. David Brody, *Labor in Crisis: The Steel Strike of 1919* (Philadelphia: J.B. Lippincott, 1965), 45–51, 59–60.

64. The most thorough accounts of the steel organizing drive and the 1919 strike are William Z. Foster, *The Great Steel Strike* (New York: B.W. Huebsch, 1920), and Brody, *Labor in Crisis*. Except where otherwise noted, I have drawn from them.

65. Freeman et al., *Who Built America?* 258–61.

66. For the strike in Gary, see Lane, *"City of the Century,"* 90–93. For a gripping portrayal of the strike from the point of view of black workers, see Attaway, *Blood on the Forge*.

67. The actual demands of the striking workers were far from radical, dealing, very concretely, with hours, wages, and union recognition. See Brody, *Labor in Crisis*, 100–101, 129. The *New York Times*, like many newspapers, gave heavy coverage to the strike. From September 23 through September 26, the *Times* ran three-line banner headlines about the strike on its front page that emphasized the strike's size and violence.

68. Foster, *The Great Steel Strike*, 1; Vorse, *Men and Steel*, 21; John Dos Passos, *The Big Money* (New York: Harcourt, Brace, 1936).

Chapter 4
"I WORSHIP FACTORIES"

1. Henry Ford, "Mass Production," in *Encyclopedia Britannica*, 13th ed. (New York: The Encyclopædia Britannica, 1926), vol. 30, 821–23; David A. Hounshell, *From the American System to Mass Production, 1800–1932* (Baltimore, MD: Johns Hopkins Press, 1984), 1, 218–19, 224; Helen Jones Earley and James R. Walkinshaw, *Setting the Pace: Oldsmobile's First 100 Years* (Lansing, MI: Public Relations Department, Oldsmobile Division, 1996), 461; The Locomobile Society of America, "List of Cars Manufactured by the Locomobile Company of America," http://www.locomobilesociety.com/cars

.cfm, and "U.S. Automobile Production Figures," https://en.wikipedia.org/wiki/U.S._Automobile_Production_Figures (both accessed Feb. 6, 2017); Joshua Freeman et al., *Who Built America? Working People and the Nation's Economy, Politics, Culture, and Society,* vol. 2 (New York: Pantheon Books, 1992), 277.

2. Hounshell, *From the American System to Mass Production,* 1, 228. My discussion of the development of the Ford system draws heavily from Hounshell's superb study.

3. Edward A. Filene, *The Way Out: A Forecast of Coming Changes in American Business and Industry* (Garden City, NY: Page & Company, 1924), 180; Vicki Goldberg, *Margaret Bourke-White: A Biography* (New York: Harper & Row, 1986), 74.

4. Hounshell, *From the American System to Mass Production,* 4–8, 15–50.

5. Eric Hobsbawm, *The Age of Capital 1848–1875* (New York: Charles Scribner's Sons, 1975), 44.

6. John A. James and Jonathan S. Skinner, "The Resolution of the Labor Scarcity Paradox," Working Paper No. 1504, National Bureau of Economic Research, Nov. 1984.

7. Hounshell, *From the American System to Mass Production,* 115–23; Alfred D. Chandler, Jr., *Scale and Scope: The Dynamics of Industrial Capitalism* (Cambridge, MA: Harvard University Press, 1994), 196.

8. Alfred D. Chandler, Jr., *The Visible Hand: The Managerial Revolution in American Business* (Cambridge, MA: Harvard University Press, 1977), 240, 249–53; Hounshell, *From the American System to Mass Production,* 240–43.

9. Until 1915, Ford partner James Couzens played a central role in the Ford Motor Company, developing many of its innovative practices and contributing greatly to its overall success. Keith Sward, *The Legend of Henry Ford* (New York: Rinehart & Company, 1948), 9–27, 43–46.

10. Sward, *The Legend of Henry Ford,* 44–45; Hounshell, *From the American System to Mass Production,* 224.

11. Stephen Meyer, *The Five Dollar Day; Labor Management and Social Control in the Ford Motor Company, 1908–1921* (Albany: State University of New York Press, 1981), 16, 18; Adam Smith, *An Inquiry into the Nature and Causes of the Wealth of Nations* ([1776] London: Oxford University Press, 1904), 6–7; Hounshell, *From the American System to Mass Production,* 227.

12. Though various accounts at the time and after, including by the Ford company, have claimed that by the time of the introduction of the assembly line complete interchangeability of parts had been achieved, apparently for several years some filing and grinding of parts on the assembly line occurred.

Sward, *The Legend of Henry Ford*, 42, 46, 68–77; *Ford Factory Facts* (Detroit, MI: Ford Motor Company, 1912), 46–47, 49; Allan Nevins and Frank Ernest Hill, *Ford: Expansion and Challenge, 1915–1933* (New York: Charles Scribner's Sons: 1957), 522; Hounshell, *From the American System to Mass Production*, 219–20, 224–25, 230–33; Meyer, *The Five Dollar Day*, 10, 22–29; Jack Russell, "The Coming of the Line; The Ford Highland Park Plant, 1910–1914," *Radical America* 12 (May–June 1978), 30–33.

13. Daniel Nelson, *Managers and Workers: The Origins of the New Factory System in the United States 1880–1920* (Madison: University of Wisconsin Press, 1975), 21–23; David Gartman, "Origins of the Assembly Line and Capitalist Control of Work at Ford," in Andrew Zimbalist, ed., *Case Studies on the Labor Process* (New York: Monthly Review Press, 1979), 197–98; Ford, "Mass Production," 822; Meyer, *The Five Dollar Day*, 29–31; Karl Marx, *Capital: A Critique of Political Economy*, vol. 1 (1867: New York: International Publishers, 1967), 380.

14. Hounshell, *From the American System to Mass Production*, 237–49; Gartman, "Origins of the Assembly Line," 201.

15. Russell, "The Coming of the Line," 33–34, 37 (includes Ford quote). Photographs of cars and trucks being assembled using the craft method at various early vehicle companies can be seen in Bryan Olsen and Joseph Cabadas, *The American Auto Factory* (St. Paul, MN: Motorbooks, 2002).

16. Hounshell, *From the American System to Mass Production*, 250–60.

17. Gartman, "Origins of the Assembly Line," 199, 201–02.

18. Hounshell, *From the American System to Mass Production*, 249–53; Russell, "The Coming of the Line," 38; Lindy Biggs, *The Rational Factory: Architecture, Technology, and Work in America's Age of Mass Production* (Baltimore, MD: Johns Hopkins University Press, 1996), 27.

19. Joyce Shaw Peterson, *American Automobile Workers, 1900–1933* (Albany: State University of New York Press, 1987), 43; Meyer, *The Five Dollar Day*, 40–41; Biggs, *The Rational Factory*, 133–34; Nevins and Hill, *Ford: Expansion and Challenge*, 534.

20. Meyer, *The Five Dollar Day*, 10, 50; Terry Smith, *Making the Modern: Industry, Art and Design in America* (Chicago: University of Chicago Press, 1993), 53; Department of Commerce, Bureau of the Census, *Abstract of the Census of Manufactures, 1919* (Washington, D.C.: Government Printing Office, 1923), 355, 374–75; Chandler, Jr., *Scale and Scope*, 27; Nelson, *Managers and Workers*, 9.

21. Nevins and Hill, *Ford: Expansion and Challenge*, 288. A selection from

the very large Ford collection of photographs documenting the Highland Park plant can be viewed online at https://www.thehenryford.org/collections-and-research/.

22. David Montgomery, *The Fall of the House of Labor; the Workplace, the State, and American Labor Activism, 1865–1925* (Cambridge: Cambridge University Press, 1987), 133–35, 238–40.

23. Meyer, *The Five Dollar Day*, 77–78, 80–85, 89–93, 156; Russell, "The Coming of the Line," 39–40.

24. In 1926, Ford reduced the workweek from six days to five, becoming one of the first major industrial companies to institute the forty-hour week. Meyer, *The Five Dollar Day*, 95–168; Peterson, *American Automobile Workers*, 156; John Reed, "Why They Hate Ford," *The Masses*, 8 (Oct. 1916), 11–12.

25. Nelson, *Managers and Workers*, 101–21; Montgomery, *The Fall of the House of Labor*, 236–38; Reed, "Why They Hate Ford"; Meyer, *The Five Dollar Day*, 114, 156–57.

26. Sward, *Legend of Henry Ford*, 107–09; Antonio Gramsci, *Selections from the Prison Notebooks of Antonio Gramsci*, ed. and trans. Quintin Hoare and Geoffrey Nowell Smith (New York: International Publishers, 1971), lxxxvi–lxxxvii, 286, 302, 305.

27. Meyer, *The Five Dollar Day*, 197–200; Sward, *Legend of Henry Ford*, 291–342. See also Harry Bennett, *We Never Called Him Henry* (Greenwich, CT: Gold Medal Books, 1951).

28. Biggs, *The Rational Factory*, 89–94. The Piquette Avenue plant is still standing. It now houses a museum and can be rented for corporate parties, weddings, and bar mitzvahs. See http://www.fordpiquetteavenueplant.org/ (accessed Sept. 8, 2015).

29. Hounshell, *From the American System to Mass Production*, 225–26; "Industry's Architect," *Time*, June 29, 1942; Grant Hildebrand, *Designing for Industry: The Architecture of Albert Kahn* (Cambridge, MA: MIT Press, 1974), 26–27. Some of Kahn's early work can be seen in W. Hawkins Ferry, *The Legacy of Albert Kahn* (Detroit, MI: Wayne State University Press, 1970).

30. George N. Pierce became the manufacturer of Pierce-Arrow automobiles. Nelson, *Managers and Workers*, 15–16; Betsy Hunter Bradley, *The Works: The Industrial Architecture of the United States* (New York: Oxford University Press, 1999), 155–58; Hildebrand, *Designing for Industry*, 28–43; Albert Kahn, "Industrial Architecture" (speech), May 25, 1939, Box 1, Albert Kahn Papers, Bentley Historical Library, University of Michigan, Ann Arbor, Michigan; Smith, *Making the Modern*, 59.

31. Biggs, *The Rational Factory*, 93–102, 110; Kahn, "Industrial Architecture."

32. Smith, *Making the Modern*, 41–42, 71; Biggs, *The Rational Factory*, 78, 109, 120–25; Hildebrand, *Designing for Industry*, 52.

33. I thank Jeffrey Trask for making this point to me. See Gillian Darley, *Factory* (London: Reaktion Books, 2003), 157–89.

34. Biggs, *The Rational Factory*, 103–4, 150; *Ford Factory Facts* (Detroit, MI: Ford Motor Company, 1915) is an expanded and updated version of the 1912 booklet.

35. Both the Lingotto plant and the New York Packard service building are still standing. The former was converted into a cultural, hotel, office, retail, and educational complex by Renzo Piano; the latter now houses a car dealership. Jean Castex, *Architecture of Italy* (Westport, CT: Greenwood Press, 2008), 47–49; Darley, *Factory*, 10–12; Christopher Gray, "The Car Is Still King on 11th Avenue," *New York Times,* July 9, 2006.

36. Photographs of all of the mentioned buildings appear in Ferry, *The Legacy of Albert Kahn*, except for the Joy house, which is in Hildebrand, *Designing for Industry*, 74. On Kahn's automobile projects and his firm organization, see Olsen and Cabadas, *The American Auto Factory*, 39, 65; George Nelson, *Industrial Architecture of Albert Kahn, Inc.* (New York: Architectural Book Publishing Company, 1939), 19–23; Smith, *Making the Modern*, 76–78, 85–87; and Hildebrand, *Designing for Industry*, 60, 124.

37. Olsen and Cabadas, *The American Auto Factory*, 39; Biggs, *The Rational Factory*, 138–40, 151. For Ford tractors, see Reynold Wik, *Henry Ford and Grassroots America* (Ann Arbor: University of Michigan Press, 1972), 82–97.

38. Biggs, *The Rational Factory*, 146, 151; Writers' Program of the Works Progress Administration, *Michigan: A Guide to the Wolverine State* (New York: Oxford University Press, 1941), 221–24; Greg Grandin, *Fordlandia: The Rise and Fall of Henry Ford's Forgotten Jungle City* (New York: Metropolitan Books, 2009). Kingsford is now owned by The Clorox Company. The Clorox Company, "A Global Portfolio of Diverse Brands" (accessed Sept., 13, 2015), https://www.thecloroxcompany.com/products/our-brands/.

39. The Rouge foundry also made parts for Fordson tractors. Biggs, *The Rational Factory*, 148–49, 152; Hounshell, *From the American System to Mass Production*, 268, 289.

40. Nelson, *Industrial Architecture of Albert Kahn, Inc.*, 132; Biggs, *The Rational Factory*, 129, 141–57; Kahn, "Industrial Architecture"; Ferry, *The Legacy of Albert Kahn*, 113–16, 120–22, 129–301; The Reminiscences of Mr. B. R. Brown Jr., Benson Ford Research Center, Dearborn, Michigan; Works Progress Administration, *Michigan*, 220–21; Hildebrand, *Designing for Indus-*

try, 91–92, 99, 102–08, 172–82. On Kahn's and Ford's antimodernism, see Albert Kahn, "Architectural Trend" (speech), April 15, 1931, Box 1, Albert Kahn Papers; Sward, *Legend of Henry Ford*, 259–75; and Smith, *Making the Modern*, 144–55 (though Smith's interpretation is very different than mine).

41. In addition to Highland Park and River Rouge, Ford built major manufacturing plants in Canada and England that built finished cars and trucks and supplied parts to foreign branch plants. As employment at the Rouge grew, it shrank at Highland Park. In 1929, when the average number of hourly employees at the Rouge was 98,337, at Highland Park it was only 13,444. After the stock market crash, employment at the Rouge fell but remained substantial. Edmund Wilson, *The American Earthquake* (Garden City, NY: Doubleday, 1958), 219–20, 234, 687; Nevins and Hill, *Ford: Expansion and Challenge*, 210, 365–66, 542–43; Chandler, Jr., *Scale and Scope*, 207–08; Bruce Pietrykowski, "Fordism at Ford: Spatial Decentralization and Labor Segmentation at the Ford Motor Company, 1920–1950," *Economic Geography* 71 (4) (Oct. 1995), 386, 389–91; Historic American Engineering Record, Mid-Atlantic Region National Park Service, "Dodge Bros. Motor Car Company Plant (Dodge Main): Photographs, Written Historical and Descriptive Data" (Philadelphia: Department of the Interior, 1980); Ronald Edsforth, *Class Conflict and Cultural Consensus: The Making of a Mass Consumer Society in Flint, Michigan* (New Brunswick, NJ: Rutgers University Press, 1987), 77; *New York Times*, May 31, 1925, Apr. 9, 1972; Hounshell, *From the American System to Mass Production*, 263–301; Biggs, *The Rational Factory*, 148; Sward, *Legend of Henry Ford*, 185–205; The Reminiscences of Mr. B. R. Brown Jr.

42. Not everyone, though, was enthralled. European carmaker André Citroen, after reporting that his visit to Dearborn left him "greatly impressed by the power of Ford's production and his marvelous industrial creations at the River Rouge plant," added "regrettably, the artistic element is absent. Nothing about Ford or his plant suggests a trace of the finer esthetic qualities." Hounshell, *From the American System to Mass Production*, 260–61; Olsen and Cabadas, *The American Auto Factory*, 61, 63, 67, 70–71; *New York Times*, Apr. 22, 1923.

43. Kahn also helped design both the General Motors and Ford exhibitions at the 1939 New York World's Fair. John E. Findling, ed., *Historical Dictionary of World's Fairs and Expositions, 1851–1988* (New York: Greenwood Press, 1990), 22; Nevins and Hill, *Ford: Expansion and Challenge*, 1–2; Grandin, *Fordlandia*, 2; Richard Guy Wilson, Dianne H. Pilgrim, and Dickran Tashjian, *The Machine Age in America 1918–1941* (New York: The Brooklyn

Museum and Harry N. Abrams, 1986), 27; Nelson, *Industrial Architecture of Albert Kahn, Inc.*, 97; Hildebrand, *Designing for Industry*, 206, 213; Works Progress Administration, *Michigan*, 286, 292–93; *New York Times*, Apr. 9, 1972; U.S. Travel Service, U.S. Department of Commerce, *USA Plant Visits 1977–1978* (Washington, D.C.: U.S. Government Printing Office, n.d.).

44. David Roediger, "Americanism and Fordism—American Style: Kate Richards O'Hare's 'Has Henry Ford Made Good?'," *Labor History* 29 (2) (Spring 1988), 241–52.

45. John Reed, "Why They Hate Ford," 11–12; Nevins and Frank Ernest Hill, *Ford: Expansion and Challenge*, 88.

46. Edmund Wilson, "The Despot of Dearborn," *Scribner's Magazine*, July 1931, 24–36; Roediger, "Americanism and Fordism—American Style," 243; Steven Fraser, *Labor Will Rule: Sidney Hillman and the Rise of American Labor* (New York: Free Press, 1991), 259–70; Filene, *The Way Out*, 199, 201, 215–17, 221. On Ford's anti-Semitism, see Sward, *Legend of Henry Ford*, 146–60.

47. John Dos Passos, *The Big Money* ([1936] New York: New American Library, 1969), 70–77, and Alfred Kazin's introduction to this edition, xi–xii. Cecelia Tichi expanded on Kazin's observation in *Shifting Gears: Technology, Literature, Culture in Modernist America* (Chapel Hill: University of North Carolina Press, 1987), 194–216.

48. Smith, *Making the Modern*, 16–18; Louis-Ferdinand Céline, *Journey to the End of the Night* ([1932] New York: New Directions, 1938); Upton Sinclair, *The Flivver King: A Story of Ford-America* (Emaus, PA: Rodale Press, 1937); Aldous Huxley's *Brave New World: A Novel* (London: Chatto & Windus, 1932).

49. Darley, *Factory*, 15–27, 34; Wilson, Pilgrim, and Tashjian, *The Machine Age in America*, 23, 29; Kim Sichel, *From Icon to Irony: German and American Industrial Photography* (Seattle: University of Washington Press, 1995); Leah Bendavid-Val, *Propaganda and Dreams: Photographing the 1930s in the U.S.S.R. and U.S.A.* (Zurich: Edition Stemmle, 1999).

50. Margaret Bourke-White, *Portrait of Myself* (New York: Simon and Schuster, Inc., 1963), quotes on 18, 33, 40, 49; Goldberg, *Margaret Bourke-White*, quote on 74. Bourke-White may have been inspired by O'Neill's play, in which one character says "I love dynamos. O love to hear them sing." Eugene O'Neill, *Dynamo* (New York: Horace Liveright, 1929), 92.

51. Wilson, Pilgrim, and Tashjian, *The Machine Age in America*, 69; Goldberg, *Margaret Bourke-White*, 87–89; *Life*, Nov. 23, 1936; William H. Young and Nancy K. Young, *The 1930s* (Westport, CT: Greenwood Press, 2002), 156. "Margaret Bourke-White Photographic Material, Itemized Listing"

is a comprehensive list of her photographs at the Margaret Bourke-White Papers, Special Collections Research Center, Syracuse University Libraries, including her factory photographs, https://library.syr.edu/digital/guides/b/bourke-white_m.htm#series7 (accessed Sept. 23, 2015). For Hine, see, for example, Jonathan L. Doherty, ed., *Women at Work: 153 Photographs by Lewis W. Hine* (New York: Dover Publications and George Eastman House, 1983).

52. Sheeler's portfolio of Rouge photographs can be seen at the Detroit Institute of Art website for the 2004 exhibition "The Photography of Charles Sheeler, American Modernist" (accessed Sept. 23, 2015), http://www.dia.org/exhibitions/sheeler/content/rouge_gallery/hydra_shear.html. Sharon Lynn Corwin, "Selling 'America': Precisionism and the Rhetoric of Industry, 1916–1939," Ph.D. dissertation, University of California, Berkeley, 2001, 17–79, 158; Carol Troyen, "Sheeler, Charles," American National Biography Online Feb. 2000 (accessed Sept. 24 2015), http://www.anb.org/articles/17/17-00795.html; Wilson, Pilgrim, and Tashjian, *The Machine Age in America*, 24, 78, 218–19; Smith, *Making the Modern*, 111–13. The Ford company returned to the strategy of selling cars through imagery of the magic and majesty of their production in a 1940 film it commissioned, *Symphony in F*, shown at the New York World's Fair. It can be seen at "Symphony in F: An Industrial Fantasia for the World of Tomorrow," The National Archives, Unwritten Record Blog, Mar. 3, 2016, https://unwritten-record.blogs.archives.gov/2016/03/03/symphony-in-f-an-industrial-fantasia-for-the-world-of-tomorrow/.

53. Leo Marx, *The Machine in the Garden: Technology and the Pastoral Ideal in America* (New York: Oxford University Press, 1964), 355–56; Nevins and Hill, *Ford: Expansion and Challenge*, 282–83. For Sheeler's photomontage "Industry," see Wilson, Pilgrim, and Tashjian, *The Machine Age in America*, 24, 218. *American Landscape* is now in the collection of the Museum of Modern Art; *Classic Landscape* in the collection of the National Gallery of Art. See also *River Rouge Plant*, Whitney Museum of American Art, and *City Interior*, Worcester Art Museum. *Amoskeag Mill Yard # 1* and *Amoskeag Canal* are in the collection of the Currier Museum of Art in Manchester, New Hampshire. *Amoskeag Mills #2* is in the collection of the Crystal Bridges Museum in Bentonville, Arkansas. Hine's Amoskeag photographs are owned by the Library of Congress and can be viewed at http://www.loc.gov/pictures/search/?q=Amoskeag%20hine (accessed Nov. 4, 2016). Bourke-White's Amoskeag photographs are in Oversize 5, folders 31–35, Margaret Bourke-White Papers.

54. Smith, *Making the Modern*, 194; Troyen, "Sheeler, Charles."

55. Carol Quirke, *Eyes on Labor: News Photography and America's Working Class* (New York: Oxford University Press, 2012), 273–74; Corwin, "Selling 'America,'" 127; *Life*, Nov. 23, 1936; Nov. 14, 1938.

56. Sharon Lynn Corwin stresses, contrary to the standard account and to Terry Smith, that workers do appear in Sheeler's Rouge photographs and are critical to their meaning. Corwin, "Selling 'America,'" 23; *Fortune*, Dec. 1940.

57. Like Bourke-White, Driggs grew up familiar with the world of industry; her father was an engineer for a steel company. Rivera and many of the Precisionists shared a past engagement with Cubism. Corwin, "Selling 'America,'" 145–48, 159–62, 165; Barbara Zabel, "Louis Lozowick and Technological Optimism of the 1920s," *Archives of American Art Journal* 14 (2) (1974), 17–21; Wilson, Pilgrim, and Tashjian, *The Machine Age in America*, 237–42, 343; Linda Bank Downs, *Diego Rivera: The Detroit Industry Murals* (New York: Norton, 1999), 21.

58. Downs, *Diego Rivera*, 22, 28.

59. Henry Ford offered a chauffeured Lincoln to Rivera and Kahlo to use in their exploration of the city, but Rivera thought it would be embarrassing for artists to be seen in such luxury, so he accepted a more modest car from Edsel instead. Mark Rosenthal, "Diego and Frida"; Juan Rafael Coronel Rivera, "April 21, 1932"; Linda Downs, "The Director and the Artist: Two Revolutionaries"; and John Dean, "'He's the Artist in the Family': The Life, Times, and Character of Edsel Ford," all in Rosenthal, *Diego Rivera and Frida Kahlo in Detroit* (Detroit, MI: Detroit Institute of Arts, 2015). On the impact of the Depression on Detroit, see Steve Babson with Ron Alpern, Dave Elsila, and John Revitte, *Working Detroit: The Making of a Union Town* (New York: Adama Books, 1984), 52–60.

60. Rosenthal, *Diego Rivera and Frida Kahlo*, 102–03, 219.

61. Rivera's depiction of the machinery and processes at the Rouge, working from sketches, photographs, and information provided by Ford engineers, is remarkably accurate. The one major exception is the giant stamping machine in the south wall panel. Rivera painted an older model machine—the one Sheeler had photographed—rather than the one then in use. (Rivera may have worked from the Sheeler photo.) Apparently Rivera preferred the anthropomorphic qualities of the older machine. For a detailed description and analysis of the murals and their relationship to actual Rouge activity, see Downs, *Diego Rivera*.

62. Rosenthal, *Diego Rivera and Frida Kahlo*, 103–07, 182; *Detroit News*, Mar. 22, 1933, and May 12, 1933. Before returning to Mexico, Rivera completed a series of murals for the leftist New Workers School in New

York City that included a portrayal of the Homestead strike. See David P. Demarest, Jr., ed., *"The River Ran Red": Homestead 1892* (Pittsburgh, PA: University of Pittsburgh Press, 1992), 218. The Rouge appears in another Detroit mural, painted in 1937 by WPA artist Walter Speck for the headquarters of United Automobile Workers Local 174. It now is in the Walter Reuther Library, Wayne State University, Detroit. See "Collection Spotlight: UAW Local 174 Mural," Oct. 20, 2016, https://reuther.wayne.edu/node/13600.

63. In another painting Kahlo began in Detroit, *Self-Portrait on the Borderline between Mexico and the United States*, the Highland Park powerhouse appears in the background. Downs, *Diego Rivera*, 58–60; Rosenthal, "Diego and Frida: High Drama in Detroit," and Solomon Grimberg, "The Lost Desire: Frida Kahlo in Detroit," in Rosenthal, *Diego Rivera and Frida Kahlo*.

64. Charles Chaplin, *Modern Times* (United Artists, 1936); Hounshell, *From the American System to Mass Production*, 319–20; Charles Musser, "Modern Times (Chaplin 1936)," (accessed Sept. 30, 2015), http://actionspeaksradio.org/chaplin-by-charles-musser-2012/); Joyce Milton, *Tramp: The Life of Charlie Chaplin* (New York: HarperCollins, 1996), 336, 348, 350; Mark Lynn Anderson, *"Modern Times"* (accessed Sept. 30, 2015), http://laborfilms.org/modern-times/; Edward Newhouse, "Charlie's Critics," *Partisan Review and Anvil*, Apr. 1936, 25–26 (includes quote from *Daily Worker* review); Stephen Kotchin, *Magic Mountain: Stalinism as a Civilization* (Berkeley: University of California Press, 1995), 184; Octavio Cortazar, *Por Primera Vez/For the First Times* (El Instituto Cubano, Lombarda Industria Cinematografia, 1967). In an odd coda, after the completion of *Modern Times*, Paulette Goddard and Chaplin ended their romantic relationship and Goddard went on to have one with Rivera. In a mural Rivera painted in San Francisco in 1940, *Unión de la Expresión Artística del Norte y Sur de este Continente* (The Marriage of the Artistic Expression of the North and of the South on This Continent), he included images of Chaplin, Kahlo, and Goddard eyeing each other suspiciously and a mashup of the Aztec goddess Coatlicue and a Detroit Motor Company stamping machine, a rare return to a theme of *Detroit Industry*. David Robinson, *Chaplin, His Life and Art* (New York: McGraw-Hill, 1985), 509; City College of San Francisco, "Pan American Unity Mural," (accessed Oct. 1, 2015), https://www.ccsf.edu/en/about-city-college/diego-rivera-mural/overview.html.

65. Ruth McKenney, *Industrial Valley* (New York: Harcourt, Brace, 1939), 261–62.

66. For an overview of this era, see Joshua B. Freeman, *American Empire, 1945–*

2000: The Rise of a Global Empire, the Democratic Revolution at Home (New York: Viking, 2012).

67. There is a large literature about labor upsurge of the 1930s, but the best single account remains Irving Bernstein, *Turbulent Years: A History of the American Worker 1933–1941* (Boston: Houghton Mifflin, 1970).

68. In addition to Bernstein, *Turbulent Years*, see, Ronald W. Schatz, *The Electrical Workers: A History of Labor at General Electric and Westinghouse, 1923–1960* (Urbana: University of Illinois Press, 1983); Daniel Nelson, *American Rubber Workers and Organized Labor, 1900–1941* (Princeton, NJ: Princeton University Press, 1988); and Sidney Fine, *The Automobile Under the Blue Eagle: Labor, Management, and the Automobile Manufacturing Code* (Ann Arbor: University of Michigan Press, 1964).

69. Bernstein, *Turbulent Years*, 509–51; Henry Kraus, *The Many and the Few: A Chronicle of the Dynamic Auto Workers* ([1947] Urbana: University of Illinois Press, 1985). See, also, Sidney A. Fine, *Sit-down: The General Motors Strike of 1936–1937* (Ann Arbor: University of Michigan Press, 1969).

70. Joshua Freeman et al., *Who Built America?* 395; Bernstein, *Turbulent Years*, 551–54, 608–09, 613; Steve Jefferys, *Management and Managed: Fifty Years of Crisis at Chrysler* (Cambridge: Cambridge University Press, 1986), 71–77; Jefferson Cowie, *Capital Moves: RCA's Seventy-Year Quest for Cheap Labor* (Ithaca, NY: Cornell University Press, 1999), 17–33.

71. Robert H. Zieger, *The CIO, 1935–1955* (Chapel Hill: University of North Carolina Press, 1995), 54–60; Bernstein, *Turbulent Years*, 432–73.

72. Bernstein, *Turbulent Years*, 478–98; Zieger, *CIO*, 79, 82.

73. Zieger, *CIO*, 121–31.

74. No similar size election again would be held until 1999, when seventy-four thousand home-care workers in Los Angeles were sent ballots to determine if they wanted union representation. John Barnard, *American Vanguard: The United Auto Workers during the Reuther Years, 1935–1970* (Detroit, MI: Wayne State University Press, 2004), 153–64; Zieger, *CIO*, 122–24; *Los Angeles Times*, Feb. 26, 1999.

75. Joshua Freeman, "Delivering the Goods: Industrial Unionism During World War II," *Labor History* 19 (4) (Fall 1978); U.S. Department of Commerce, *Historical Statistics of the United States, 1789–1945* (Washington, D.C., U.S. Government Printing Office, 1949), 72. See also Nelson Lichtenstein, *Labor's War at Home: The CIO in World War II* ([1982] Philadelphia: Temple University Press, 2003).

Chapter 5
"COMMUNISM IS SOVIET POWER
PLUS THE ELECTRIFICATION OF THE WHOLE COUNTRY"

1. *Detroit Sunday News*, Dec. 15, 1929. Photographs of the plant site and construction are in box 10, Albert Kahn Papers, Bentley Historical Library, University of Michigan, Ann Arbor, Michigan. See also "Agenda for Meeting with Russian Visitors—Saturday, June 13, 1964," Russian Scrapbooks, vol. II, box 13, Kahn Papers; *Those Who Built Stalingrad, As Told by Themselves* (New York: International Publishers, 1934), 29; Alan M. Ball, *Imagining America: Influence and Images in Twentieth-Century Russia* (Lanham, MD: Rowman & Littlefield, 2003), 124; *New York Times*, Mar. 29, 1930, May 18, 1930; Margaret Bourke-White, *Eyes on Russia* (New York: Simon and Schuster, 1931), 118–27.

2. V. I. Lenin, "Our Foreign and Domestic Position and Party Tasks," Speech Delivered to the Moscow Gubernia Conference of the R.C.P.(B.), Nov. 21, 1920, *Lenin's Collected Works, Volume 31* (Moscow: Progress Publishers, Moscow, 1966), 419–20.

3. Edward Hallett Carr and R. W. Davies, *Foundations of a Planned Economy, 1926–1929,* Vol. I–II (London: Macmillan, 1969), 844, 898–902; Alexander Erlich, *The Soviet Industrialization Debate, 1924–1928* (Cambridge, MA: Harvard University Press, 1967), 164–65; J. V. Stalin, "A Year of Great Change, On the Occasion of the Twelfth Anniversary of the October Revolution," *Pravda* 259 (Nov. 7, 1929), https://www.marxists.org/reference/archive/stalin/works/1929/11/03.htm; Stephen Kotkin, *Magnetic Mountain: Stalinism as a Civilization* (Berkeley: University of California Press, 1995), 32 (quoted passage), 69–70, 363, 366.

4. Arens went on to become a leading industrial designer, working for some of the best-known American corporations. Barnaby Haran cites Arens's comments in his article "Tractor Factory Facts: Margaret Bourke-White's *Eyes on Russia* and the Romance of Industry in the Five-Year Plan," *Oxford Art Journal* 38 (1) (2015), 82. The full text is in *New Masses* 3 (7) (Nov. 1927), 3. On Arens, see "Biographical History," Egmont Arens Papers Special Collections Research Center, Syracuse University Libraries, accessed Feb. 23, 2016, http://library.syr.edu/digital/guides/a/arens_e.htm#d2e97.

5. Of course, there always were some government-owned factories, particularly

to produce armaments. As discussed in Chapter 4, at times these played an important role in the development of production techniques.

6. On the impact of scientific management and mass production in Europe, see Thomas P. Hughes, *American Genesis: A Century of Invention and Technological Enthusiasm, 1870–1970* (New York: Viking, 1989), 285–323; Judith A. Merkle, *Management and Ideology: The Legacy of the International Scientific Management Movement* (Berkeley: University of California Press, 1980), esp. 105, 136–223; Charles S. Maier, "Between Taylorism and Technocracy: European Ideologies and the Vision of Industrial Productivity in the 1920s," *Journal of Contemporary History* 5 (2) (1970), pp. 27–61; and Antonio Gramsci, "Americanism and Fordism," in *Selections from the Prison Notebooks of Antonio Gramsci*, ed. and trans. Quintin Hoare and Geoffrey Nowell Smith (New York: International Publishers, 1971).

7. Lenin was particularly attracted to Gilbreth's work (as other Russian communists would be) because, by simplifying motions for completing tasks, it claimed to increase productivity without increasing the exploitation of workers as speedup did. S. A. Smith, *Red Petrograd: Revolution in the Factories 1917–1918* (Cambridge: Cambridge University Press, 1983), 7–12; Merkle, *Management and Ideology*, 105–06, 179; Daniel A. Wren and Arthur G. Bedeian, "The Taylorization of Lenin: Rhetoric or Reality?" *International Journal of Social Economics* 31 (3) (2004), 287–99 (quote from Lenin on 288); V. I. Lenin, *Imperialism: The Highest Stage of Capitalism; A Popular Outline* ([1917] New York: International Publishers, 1939).

8. Lenin's remarks about Taylor were soon translated into English, circulated in the United States, and frequently quoted in business circles. Wren and Bedeian, "Taylorization of Lenin," 288–89; Merkle, *Management and Ideology*, 111–15 (quote on 113).

9. Isaac Deutscher, *The Prophet Armed; Trotsky: 1879–1921* (New York: Oxford University Press, 1954), 499–502; Merkle, *Management and Ideology*, 118–19; Kendall E. Bailes, "Alexei Gastev and the Soviet Controversy over Taylorism, 1918–24," *Soviet Studies* 29 (3) (July 1977), 374, 380–83.

10. Merkle, *Management and Ideology*, 114–20; Bailes, "Alexei Gastev"; Vladimir Andrle, *Workers in Stalin's Russia: Industrialization and Social Change in a Planned Economy* (New York: St. Martin's Press, 1988), 101–02; Wren and Bedeian, "Taylorization of Lenin," 290–91; Deutscher, *The Prophet Armed*, 498–501.

11. An earlier All-Russian Conference on Scientific Management had been organized by Trotsky in 1921, but failed to resolve the differences between the two sides of the debate. Bailes, "Alexei Gastev," 387–93; Kendall E. Bailes, "The

American Connection: Ideology and the Transfer of American Technology to the Soviet Union, 1917–1941," *Comparative Studies in Society and History* 23 (3) (July 1981), 437; Wren and Bedeian, "Taylorization of Lenin," 291.

12. When a delegation from the Ford Motor Company visited Gastev's institute in 1926, they deemed it "a circus, a comedy, a crazy house," "a pitiful waste of young people's time." Merkle, *Management and Ideology*, 123; Andrle, *Workers in Stalin's Russia*, 93–94; Bailes, "Alexei Gastev," 391, 393; Timothy W. Luke, *Ideology and Soviet Industrialization* (Westport, CT: Greenwood Press, 1985), 165–66; Wren and Bedeian, "Taylorization of Lenin" 291–96; Ball, *Imagining America*, 28–29.

13. My discussion of RAIC is based on Steve Fraser, "The 'New Unionism' and the 'New Economic Policy'," in James E. Cronin and Carmen Sirianni, eds., *Work, Community and Power: The Experience of Labor in Europe and America, 1900–1925* (Philadelphia: Temple University Press, 1983).

14. William Z. Foster, *Russian Workers and Workshops in 1926* (Chicago: Trade Union Educational League, 1926), 52; Erlich, *Soviet Industrialization Debate*, 24–25, 105–06, 114.

15. Erlich, *Soviet Industrialization Debate*, xvii–xviii, 140, 147, 161; Smith, *Red Petrograd*, 7–8, 10–12; Orlando Figes, *Revolutionary Russia, 1891–1991: A History* (New York: Metropolitan Books, 2014), 112.

16. Bailes, "The American Connection," 430–31; Hans Rogger, "Amerikanizm and the Economic Development of Russia," *Comparative Studies in Society and History* 23 (3) (July 1981); Hughes, *American Genesis*, 269; Dana G. Dalrymple, "The American Tractor Comes to Soviet Agriculture: The Transfer of a Technology," *Technology and Culture* 5 (2) (Spring 1964), 192–94, 198; Allan Nevins and Frank Ernest Hill, *Ford: Expansion and Challenge, 1915–1933* (New York: Charles Scribner's Sons: 1957), 255, 673–77.

17. Foster also claimed that Soviet workers accepted piecework and Taylorism because "The benefits of increased production flow to the workers, not to greedy capitalists." William Z. Foster, *Russian Workers*, 13, 54; *New York Times*, Feb. 17, 1928 ("Fordizatsia").

18. In seeing socialism as the outcome of a combination of Soviet rule with American methods, Trotsky was not only echoing Lenin but also voicing a common Bolshevik belief. In 1923, for example, Nikolai Bukharin declared "We need Marxism plus Americanism." Rogger, "Amerikanizm," 384. The Trotsky quotes come from his essay "Culture and Socialism," *Krasnaya Nov*, 6 (Feb. 3, 1926), translated by Brian Pearce, in Leon Trotsky, *Problems of Everyday Life and Other Writings on Culture and Science* (New York: Monad Press, 1973).

19. The Five-Year Plan was a highly detailed document, running more than 1,700 pages long. Erlich, *Soviet Industrialization Debate*; Carr and Davies, *Foundations of a Planned Economy*, 894, 896; Figes, *Revolutionary Russia*, 4, 139, 146–48.

20. There perhaps was a cultural element in the Soviet embrace of industrial giantism as well; Russia, before and after the revolution, had a general predilection to monumentality, evident, for example, in buildings from the Hermitage to the never completed Moscow Palace of Soviets. My thanks to Kate Brown for this point. Carr and Davies, *Foundations of a Planned Economy*, 844, 898–902; Erlich, *Soviet Industrialization Debate*, 67–68, 107–08, 140; Andrle, *Workers in Stalin's Russia*, 27; *Those Who Built Stalingrad*, 33–38.

21. Bailes, "The American Connection," 431; Merkle, *Management and Ideology*, 125; Rogger, "Amerikanizm," 416–17.

22. Another American, Bill Shatov, supervised a second, large early Soviet project, the Turksib railway, but that was a very different story; Shatov was a Russian-born anarchist, active in the United States in the Industrial Workers of the World, who returned to Russia in 1917. Hughes, *American Genesis*, 264–69; Carr and Davies, *Foundations of a Planned Economy*, 900–901; Bourke-White, *Eyes on Russia*, 76–88; Sonia Melnikova-Raich, "The Soviet Problem with Two 'Unknowns': How an American Architect and a Soviet Negotiator Jump-Started the Industrialization of Russia: Part II: Saul Bron," *Industrial Archeology* 37 (1/2) (2011), 8–9. Melnikova-Raich's article is the second part of her revelatory examination of the role of American companies and experts in Soviet industrialization based on extensive research in both U.S. and Soviet archives. On Shatov, see the Emma Goldman Papers, Editors' Notes (accessed Jan. 11, 2016), http://editorsnotes.org/projects/emma/topics/286/.

23. Adler, "Russia 'Arming' with Tractor"; Maurice Hindus, "Preface," in Bourke-White, *Eyes on Russia*, 14–15; Dalrymple, "The American Tractor Comes to Soviet Agriculture," 210; Andrle, *Workers in Stalin's Russia*, 3.

24. The Soviets planned to produce a tractor based on an International Harvester model, receiving cooperation from the company without paying it royalties. *New York Times*, Nov. 5, 1928, May 5, 1929, and May 7, 1929; Sonia Melnikova-Raich, "The Soviet Problem with Two 'Unknowns': How an American Architect and a Soviet Negotiator Jump-Started the Industrialization of Russia: Part I: Albert Kahn," *Industrial Archeology* 36 (2) (2010), 60–61, 66; *Economic Review of the Soviet Union*, Apr. 1, 1930.

25. *Detroit Free Press*, May 14, 1929, and June 1, 1929.

26. Melnikova-Raich, "The Soviet Problem with Two 'Unknowns,' Part I," 61,

66–68; *New York Times*, July 1, 1929, Mar. 29, 1930, May 18, 1930, and Mar. 27, 1932; *Those Who Built Stalingrad,* 38–45, 50–56 (Ivanov quote on 52), 206; Andrle, *Workers in Stalin's Russia*, 84–85; Rogger, "Amerikanizm," 383–84.

27. *New York Times*, June 19, 1930; *Those Who Built Stalingrad,* 13, 62.

28. Melnikova-Raich, "The Soviet Problem with Two 'Unknowns,' Part II," 9–11, 23–24; *New York Times*, May 5, 1929, May 7, 1929, June 1, 1929; Nevins and Hill, *Ford: Expansion and Challenge,* 677–78, 683; Richard Cartwright Austin, *Building Utopia: Erecting Russia's First Modern City, 1930* (Kent, OH: Kent State University Press, 2004), 12.

29. Melnikova-Raich, "The Soviet Problem with Two 'Unknowns,' Part II," 11–12; *Michigan Manufacturer and Financial Record*, Apr. 19, 1930; Lewis H. Siegelbaum, *Cars for Comrades: The Life of the Soviet Automobile* (Ithaca, NY: Cornell University Press, 2008), 40; Betsy Hunter Bradley, *The Works: The Industrial Architecture of the United States* (New York: Oxford University Press, 1999), 22; Austin, *Building Utopia*, 5–6, 13–19.

30. Austin, *Building Utopia*, 31–43, 59–101, 121–39; *New York Times*, Dec. 2, 1931.

31. In April 1930, the Soviet Automobile Construction Trust decided it had been a mistake to ask Austin to design the autoworkers' city: "If Americans are specialists in automobile construction, they are certainly far from specialists in designing Socialist town [sic] for the Soviet Republics." Nonetheless, even in the radical socialist vision for the city there was some American influence. One of the key figures involved, architect and educator Alexander Zelenko, had spent time in the United States, including visits to Hull House in Chicago and the University Settlement in New York, where he was influenced by the ideas of John Dewey. *New York Times*, Dec. 16, 1929, Apr. 11, 1931, Mar. 27, 1932; Yordanka Valkanova, "The Passion for Educating the 'New Man': Debates about Preschooling in Soviet Russia, 1917–1925," *History of Education Quarterly* 49 (2) (May 2009), 218; Austin, *Building Utopia*, 45–53, 84–85, 161–68; Kotkin, *Magnetic Mountain*, 366.

32. The very popular Soviet novel *Cement*, by Fyodo Vasilievich Gladkov ([1925] New York: Frederick Ungar Publishing Co., 1973), vividly portrays the huge obstacles and heroic efforts involved in Soviet industrialization. For first-person accounts in English of work on First-Year Plan projects, see *Those Who Built Stalingrad* and John Scott, *Behind the Urals: An American Worker in Russia's City of Steel* (Cambridge, MA: Houghton Mifflin, 1942).

33. On-site reports include *The Detroit Sunday News*, Dec. 15, 1929, and *New York Times*, Nov. 21, 1930. *Time* coverage includes "Great Kahn," May 20,

1929, "Austin's Austingrad," Sept. 16, 1929, and "Architects to Russia," Jan. 20, 1930.

34. Saul G. Bron, *Soviet Economic Development and American Business* (New York: Horace Liveright, 1930), 76, 144–46.

35. Melnikova-Raich, "The Soviet Problem with Two 'Unknowns,' Part I," 60–63; *New York Times*, Jan. 11, 1930; "Architects to Russia," *Time*, Jan. 20, 1930; Terry Smith, *Making the Modern: Industry, Art and Design in America* (Chicago: University of Chicago Press, 1993), 85; *Detroit Free Press*, Jan. 18, 1930; *Detroit Times*, Mar. 17, 1930.

36. "Industry's Architect," *Time*, June 29, 1942; Melnikova-Raich, "The Soviet Problem with Two 'Unknowns,' Part I," 62–66, 75.

37. The design of the tractor to be produced in Chelyabinsk and much of the engineering for its manufacture was done at a Detroit office that had twelve U.S. and forty Soviet engineers. Melnikova-Raich, "The Soviet Problem with Two 'Unknowns,' Part I," 69–71.

38. *Those Who Built Stalingrad*, 56-58, 261; Bourke-White, *Eyes on Russia*, 188. Of course, not speaking Russian and unfamiliar with the circumstances, it is quite possible that Bourke-White and other American observers failed to fully understand what they were seeing and its causes.

39. *New York Times*, Nov. 7, 1930, Nov. 24, 1930, Dec. 27, 1930, Sept. 28, 1931, Oct. 4, 1931, Apr. 14, 1934; Nevins and Hill, *Ford: Expansion and Challenge*, 522; Meredith Roman, "Racism in a 'Raceless' Society: The Soviet Press and Representations of American Racial Violence at Stalingrad in 1930," *International Labor and Working-Class History* 71 (Spring 2007), 187; *Those Who Built Stalingrad*, 64–66, 161, 164, 228–29, 261, 263.

40. *New York Times*, July 20, 1930; Austin, *Building Utopia*, 190–91; Victor Reuther, *The Brothers Reuther and the Story of the UAW* (Boston: Houghton Mifflin, 1976), 93, 101.

41. *New York Times*, July 20, 1930; May 11, 1931; May 14, 1931; May 18, 1931; Dec. 2, 1931 (Duranty), May 18, 1932; Austin, *Building Utopia*, 190–91, 197; Andrle, *Workers in Stalin's Russia*, 35; Reuther, *Brothers Reuther*, 88, 93, 101, 110.

42. Melnikova-Raich, "The Soviet Problem with Two 'Unknowns,' Part I," 69; *Those Who Built Stalingrad*, 158; *New York Times*, Dec. 2, 1931.

43. The following account of Magnitogorsk is based primarily on Stephen Kotkin's brilliant history, *Magnetic Mountain*, and the first-person account by American John Scott, who worked in the plant, *Behind the Urals*.

44. "Mighty Giant" from *USSR in Construction*, 1930, no. 9, p. 14. The Nizhny Tagil plant looked very much like a Kahn factory, but apparently only Soviet

specialists were involved in designing, building, and starting it up, including many veterans of First Five-Year Plan projects. See *USSR in Construction*, 1936, no. 7 (July).

45. "Super-American tempo" from *USSR in Construction*, 1930, no. 9, p.14. On the weather, see http://www.weatherbase.com/weather/weather.php3? s=83882&cityname=Magnitogorsk-Chelyabinsk-Russia (accessed Jan. 26, 2016) and Scott, *Behind the Urals*, 9–10, 15. For many Americans besides Scott, cold was a defining feature of their experience in the Soviet Union. When Victor Herman, who accompanied his father to the Gorky auto plant, attended a Kremlin celebration of the first vehicles to come off the line, the first thing he noticed was the warmth in the banquet hall, realizing that he had not been "really all-over warm" since arriving in the country. Victor Herman, *Coming Out of the Ice* (New York: Harcourt Brace Jovanovich, 1979), 53.

46. Kotkin and Scott both extensively discuss the use of unfree labor. See, also, William Henry Chamberlin, *Russia's Iron Age* (Boston: Little, Brown, 1934), 51–53; Lynne Viola, *The Unknown Gulag: The Lost World of Stalin's Special Settlements* (New York: Oxford University Press, 2007), 101.

47. In addition to Kotkin and Scott (quoted passage on 159), see Melnikova-Raich, "The Soviet Problem with Two 'Unknowns,'" Part II, 19; Herman, *Coming Out of the Ice*; and Siegelbaum, *Cars for Comrades*, 58–59.

48. Scott, *Behind the Urals*, 204–05, 277–79.

49. Siegelbaum, *Cars for Comrades*, 45; Andrle, *Workers in Stalin's Russia*, 16; Robert C. Allen, *Farm to Factory: A Reinterpretation of the Soviet Industrial Revolution* (Princeton, NJ: Princeton University Press, 2003), 92–93, 102–06. On the difficulty of obtaining accurate Soviet economic data, see Oscar Sanchez-Sibony, *Red Globalization: The Political Economy of the Soviet Cold War from Stalin to Khrushchev* (Cambridge: Cambridge University Press, 2014), 12–19.

50. Kotkin, *Magnetic Mountain*, 70, 363; Sheila Fitzpatrick, *Everyday Stalinism: Ordinary Life in Extraordinary Times: Soviet Russia in the 1930s* (New York: Oxford University Press, 1999), 79–83.

51. Scott, *Behind the Urals*, 16; *Those Who Built Stalingrad*, 168–73.

52. Andrle, *Workers in Stalin's Russia*, 35; Scott, *Behind the Urals*, 144; Kotkin, *Magnetic Mountain*, 189. Tensions about shifting gender roles are a major theme in *Cement*, Gladkov's widely read novel about the struggle to reopen a huge, prerevolution cement factory.

53. Nelson Lichtenstein, *The Most Dangerous Man in Detroit: Walter Reuther and the Fate of American Labor* (New York: Basic Books, 1995), 39; *Those Who Built Stalingrad*, 98.

54. Scott, *Behind the Urals*, 138, 152, 212–19; Fitzpatrick, *Everyday Stalinism*, 87; Kotkin, *Magnetic Mountain*, 214–15.

55. Scott, *Behind the Urals*, 40; Katerina Clark, "Little Heroes and Big Deeds: Literature Responds to the First Five-Year Plan," in Sheila Fitzpatrick, ed., *Cultural Revolution in Russia, 1928–1931* (Bloomington: Indiana University Press, 1978), 197; Reuther, *Brothers Reuther*, 98–99; *Those Who Built Stalingrad*, 52–53. Oddly, artificial palm trees seemed to have been something of a rage in the Soviet Union; when Ernst May and a team of German architects entered the country in 1929 to design new industrial cities, they found artificial palms common in railway waiting rooms. Ernst May, "Cities of the Future," in Walter Laqueur and Leopold Labedz, eds., *Future of Communist Society* (New York: Praeger, 1962), 177.

56. Fitzpatrick, *Everyday Stalinism*, 49, 55–56, 95–103; Andrle, *Workers in Stalin's Russia*, 37; A. Baikov, *Magnitogorsk* (Moscow: Foreign Language Publishing House, 1939), 19, 30–31; Kotkin, *Magnetic Mountain*, 67, 182–92, 290–91; Scott, *Behind the Urals*, 235–36.

57. Herman, *Coming Out of the Ice*, 38; Scott, *Behind the Urals*, 234; Kotkin, *Magnetic Mountain*, 108–23.

58. Fitzpatrick, *Everyday Stalinism*, 80–82; *Those Who Built Stalingrad*, 212–19.

59. Clark, "Little Heroes and Big Deeds," 190–92; Susan Tumarkin Goodman, "Avant-garde and After: Photography in the Early Soviet Union," in Goodman and Jens Hoffman, eds., *The Power of Pictures: Early Soviet Photography, Early Soviet Film* (New Haven, CT: Yale University Press, 2015), 23, 31–32; Lydia Chukovskaya, *Sofia Petrovna* (1962; Evanston, IL: Northwestern University Press, 1988), 4. Chukovskaya's novella was not published in Russian until 1962 and in English until 1967.

60. For a comparison of documentary photography in the United States and the Soviet Union, see Bendavid-Val, *Propaganda and Dreams: Photographing the 1930s in the U.S.S.R. and U.S.A.* (Zurich: Edition Stemmle, 1999).

61. Over time, the magazine began covering more varied topics, including political events, the army, ethnic groups, distant regions of the country, and sports. *USSR in Construction*, 1930–1941; *USSR in Construction: An Illustrated Exhibition Magazine* (Sundsvall, Sweden: Fotomuseet Sundsvall, 2006); University of Saskatchewan Library, Digital Collections, USSR in Construction, "About" (accessed Feb. 5, 2016), http://library2.usask.ca/USSRConst/about; Goodman, "Avant-garde and After," 27–28; Bendavid-Val, *Propaganda and Dreams*, 62–65.

62. *SSSR stroit sotsializm* (Moskova: Izogiz, 1933); *USSR in Construction: An Illustrated Exhibition Magazine* (press run data); B. M. Tal, *Industriia*

sotsializma. Tiazhelaia promyshlennost'k VII vsesoiuznomu s'ezdy sovetov [*Industry of Socialism. Heavy Industry for the Seventh Congress of Soviets*] (Moscow: Stroim, 1935).

63. Goodman, "Avant-garde and After," 15, 17; *USSR in Construction*, 1930, no. 1.

64. Goodman, "Avant-garde and After," 22–27, 38. Leah Bendavid-Val stresses similarities between Soviet and U.S. photographers in *Propaganda and Dreams*, which includes photographs of Magnitogorsk by Debabov, Albert, and Petrusov. For more extensive collections of Petrusov's work, see *Georgij Petrussow, Pioneer Sowjetischer Photographie* (Köln, Germany: Galerie Alex Lachmann, n.d.) and *Georgy Petrusov: Retrospective/Point of View* (Moscow: GBUK "Multimedia Complex of Actual Arts," Museum "Moscow House of Photography," 2010).

65. *Entuziazm (Simfonija Donbassa)*, Ukrainfilm, 1931. Filmmakers like Vertov and Sergei Eisenstein, who used avant-garde techniques to pursue revolutionary themes, drew considerable attention outside of the Soviet Union, but domestic audiences preferred more conventional entertainment. Jens Hoffman, "Film in Conflict," in Goodman and Hoffman, *The Power of Pictures*.

66. The Soviets also published in English a collection of letters from foreigners who worked in the Soviet Union. Melnikova-Raich, "The Soviet Problem with Two 'Unknowns,' Part II," 17–18; Cynthia A. Ruder, *Making History for Stalin; The Story of the Belomor Canal* (Gainesville: University Press of Florida, 1998); *Those Who Built Stalingrad*; Baikov, *Magnitogorsk*; Garrison House Ephemera (accessed Nov. 13, 2016), http://www.garrisonhouseephemera. com/?page=shop/flypage&product_id=546; *Sixty Letters: Foreign Workers Write of Their Life and Work in the U.S.S.R.* (Moscow: Co-operative Publishing Society of Foreign Workers in the U.S.S.R., 1936).

67. Duranty's articles on Soviet industry are too numerous to individually cite. For Chamberlin, see *Russia's Iron Age*. On American academic experts and intellectuals, see David C. Engerman, *Modernization from the Other Shore: American Intellectuals and the Romance of Russian Development* (Cambridge, MA: Harvard University Press, 2003), esp. 5–6, 9, 156–57, 166, 237 (Fischer quote).

68. Hans Schoots, *Living Dangerously: A Biography of Joris Ivens* (Amsterdam: Amsterdam University Press, 2000), 74–81.

69. Bourke-White returned to the U.S.S.R. in 1941, when she photographed Moscow during German bombing raids, Stalin in the Kremlin, and the front line. Bourke-White, *Eyes on Russia* (quotes on 23 and 42); Margaret Bourke-White, *Portrait of Myself* (New York: Simon and Schuster, Inc., 1963), 90–

104, 174–88; Vicki Goldberg, *Margaret Bourke-White: A Biography* (New York: Harper & Row, 1986), 128– 32; Haran, "Tractor Factory Facts."

70. To compare Bourke-White's Soviet and U.S. textile mill photographs, see Bourke-White, *Eyes on Russia* and Bourke-White, "Amoskeag" (1932), reproduced in Richard Guy Wilson, Dianne H. Pilgrim, and Dickran Tashjian, *The Machine Age in America 1918–1941* (New York: Brooklyn Museum and Harry N. Abrams, 1986), 234. Bourke-White also took similar photographs at the American Woolen Company in Lawrence, Massachusetts. For an interesting discussion of her Soviet work, see Haran, "Tractor Factory Facts."

71. A drop in grain production during the first years of collectivization, combined with the export of grain, exacerbated the food crisis. Sanchez-Sibony, *Red Globalization*, 36–53 (Stalin quote on 51); Bailes, "The American Connection," 433, 442–43; *Those Who Built Stalingrad,* 150, 198; Scott, *Behind the Urals*, 86–87, 174; Melnikova-Raich, "The Soviet Problem with Two 'Unknowns,' Part I," 74–75; *New York Times*, Mar. 26, 1932; *Detroit Free Press*, Mar. 29, 1932; *Daily Express*, Apr. 19, 1932; *Detroit News*, Apr. 24, 1932; Nevins and Hill, *Ford: Expansion and Challenge*, 682.

72. Merkle, *Management and Ideology*, 132; Bailes, "The American Connection," 442–44; *Those Who Built Stalingrad,* 54, 198; Michael David-Fox, *Showcasing the Great Experiment: Cultural Diplomacy and Western Visitors to the Soviet Union, 1921–1941* (New York: Oxford University Press, 2012), 285–86, 297–99; Melnikova-Raich, "The Soviet Problem with Two 'Unknowns,' Part I," 75–76; Scott, *Behind the Urals*, 230–31.

73. Bailes, "The American Connection," 445; Chamberlin, *Russia's Iron Age*, 61–65; R. W. Davies, Mark Harrison, and S. G. Wheatcroft, eds., *The Economic Transformation of the Soviet Union, 1913–1945* (New York: Cambridge University Press, 1993), 95, 155; Figes, *Revolutionary Russia*, 5, 178.

74. Wikipedia, "Alexei Gastev" (accessed Nov. 12, 2016), https://en.wikipedia.org/wiki/Aleksei_Gastev; Melnikova-Raich, "The Soviet Problem with Two 'Unknowns,' Part II," 17–20; Patrick Flaherty, "Stalinism in Transition, 1932–1937," *Radical History Review*, 37 (Winter 1987). Bill Shatov, who had returned home from the United States and supervised the Turksib railway project, was exiled to Siberia in 1937 and executed the following year. Emma Goldman Papers, Editors' Notes (accessed Jan. 11, 2016), http://editorsnotes.org/projects/emma/topics/286/. For an account of the long imprisonment, Siberian exile, and eventual return to the United States of a young American worker at the Gorky auto plant, see Herman, *Coming Out of the Ice.*

75. Donald Filtzer, *Soviet Workers and Stalinist Industrialization: The Formation of Modern Soviet Production Relations, 1928–1941* (Armonk, NY: M.E. Sharpe, 1986), 126–27, 261–66; Erlich, *Soviet Industrialization Debate*, 182–83; Allen, *Farm to Factory*, 152, 170–71; Flaherty, "Stalinism in Transition," 48–49.

76. After their revolution, the Soviets (like the French) introduced a new organization of time, replacing the weekend with a system of one day off work during every five days (four days in the metallurgy industry), later switching to one day off every six days, before ultimately returning to more conventional timekeeping. Filtzer, *Soviet Workers*, 91–96, 156; Kate Brown, *A Biography of No Place: From Ethnic Borderland to Soviet Heartland* (Cambridge, MA: Harvard University Press, 2003), 92–117; Fitzpatrick, *Everyday Stalinism*, 4, 42–45.

77. Most of the twenty thousand Stalingrad Tractor Factory workers were evacuated as the battle broke out. The factory was rebuilt after the war. The Nizhny Tagil Railroad Car Factory also was converted to military production, and, like the Chelyabinsk plant, continues to produce both military and civilian equipment, employing thirty thousand workers in 2016. Melnikova-Raich, "The Soviet Problem with Two 'Unknowns,' Part I," 68–69, 71–73; Reuther, *Brothers Reuther*, 102–03; Siegelbaum, *Cars for Comrades*, 61–62; Jochen Hellbeck, *Stalingrad: The City that Defeated the Third Reich* (New York: Public Affairs Press, 2015), 89; "History—Chelyabinsk tractor plant (ChTZ)" (accessed Jan. 18, 2016), http://chtz-uraltrac.ru/articles/categories/24.php; *New York Times*, Feb. 25, 2016; Scott, *Behind the Urals*, vii–viii, 63–65, 103.

78. John P. Diggins, *Up from Communism* ([1975] New York: Columbia University Press, 1994), 189–98; Christopher Phelps, "C.L.R. James and the Theory of State Capitalism," in Nelson Lichtenstein, ed., *American Capitalism: Social Thought and Political Economy in the Twentieth Century* (Philadelphia: University of Pennsylvania Press, 2006); Filtzer, *Soviet Workers*, 270–71.

79. Andrle, *Workers in Stalin's Russia*, 126–76, 198–201; Kotkin, *Magnetic Mountain*, 206–07, 318–19; Filtzer, *Soviet Workers*, 233–36; Federico Bucci, *Albert Kahn: Architect of Ford* (New York: Princeton Architectural Press, 1993), 92.

80. If anything, Freyn thought the Soviets were a bit too democratic; it would be better if "more decisions might be made by responsible individuals rather than by committees and commissions." Edmund Wilson, "A Senator and an Engineer," *New Republic*, May 27, 1931; "An American Engineer Looks at the Five Year Plan," *New Republic*, May 6, 1931; *Detroit News* Apr. 24, 1932.

Chapter 6
"COMMON REQUIREMENTS OF INDUSTRIALIZATION"

1. Many of Burnham's arguments had been put forth earlier by Bruno Rizzi, but received little notice outside of small, left-wing circles. At roughly the same time, C. L. R. James broke with Trotsky to describe the Soviet Union as "state capitalist," with productive enterprises collectively owned by a reemerging capitalist class through the government. Ultimately, the United States, too, James argued, would become state capitalist. James Burnham, *The Managerial Revolution* ([1941] Bloomington: Indiana University Press, 1960); Isaac Deutscher, *The Prophet Outcast; Trotsky: 1929–1940* (New York: Oxford University Press, 1963), 459–77; Christopher Phelps, "C.L.R. James and the Theory of State Capitalism," in Nelson Lichtenstein, ed., *American Capitalism: Social Thought and Political Economy in the Twentieth Century* (Philadelphia: University of Pennsylvania Press, 2006).

2. These paragraphs draw substantially from David C. Engerman, "To Moscow and Back: American Social Scientists and the Concept of Convergence," in Lichtenstein, ed., *American Capitalism.*

3. http://brooklynnavyyard.org/the-navy-yard/history/ (accessed Mar. 29, 2016). For a popular overview of the role of private business in wartime defense production, see Arthur Herman, *Freedom's Forge: How American Business Produced Victory in World War II* (New York: Random House, 2012).

4. To undertake the war work, Kahn's firm grew from four hundred to six hundred employees. Hawkins Ferry, *The Legacy of Albert Kahn* (Detroit, MI: Wayne State University Press, 1970), 25–26.

5. To expand the labor pool for Willow Run, Ford opened up jobs to women, who eventually made up 35 percent of the workforce. However, in a departure from its policy at Highland Park and the Rouge, the company all but spurned African Americans. Willow Run workers eventually achieved productivity far above the airplane industry norm. The latest use of the Willow Run factory grounds has been as a test site for driverless cars. Sarah Jo Peterson, *Planning the Home Front: Building Bombers and Communities at Willow Run* (Chicago: University of Chicago Press, 2013); Allan Nevins and Frank Ernest Hill, *Ford: Expansion and Challenge, 1915–1933* (New York: Charles Scribner's Sons: 1957), 242–47; Nelson Lichtenstein, *The Most Dangerous Man in Detroit: Walter Reuther and the Fate of American Labor* (New York: Basic Books, 1995), 160–74; Gail Radford, *Modern Housing*

for America: Policy Struggles in the New Deal Era (Chicago: University of Chicago Press, 1996), 121–32; *New York Times,* June 6, 2016.

6. Not all the workers at other airplane manufacturers were housed in single plants; Republic and Grumman built auxiliary factories near their main plants to bring work nearer to where workers lived, reducing problems with commuting and housing. T. P. Wright Memorandum for Charles E. Wilson, Mar. 21, 1943, box 7, National Aircraft War Production Council, Harry S. Truman Library, Independence, MO; Ferry, *Legacy of Albert Kahn,* 25, 127–28; Tim Keogh, "Suburbs in Black and White: Race, Jobs and Poverty in Twentieth-Century Long Island," Ph.D. dissertation, City University of New York, 2016, 53–56, 77; T. M. Sell, *Wings of Power: Boeing and the Politics of Growth in the Northwest* (Seattle: University of Washington Press, 2001), 19; John Gunther, *Inside U.S.A.* (New York: Harper & Brothers, 1947), 142–43.

7. "Bethlehem Ship," *Fortune,* Aug. 1945, 220; Bernard Matthew Mergen, "A History of the Industrial Union of Marine and Shipbuilding Workers of America, 1933–1951," Ph.D. dissertation, University of Pennsylvania, 1968, 2–3, 103–04, 134–37, 142; [Baltimore] *Evening Sun,* Dec. 8, 1943; Apr. 5, 1944; Apr. 20, 1944; May 15, 1944; July 1, 1944; Karen Beck Skold, "The Job He Left Behind: American Women in Shipyards During World War II," in Carol R. Berkin and Clara M. Lovett, eds., *Women, War, and Revolution* (New York: Holmes & Meier, 1980), esp. 56–58; Eric Arnesen and Alex Lichtenstein, "Introduction: 'All Kinds of People,'" in Katherine Archibald, *Wartime Shipyard: A Study in Social Disunity* ([1947] Urbana: University of Illinois Press, 2006), xvi, xxxi–xxxv; Joshua B. Freeman, *American Empire, 1945–2000: The Rise of a Global Empire, the Democratic Revolution at Home* (New York: Viking, 2012), 21; Peterson, *Planning the Home Front,* 279.

8. For the impact of World War II on the American working class, see Joshua Freeman, "Delivering the Goods: Industrial Unionism during World War II," *Labor History* 19 (4) (Fall 1978); Nelson Lichtenstein, "The Making of the Postwar Working Class: Cultural Pluralism and Social Structure in World War II," *The Historian* 51 (1) (Nov. 1988), 42–63; Gary Gerstle, "The Working Class Goes to War," *Mid-America* 75 (3) (1993), 303–22. Dorothea Lange and Charles Wollenberg, *Photographing the Second Gold Rush: Dorothea Lange and the East Bay at War, 1941–1945* (Berkeley, CA: Heyday Books, 1995).

9. Jack Metzgar, "The 1945–1946 Strike Wave," in Aaron Brenner, Benjamin Day, and Immanuel Ness, eds., *The Encyclopedia of Strikes in American*

History (Armonk, NY: M.E. Sharpe, 2009); Art Preis, *Labor's Giant Step: Twenty years of the CIO* (New York: Pioneer Press, 1965), 257–83.

10. Ronald W. Schatz, *The Electrical Workers: A History of Labor at General Electric and Westinghouse, 1923–1960* (Urbana: University of Illinois Press, 1983), 105–64; Nelson Lichtenstein, *The Most Dangerous Man in Detroit*, 282–98; Freeman, *American Empire*, 119–24; *Labor's Heritage: Quarterly of the George Meany Memorial Archives*, 4 (1992), 28; Joshua Freeman, "Labor During the American Century: Work, Workers, and Unions Since 1945," in Jean-Christophe Agnew and Roy Rosenzweig, eds., *A Companion to Post-1945 America* (Malden, MA: Blackwell, 2002); Ruth Milkman, *Farewell to the Factory: Auto Workers in the Late Twentieth Century* (Berkeley: University of California Press, 1997); Charles Corwin in *New York Daily Worker*, Feb. 4, 1949, quoted in Karen Lucic, *Charles Sheeler and the Cult of the Machine* (Cambridge, MA: Harvard University Press, 1991), 114; Jack Metzgar, *Striking Steel: Solidarity Remembered* (Philadelphia: Temple University Press, 2000), 30–45 (quote on 39).

11. Daniel Nelson, *American Rubber Workers and Organized Labor, 1900–1941* (Princeton, NJ: Princeton University Press, 1988), 82–83, 234–45, 257–64, 271, 307–09, 315–17; Charles A. Jeszeck, "Plant Dispersion and Collective Bargaining in the Rubber Tire Industry," Ph.D. dissertation, University of California, Berkeley, 1982, 31, 47–54, 106–08.

12. The Bloomington plant swelled to more than eight thousand employees after RCA began producing televisions there, but the company eventually shifted much of the production first to Memphis and then Ciudad Juárez, Mexico. Jefferson Cowie, *Capital Moves: RCA's Seventy-Year Quest for Cheap Labor* (Ithaca, NY: Cornell University Press, 1999), 10, 15, 17, 22–35, 42–43.

13. In a further effort to avoid interruptions in production, General Motors, unlike Ford, made it a policy to use outside suppliers for a majority of the parts and accessories that went into its vehicles. Douglas Reynolds, "Engines of Struggle: Technology, Skill and Unionization at General Motors, 1930–1940," *Michigan Historical Review* 15 (Spring 1989), 79–80; *New York Times*, Aug. 12, 1935; Alfred D. Chandler, Jr., *Scale and Scope: The Dynamics of Industrial Capitalism* (Cambridge, MA: Harvard University Press, 1994), 208.

14. Jeszeck, "Plant Dispersion," 33–35; "Flying High," Kansas City Public Library, http://www.kclibrary.org/blog/week-kansas-city-history/flying-high, and "Fairfax Assembly Plant," GM Corporate Newsroom, http://media.gm.com/media/us/en/gm/company_info/facilities/assembly/fairfax.html (both accessed Apr. 5, 2016); Schatz, *Electrical Workers*, 233. On

war-related industrial development in the Southwest, see Elizabeth Tandy Shermer, *Sunbelt Capitalism: Phoenix and the Transformation of American Politics* (Philadelphia: University of Pennsylvania Press, 2013).

15. Metzgar, "The 1945–1946 Strike Wave"; Freeman, *American Empire*, 39–41; Kim Phillips-Fein, *Invisible Hands: The Making of the Conservative Movement from the New Deal to Reagan* (New York: W. W. Norton, 2009), 93–97; Elizabeth A. Fones-Wolf, *Selling Free Enterprise: The Business Assault on Labor and Liberalism, 1945–60* (Urbana: University of Illinois Press, 1994), 138–39.

16. Kim Phillips-Fein, "Top-Down Revolution: Businessmen, Intellectuals and Politicians Against the New Deal, 1945–1964," Ph.D. dissertation, Columbia University, 2004, 220; Joshua B. Freeman, *Working-Class New York: Life and Labor since World War II* (New York: New Press, 2000), 60–71; Tami J. Friedman, "Communities in Competition: Capital migration and plant relocation in the United States carpet industry, 1929–1975," Ph.D. dissertation, Columbia University, 2001, 22, 70–76, 201–04.

17. Schatz, *Electrical Workers*, 170–75; Phillips-Fein, *Invisible Hands*, 97–114.

18. Schatz, *Electrical Workers*, 233–34.

19. Schatz, *Electrical Workers*, 234–36; Freeman, *American Empire*, 303–06; Thomas J. Sugrue, *The Origins of the Urban Crisis: Race and Inequality in Postwar Detroit* (Princeton, NJ: Princeton University Press, 1996), 128–29; James C. Cobb, *The Selling of the South: The Southern Crusade for Industrial Development, 1936–1980* (Baton Rouge: Louisiana State University Press, 1982); Friedman, "Communities in Competition," 111–66.

20. See, for example, Martin Beckman, *Location Theory* (New York: Random House, 1968); Gerald J. Karaska and David F. Bramhall, *Locational Analysis for Manufacturing: A Selection of Readings* (Cambridge, MA: MIT Press, 1969); and Paul Krugman, *Geography and Trade* (Leuven, Belgium: Leuven University Press and Cambridge, MA: MIT Press, 1989), esp. 62–63 for discussion of Akron.

21. Counter to the common management view, the productivity of unionized workers often exceeded that of nonunion workers. Roger W. Schmenner, *Making Business Location Decisions* (Englewood Cliffs, NJ: Prentice-Hall, 1982), vii, 10–11, 124–26, 154–57, 239; Phillips-Fein, *Invisible Hands*, 104; Lawrence Mishel and Paula B. Voos, eds., *Unions and Economic Competitiveness* (New York: M.E. Sharpe, 1992).

22. Kimberly Phillips-Fein, "American Counterrevolutionary: Lemuel Ricketts Boulware and General Electric, 1950–1960," in Lichtenstein, ed., *American Capitalism*, 266–67; John Barnard, *American Vanguard: The United Auto*

Workers during the Reuther Years, 1935–1970 (Detroit, MI: Wayne State University Press, 2004), 483; Cowie, *Capital Moves*, 53–58. See also Friedman, "Communities in Competition," 380–81, 403–21.

23. Sugrue, *Origins of the Urban Crisis*, 130–35.

24. Steve Jefferys, *Management and Managed: Fifty Years of Crisis at Chrysler* (Cambridge: Cambridge University Press, 1986), 155; Historic American Engineering Record, Mid-Atlantic Region National Park Service, "Dodge Bros. Motor Car Company Plant (Dodge Main): Photographs, Written Historical and Descriptive Data" (Philadelphia: Department of the Interior, 1980), 20.

25. Freeman, *American Empire*, 115; U.S. Bureau of the Census, *1967 Census of Manufactures*, vol. 1: *Summary and Subject Statistics* (Washington, D.C.: U.S. Government Printing Office, 1971), table 1 (pages 2–4).

26. Charles Fishman, "The Insourcing Boom," *The Atlantic*, Dec. 2012; Mark Reilly, "General Electric Appliance Park," in John E. Kleber, ed., *The Encyclopedia of Louisville* (Lexington, KY: University Press of Kentucky, 2000), 333–34.

27. "The Rebirth of Ford," *Fortune*, May 1947, 81–89. The Evans photographs are now held by the Metropolitan Museum of Art and can be seen at http://www.metmuseum.org/art/collection/search/281891 and http://www.metmuseum.org/art/collection/search/279282 (accessed Apr. 11, 2016).

28. Warren Bareiss, "The Life of Riley," Museum of Broadcast Communications—Encyclopedia of Television (accessed Apr. 11, 2016), http://www.museum.tv/eotv/lifeofriley.htm. See also George Lipsitz, "The Meaning of Memory: Family, Class, and Ethnicity in Early Network Television Programs," *Cultural Anthropology* 1 (4) (Nov. 1986), 355–87.

29. Nelson Lichtenstein, *State of the Union: A Century of American Labor* ([2002] Princeton, NJ: Princeton University Press, 2013), 148–62, 215–18; U.S. Bureau of the Census, *Census of Manufactures, 1972*, vol. 1, *Subject and Special Statistics* (Washington, D.C.: U.S. Government Printing Office, 1976), 68; Daniel Bell, *The Coming of Post-Industrial Society; A Venture in Social Forecasting* (New York: Basic Books, 1973); Freeman, *American Empire*, 303–06, 344–49; Metzgar, *Striking Steel*, 210–23.

30. Anders Åman, *Architecture and Ideology in Eastern Europe during the Stalin Era; An Aspect of Cold War History* (Cambridge, MA: MIT Press, 1992), 76; Sonia Melnikova-Raich, "The Soviet Problem with Two 'Unknowns': How an American Architect and a Soviet Negotiator Jump-Started the Industrialization of Russia: Part II: Saul Bron," *Industrial Archeology* 37 (1/2) (2011), 21–22; "History—Chelyabinsk tractor plant (ChTZ)" (accessed Jan. 18,

2016), http://chtz-uraltrac.ru/articles/categories/24.php; *New York Times*, Feb. 25, 2016; Stephen Kotkin, *Steeltown, USSR: Soviet Society in the Gorbachev Era* (Berkeley: University of California Press, 1991), xii–xiii, 2, 5.

31. Kate Brown, *Plutopia: Nuclear Families, Atomic Cities, and the Great Soviet and American Plutonium Disasters* (New York: Oxford University Press, 2013); Lewis H. Siegelbaum, *Cars for Comrades: The Life of the Soviet Automobile* (Ithaca, NY: Cornell University Press, 2008), 80–81.

32. Alan M. Ball, *Imagining America: Influence and Images in Twentieth-Century Russia* (Lanham, MD: Rowman & Littlefield, 2003), 162.

33. Most of the housing in Avtograd consisted of apartments for individual families in five to sixteen story buildings. Siegelbaum, *Cars for Comrades*, 81–109; *Wall Street Journal*, Apr. 11, 2016.

34. KAMAZ, "History," https://kamaz.ru/en/about/history/ (accessed May 2, 2017).

35. Siegelbaum, *Cars for Comrades*, 112–24; *Wall Street Journal*, Apr. 11, 2016; KAMAZ, "History"; KAMAZ, "General Information" (accessed May 2, 2017), https://kamaz.ru/en/about/general-information/.

36. Czechoslovakia was exceptional in having a large communist party with substantial popular support. Tony Judt, *Postwar: A History of Europe Since 1945* (New York: Penguin, 2005), 129–39, 165–96; Åman, *Architecture and Ideology*, 12, 28–30, 147; Mark Pittaway, "Creating and Domesticating Hungary's Socialist Industrial Landscape: From Dunapentele to Sztálinváros, 1950–1958," *Historical Archaeology* 39 (3) (2005), 76, 79–80.

37. Romania never had a "first socialist city" of the sort found elsewhere in Eastern Europe. Åman, *Architecture and Ideology*, 77 ("cult of steel"), 81, 147, 157–61; Ulf Brunnbauer, "'The Town of the Youth': Dimitrovgrad and Bulgarian Socialism," *Ethnologica Balkanica* 9 (2005), 92–95. See also Paul R. Josephson, *Would Trotsky Wear a Bluetooth? Technological Utopianism under Socialism, 1917–1989* (Baltimore, MD: Johns Hopkins University Press, 2012), 65–119.

38. Åman, *Architecture and Ideology*, esp. 33–39, 102–03, 158, 162; Pittaway, "Hungary's Socialist Industrial Landscape," 78–81, 85–87; Brunnbauer, "'The Town of the Youth,'" 94, 98–111; Katherine Lebow, *Unfinished Utopia: Nowa Huta, Stalinism, and Polish Society, 1949–56* (Ithaca, NY: Cornell University Press, 2013), 46, 52–56.

39. Paweł Jagło, "Steelworks," in *Nowa Huta 1949+* [English version] (Kraków: Muzeum Historyczne Miasta Krakowa, 2013), quote on 18; Lebow, *Unfinished Utopia*, 19–26, 36–40, 69; Alison Stenning, "Placing (Post-)Socialism: The Making and Remaking of Nowa Huta, Poland," *European Urban and*

Regional Studies 7 (Apr. 2000), 100–01; Boleslaw Janus, "Labor's Paradise: Family, Work, and Home in Nowa Huta, Poland, 1950–1960," *East European Quarterly* XXXIII (4) (Jan. 2000), 469; H. G. J. Pounds, "Nowa Huta: A New Polish Iron and Steel Plant," *Geography* 43 (1) (Jan. 1958), 54–56; interview with Stanisław Lebiest, Roman Natkonski, and Krysztof Pfister, Nowa Huta, Poland, May 19, 2015. The largest U.S. Steel mill, in terms of employment, the Bethlehem Steel Sparrows Point complex, had 28,600 workers in 1957 and a capacity of 8.2 million tons a year. The U.S. Steel mill in Gary, Indiana, peaked at an estimated 25,000 workers in 1976. In 1996, with only 7,800 workers remaining, it produced 12.8 million tons of steel. Mark Reutter, *Sparrows Point; Making Steel—The Rise and Ruin of American Industrial Might* (New York: Summit Books, 1988), 10, 413; *Chicago Tribune*, Feb. 26, 1996.

40. Lebow, *Unfinished Utopia*, 37–40; 61–62, 74–77, 82–88, 92–93, 97–98, 103; Janus, "Labor's Paradise," 455–56; *Poland Today* 6 (7–8) (July–Aug. 1951), 14. Photographs of the construction of Nowa Huta, including of female plasterers, can be seen in Henryk Makarewicz and Wiktor Pental, *802 Procent Normy; pierwsze lata Nowej Huty* [*802% Above the Norm: The Early Years of Nowa Huta*] (Kraków: Fundacja Imago Mundi: Vis-à-vis/etiuda, [2007]).

41. Lebow, *Unfinished Utopia*, 65, 71, 157–58; Paweł Jagło, "Architecture of Nowa Huta," in *Nowa Huta 1949+, 26.*

42. Leszek J. Sibila, *Nowa Huta Ecomuseum: A Guidebook* (Kraków: The Historical Museum of the City of Kraków, 2007); Jagło, "Architecture of Nowa Huta"; Lebow, *Unfinished Utopia*, 29–35, 41–42, 71–73; Åman, *Architecture and Ideology*, 102–103, 151–53; *Nowa przestrzeń; Modernizm w Nowej Hucie* (Kraków: Muzeum Historyczne Miasta Krakowa, 2012). For U.S. comparison, see Freeman, *American Empire*, 12–27, 136–39.

43. Lebow, *Unfinished Utopia*, 3, 146–49; Åman, *Architecture and Ideology*, 151; stamps: https://www.stampworld.com/en_US/stamps/Poland/Postage%20stamps/?year=1951 and http://colnect.com/en/stamps/list/country/4365-Poland/theme/3059-Cranes_Machines (accessed Nov. 25, 2016); Anne Applebaum, *Iron Curtain: The Crushing of Eastern Europe, 1944–56* (New York: Doubleday, 2012), 360, 372, 377–78, 384–85 (quotes from Ważyk in her translation on 384); Andrzej Wajda, *Man of Marble* (Warsaw: Zespól Filmowy X, 1977). See also Marci Shore, "Some Words for Grown-Up Marxists: 'A Poem for Adults' and the Revolution from Within," *Polish Review* 42 (2) (1997), 131–54.

44. Brunnbauer, "'The Town of the Youth,'" 96–97, 105; Janus, "Labor's Paradise," 454–55; Pittaway, "Hungary's Socialist Industrial Landscape," 75–76, 82–85.

45. Judt, *Postwar*, 172; Janus, "Labor's Paradise," 464–65; Lebow, *Unfinished Utopia*, 45, 47, 50–51, 56; interview with Lebiest et al.

46. Janus, "Labor's Paradise," 459–64; Brunnbauer, "'The Town of the Youth,'" 105; Lebow, *Unfinished Utopia*, 124–25, 138–45; Pittaway, "Hungary's Socialist Industrial Landscape," 87.

47. Sztálinváros was renamed Dunaújváros in 1961. Josephson, *Would Trotsky Wear a Bluetooth?*, 85–86; Applebaum, *Iron Curtain*, 459; Pittaway, "Hungary's Socialist Industrial Landscape," 88–89.

48. Paweł Jagło, "Defense of the Cross," in *Nowa Huta 1949+*, 39–40; Lebow, *Unfinished Utopia*, 161–69.

49. Paweł Jagło, "Anti-Communist Opposition," in *Nowa Huta 1949+*; Stenning, "Placing (Post-) Socialism," 105–06; *Chicago Tribune*, June 10, 1979.

50. Kraków environmentalists often blamed the steel mill for the severe air pollution in the city, but prevailing winds took emissions from Nowa Huta eastward, away from the city, not toward it. Local plants, industry west of Kraków, coal-burning furnaces, and growing traffic were more responsible. Maria Lempart, "Myths and facts about Nowa Huta," in *Nowa Huta 1949+*, 50.

51. Judt, *Postwar*, 587–89; Stenning, "Placing (Post-)Socialism," 106; Jagło, "Anti-Communist Opposition."

52. The official government-recognized union tacitly supported the 1988 strike, though with its own, more modest demands. The discussion of Solidarity in Nowa Huta is drawn primarily from Lebow, *Unfinished Utopia*, 169–76, and my interview with Lebiest et al. See also Jagło, "Anti-Communist Opposition"; *New York Times*, Nov. 11, 1982, Apr. 29, 1988, May 3, 1988, and May 6, 1988; and Judt, *Postwar*, 605–08.

53. Interview with Lebiest et al.

54. "Poland Fights for Gdansk Shipyard," BBC News, Aug. 21, 2007, http://news .bbc.co.uk/2/hi/business/6956549.stm; "Gdansk Shipyard Sinking from Freedom to Failure," *Toronto Star* (accessed May 6, 2016), https://www. thestar.com/news/world/2014/01/27/gdansk_shipyard_sinking_from_ freedom_to_failure.html).

55. *New York Times*, Nov. 27, 1989; interview with Lebiest et al.; Jagło, "Steelworks,"19–20; Stenning, "Placing (Post-)Socialism," 108–10, 116.

56. *New York Times*, Oct. 6, 2015, and Oct. 7, 2015.

57. Harold James, *Krupp: A History of the Legendary German Firm* (Princeton, NJ: Princeton University Press, 2012), 39; Werner Abelshauser, *The Dynamics of German Industry: Germany's Path toward the New Economy and the American Challenge* (New York: Berghahn Books, 2005), 3, 85–86, 89.

58. Though in some respects the Wolfsburg plant was modeled on River Rouge, Volkswagen did not integrate backward to make all its parts, instead purchasing many from a network of closely connected suppliers. Abelshauser, *Dynamics of German Industry*, 91–104, 108–09; Volker R. Berghahn, *The Americanization of West German Industry 1945–1973* (Lemington Spa, NY: Berg, 1986), 304–09.

59. Werner Abelshauser, Wolfgang Von Hippel, Jeffrey Allan Johnson, and Raymond G. Stokes, *German Industry and Global Enterprise; BASF: The History of a Company* (Cambridge: Cambridge University Press, 2004), 487–99 (quote on 488); *New York Times*, Oct. 27, 2014; "BASF Headquarters" (accessed May 16, 2016), https://www.basf.com/us/en/company/career/why-join-basf/basf-at-a-glance/basf-headquarters.html.

60. *New York Times*, Oct. 6, 2015; Gillian Darley, *Factory* (London: Reaktion Books, 2003), 187–89.

61. Joel Beinin, *Workers and Peasants in the Modern Middle East* (Cambridge: Cambridge University Press, 2001), 99–113 ("citadels" on 109), 127, 158; Beinin, "Egyptian Textile Workers Confront the New Economic Order," Middle East Research and Information Project, Mar. 25, 2007, http://www.merip.org/mero/mero032507; Beinin, "The Militancy of Mahalla al-Kubra," Middle East Research and Information Project, Sept. 29, 2007, http://www.merip.org/mero/mero092907; "The Factory," *Al Jazeera*, Feb. 22, 2012, http://www.aljazeera.com/programmes/revolutionthrougharabeyes/2012/01/201213013135991429.html; "Mahalla textile workers' strike enters eighth day," *Daily News Egypt*, Feb. 17, 2014, http://www.dailynewsegypt.com/2014/02/17/mahalla-textile-workers-strike-enters-eighth-day/; Alex MacDonald and Tom Rollins, "Egypt's Mahalla textile factory workers end four-day strike after deal reached," *Middle East Eye*, Jan. 17, 2015, http://www.middleeasteye.net/news/egypts-mahalla-textile-factory-workers-end-four-day-strike-after-management-agreement-260129749.

Chapter 7
"FOXCONN CITY"

1. Pun Ngai, Shen Yuan, Guo Yuhua, Lu Huilin, Jenny Chan, and Mark Selden, "Apple, Foxconn, and Chinese Workers' Struggles from a Global Labor Perspective," *Inter-Asia Cultural Studies* 17 (2) (2016), 166; Jason Dean, "The Forbidden City of Terry Gou," *Wall Street Journal*, Aug. 11, 2007. Ngai, Chan, and Selden have written the most important study of Foxconn and of Apple in China, *Dying for an iPhone*, from which I have greatly benefited. It

is forthcoming in English but available in Spanish and Italian editions, *Morir por un iPhone* (Bueno Aires: Ediciones Continente S.R.L., 2014) and *Moirire per un iPhone* (Milan: Jaca Books, 2015).

2. To offset the wage hikes, Foxconn also raised its prices. *New York Times*, May 25, 2010, June 2, 2010; Elizabeth Woyke, *The Smartphone: Anatomy of an Industry* (New York: New Press, 2014), 135–36; *Bloomberg Businessweek*, June 7, 2010, Sept. 13, 2010; "Foxconn's Business Partners Respond to Suicides," CCTV Com English, May 20, 2010, http://english.cntv.cn/program/china24/20100520/101588 .shtml; "Foxconn Shares Dive on Suicides," CCTV Com English, June 29, 2010, http://english.cntv.cn/program/bizasia/20100528/102843.shtml; "Foxconn to Hike Prices to Offset Pay Increase," CCTV Com English, July 22, 2010, http://english.cntv.cn/20100722/104196.shtml; "Foxconn Hikes Salaries Again in South China Factory After Suicides," CCTV Com English, Oct. 1, 2010, http://english.cntv.cn/program/20101001/101698.shtml.

3. Ngai, Chan, and Selden, *Dying for an iPhone*; *Bloomberg Businessweek*, Sept. 13, 2010; James Fallows, "Mr. China Comes to America," *The Atlantic,* Dec. 2012.

4. For various statements of the number of employees at the Foxconn Shenzhen factories in 2010, see "Foxconn Hikes Salaries Again in South China Factory After Suicides," CCTV Com English, Oct. 1, 2010; *Bloomberg Businessweek*, June 7, 2010, Sept. 13, 2010; *New York Times*, May 25, 2010; Pun Ngai, *Migrant Labor in China: Post-Socialist Transformations* (Cambridge: Polity Press, 2016), 101, 119. See also Charles Duhigg and Keith Bradsher, "How the U.S. Lost Out on iPhone Work," *New York Times*, Jan. 21, 2012 ("unimaginable").

5. Foxconn factories outside of China are generally much smaller, in some cases modest-sized assembly plants serving local markets, built to circumvent tariffs. Some Foxconn factories make parts or finished products for multiple clients, including Microsoft, IBM, Intel, Cisco, GE, Amazon, HP, Dell, Motorola, Panasonic, Sony, Toshiba, Nintendo, Samsung, LG, Nokia, Acer, and Lenovo. Others serve just one client or even make only one product. Ngai, *Migrant Labor*, 105; Rutvica Andrijasevic and Devi Sacchetto, "Made in the EU: Foxconn in the Czech Republic," *WorkingUSA*, Sept. 2014; Devi Sacchetto and Martin Cecchi, "On the Border: Foxconn in Mexico," *openDemocracy,* Jan. 16, 2015, https://www .opendemocracy.net/devi-sacchetto-mart%C3%ACn-cecchi/on-border-foxconn -in-mexico; Ngai, Chan, and Selden, *Dying for an iPhone*; *New York Times*, Mar. 29, 2012; David Barboza, "China's 'iPhone City,' Built on Billions in Perks," *New York Times*, Dec. 29, 2016.

6. Ngai, Chan, and Selden, *Dying for an iPhone*; New York Times, Dec. 11, 2013; "BBC Documentary Highlights Conditions at a Chinese iPhone Factory, But Is It All Apple's Fault?" *MacWorld*, Dec. 19, 2014, http://www.macworld.com/article/2861381/bbc-documentary-highlights-conditions-at-a-chinese-iphone-factory-but-is-it-all-apples-fault.html.

7. Ngai, *Migrant Labor*, 102; Boy Lüthje, Siqi Luo, and Hao Zhang, *Beyond the Iron Rice Bowl: Regimes of Production and Industrial Relations in China* (Frankfurt: Campus Verlag, 2013), 195, 198; Hao Ren, ed., *China on Strike: Narratives of Workers' Resistance*, English edition edited by Zhongjin Li and Eli Friedman (Chicago: Haymarket Books, 2016), 11, 201–03; Jennifer Baichwal, *Manufactured Landscapes* (Foundry Films and National Film Board of Canada, 2006).

8. David Barboza, "In Roaring China, Sweaters Are West of Socks City," *New York Times*, Dec. 24, 2004; Ngai, *Migrant Labor*, 102.

9. *New York Times*, Nov. 8, 1997, Mar. 28, 2000; Nelson Lichtenstein, *The Retail Revolution: How Wal-Mart Created a Brave New World of Business* (New York: Metropolitan Books, 2009), 173; Richard P. Appelbaum, "Giant Transnational Contractors in East Asia: Emergent Trends in Global Supply Chains," *Competition & Change* 12 (Mar. 2008), 74; "About PCG," http://www.pouchen.com/index.php/en/about/locations, and "Yue Yuen Announces Audited Results for the Year 2015," http://www.yueyuen.com/index.php/en/news-pr/1147-2016-03-23-yue-yuen-announces-audited-results-for-the-year-2015 (both accessed June 3, 2016); International Trade Union Confederation, *2012 Annual Survey of Violations of Trade Union Rights—Vietnam*, June 6, 2012, http://www.refworld.org/docid/4fd889193.html.

10. Some ten thousand Soviet technicians were posted to China to help with the industrialization drive, while nearly three times that many Chinese went to the Soviet Union for training. Carl Riskin, *China's Political Economy: The Quest for Development Since 1949* (Oxford: Oxford University Press, 1987), 53–63, 74; Nicholas R. Lardy, "Economic Recovery and the 1st Five-Year Plan," in Roderick MacFarquhar and John K. Fairbank, eds., *The Cambridge History of China,* vol. 14: *The People's Republic,* part 1: *The Emergence of Revolutionary China, 1949–1965* (Cambridge: Cambridge University Press, 2008), 157–60, 177–78.

11. Riskin, *China's Political Economy*, 64, 117–18, 125–27, 133, 139, 161–65; Kenneth Lieberthal, "The Great Leap Forward and the Split in the Yenan Leadership," in MacFarquhar and Fairbank, eds., *The Cambridge History of China,* vol. 14; Stephen Andors, *China's Industrial Revolution: Politics, Plan-*

ning, and Management, 1949 to the Present (New York: Pantheon, 1977), 68–134. By the 1990s, Anshan had become China's largest industrial enterprise, employing some 220,000 workers. "Anshan Iron and Steel Corporation," in Lawrence R. Sullivan, *Historical Dictionary of the People's Republic of China*, second edition (Plymouth, UK: Scarecrow Press, 2007), 24–26. See, also, Cheng Tsu-yuan, *Ashan Steel Factory in Communist China* (Hong Kong: The Urban Research Institute, 1955).

12. Andors, *China's Industrial Revolution*, 144–47, 158–59.

13. Andors, *China's Industrial Revolution*, 135–42.

14. While there was a push during the Cultural Revolution for the despecialization of factories, there apparently was not an effort to despecialize the work of individual workers in the production process, even as they were given expanded roles in management and other aspects of factory function. Andors, *China's Industrial Revolution*, 160–240.

15. Ngai, *Migrant Labor*, 11, 15.

16. Henry Yuhuai He, *Dictionary of the Political Thought of the People's Republic of China* (London: Routledge, 2015), 287; Michael J. Enright, Edith E. Scott, and Ka-mun Chang, *Regional Powerhouse: The Greater Pearl River Delta and the Rise of China* (Singapore: John Wiley & Sons, 2005), 6, 36–38.

17. Pun Ngai, *Made in China: Women Factory Workers in a Global Workplace* (Durham, NC: Duke University Press, 2005), 1, 7.

18. Gabriel Kolko, *Vietnam: Anatomy of a Peace* (London: Routledge, 1997); The World Bank, "Vietnam, Overview," Apr. 11, 2016, http://www .worldbank.org/en/country/vietnam/overview; Nguyen Thi Tue Anh, Luu Minh Duc, and Trinh Doc Chieu, "The Evolution of Vietnamese Industry," Learning to Compete Working Paper No. 19, Brookings Institution (accessed Aug. 13, 2016), https://www.brookings.edu/wp-content/uploads/2016/07/ L2C_WP19_Nguyen-Luu-and-Trinh-1.pdf.

19. Enright, Scott, and Chang, *Regional Powerhouse*, 6, 12, 16, 36, 38–39, 67–68, 74, 98, 101–02, 117.

20. Enright, Scott, and Chang, *Regional Powerhouse*, 75, 98, 108; Andrew Ross, *Fast Boat to China: Corporate Flight and the Consequences of Free Trade— Lessons from Shanghai* (New York: Pantheon, 2006), 24–26; *Bloomberg Businessweek*, Sept. 13, 2010.

21. Enright, Scott, and Chang, *Regional Powerhouse*, 47.

22. Ngai, *Migrant Labor*, 2, 20–21, 25, 32, 76–78.

23. *The Guardian*, July 31, 2014, https://www.theguardian.com/world/2014/ jul/31/china-reform-hukou-migrant-workers; Ren, ed., *China on Strike*, 4–5; Ngai, *Made in China*, 36, 43–46.

24. For an interesting portrait of life in a state-owned factory during the 1980s, see Lijoa Zhang, *"Socialism Is Great!" A Worker's Memoir of the New China* (New York: Atlas & Co., 2008). See, also, Ching Kwan Lee, *Against the Law: Labor Protests in China's Rustbelt and Sunbelt* (Berkeley: University of California Press, 2007), 35–36; Ross, *Fast Boat to China*, 57.

25. Ngai, *Migrant Labor*, 35, 93, 128–29; "Workers Strike at China Footwear Plant Over Welfare Payments," *Wall Street Journal*, Apr. 16, 2014, http://www.wsj.com/articles/SB10001424052702304626304579505451938007332; Ren, ed., *China on Strike*, 186.

26. Ngai, *Migrant Labor*, 31. For an in-depth comparison of systems of manufacturing in China, see Lüthje, Luo, and Zhang, *Beyond the Iron Rice Bowl*.

27. The wages and benefit contributions of export factories in high-cost regions would not have been enough to support locally-living families and the services provided to them, the cost of "social reproduction." Enright, Scott, and Chang, *Regional Powerhouse*, 192, 250; Ngai, *Migrant Labor*, 32–35.

28. Ngai, *Migrant Labor*, 83–104, 123; Hong Xue, "Local Strategies of Labor Control: A Case Study of Three Electronics Factories in China," *International Labor and Working-Class History* 73 (Spring 2008), 92; Anita Chen, *China's Workers Under Assault: The Exploitation of Labor in a Globalizing Economy* (Armonk, NY: M.E. Sharpe, 2001), 12; Ngai, Chan, and Selden, *Dying for an iPhone*; Duhigg and Bradsher, "How the U.S. Lost Out on iPhone Work"; Ren, ed., *China on Strike*, 7, 184.

29. For a fine, painful portrait of a migrant worker family and their trips back home, see the documentary film *Last Train Home,* directed by Lixin Fan (EyeSteel Films, 2009). Ngai, *Migrant Labor*, 30–32; Xue, "Local Strategies of Labor Control," 85, 98–99; U.S. Department of Labor, Bureau of Labor Statistics, "The Employment Situation—May 2014," http://www.bls.gov/news.release/archives/empsit_06062014.pdf (accessed July 16, 2016); Michael Bristow, "China's holiday rush begins early," BBC News, Jan. 7, 2009, http://news.bbc.co.uk/2/hi/asia-pacific/7813267.stm; Ross, *Fast Boat to China*, 16; *New York Times*, Jan. 26, 2017.

30. As in China, in Vietnam migrant workers form a large part of the workforce in foreign-owned factories, especially near Ho Chi Minh City. See Anita Chan, "Introduction," in Chan, ed., *Labour in Vietnam* (Singapore: Institute of Southeast Asian Studies, 2011), 4.

31. In addition to the Peter Charlesworth photography of workers making Reebok shoes, see, for example, Dong Hung Group, "Shoe Manufacturers in Vietnam" (2012), http://www.donghungfootwear.com/en/phong-su-ve-dong-hung-group

.html, which includes factory photographs and a video showing the processes used for making sneakers. See, also, Tom Vanderbilt, *The Sneaker Book: Anatomy of an Industry and an Icon* (New York: New Press, 1998), 78–80.

32. For EUPA, see the documentary film *Factory City* (Discovery Channel, 2009). Ngai, Chan, and Selden, *Dying for an iPhone*; Dean, "The Forbidden City of Terry Gou"; Alfred Marshall, *Principles of Economics* ([1890] London: Macmillan and Co., Ltd., 1920), 8th ed., IV.XI.7, http://www.econlib.org/library/Marshall/marP25.html#Bk.IV,Ch.XI (accessed Sept. 22, 2014); Alfred D. Chandler, Jr., *Scale and Scope: The Dynamics of Industrial Capitalism* (Cambridge, MA: Harvard University Press, 1994), 25.

33. For Appelbaum's analysis, on which I lean heavily, see Appelbaum, "Giant Transnational Contractors."

34. The classic discussion of the importance of the link between manufacturing and distribution is Alfred D. Chandler, *The Visible Hand: The Managerial Revolution in American Business* (Cambridge, MA: Harvard University Press, 1977). See also, Nelson Lichtenstein, "The Return of Merchant Capitalism," *International Labor and Working-Class History* 81 (2012), 8–27; http://www.clarksusa.com/us/about-clarks/heritage (accessed July 19, 2016).

35. Joshua B. Freeman, *American Empire, 1945–2000: The Rise of a Global Empire, the Democratic Revolution at Home* (New York: Viking, 2012), 343–54.

36. Vanderbilt, *Sneaker Book*, 8–25, 76–88.

37. Boy Lüthje, "Electronics Contract Manufacturing: Global Production and the International Division of Labor in the Age of the Internet," *Industry and Innovation* 9 (3) (Dec. 2002), 227–47.

38. There is a large literature on changes in retailing. In addition to Appelbaum, "Giant Transnational Contractors," particularly useful works include Charles Fishman, *The Wal-Mart Effect: How the World's Most Powerful Company Really Works—and How It's Transforming the American Economy* (New York: Penguin, 2006); Lichtenstein, *The Retail Revolution*; and Xue Hong, "Outsourcing in China: Walmart and Chinese Manufacturers," in Anita Chan, ed., *Walmart in China* (Ithaca, NY: Cornell University Press, 2011).

39. For a pioneering critical look at modern branding, see Naomi Klein, *No Logo: Taking Aim at the Brand Bullies* (New York: Picador, 1999). Lüthje, "Electronics Contract Manufacturing," 230 (Nishimura quote); Marcelo Prince and Willa Plank, "A Short History of Apple's Manufacturing in the U.S.," *The Wall Street Journal*, Dec. 6, 2012, http://blogs.wsj.com/digits/2012/12/06/a-short-history-of-apples-manufacturing-in-the-u-s/; Peter Burrows, "Apple's Cook Kicks Off 'Made

in USA' Push with Mac Pro," Dec. 19, 2013, http://www.bloomberg.com/news/articles/2013-12-18/apple-s-cook-kicks-off-made-in-usa-push-with-mac-pro; G. Clay Whittaker, "Why Trump's Idea to Move Apple Product Manufacturing to the U.S. Makes No Sense," *Popular Science*, Jan. 26, 2016, http://www.popsci.com/why-trumps-idea-to-move-apple-product-manufacturing-to-us-makes-no-sense; Klein, *No Logo*, 198–99.

40. Vanderbilt, *Sneaker Book*, 90–99; *New York Times*, Nov. 8, 1997; Klein, *No Logo*, 197–98, 365–79; Donald L. Barlett and James B. Steele, "As Apple Grew, American Workers Left Behind," Nov. 16, 2011, http://americawhatwentwrong.org/story/as-apple-grew-american-workers-left-behind/; David Pogue, "What Cameras Inside Foxconn Found," Feb. 23, 2012, http://pogue.blogs.nytimes.com/2012/02/23/what-cameras-inside-foxconn-found/.

41. Lüthje, "Electronics Contract Manufacturing," 231, 234, 236–37; Boy Lüthje, Stefanie Hürtgen, Peter Pawlicki, and Martina Sproll, *From Silicon Valley to Shenzhen: Global Production and Work in the IT Industry* (Lanham, MD: Rowman & Littlefield, 2013), 69–149; Appelbaum, "Giant Transnational Contractors," 71–72.

42. For the container revolution, see Marc Levinson, *The Box: How the Shipping Container Made the World Smaller and the World Economy Bigger* (Princeton, NJ: Princeton University Press, 2006).

43. David Barboza, "In Roaring China, Sweaters Are West of Socks City"; Oliver Wainwright, "Santa's Real Workshop: The Town in China That Makes the World's Christmas Decorations," *The Guardian*, Dec. 19, 2014, https://www.theguardian.com/artanddesign/architecture-design-blog/2014/dec/19/santas-real-workshop-the-town-in-china-that-makes-the-worlds-christmas-decorations.

44. Ngai et al., "Apple, Foxconn, and Chinese Workers' Struggles," 169; *Wall Street Journal*, July 22, 2014; Adam Starariano and Peter Burrows, "Apple's Supply-Chain Secret? Hoard Lasers," *Bloomberg Businessweek*, Nov. 3, 2011, http://www.bloomberg.com/news/articles/2011-11-03/apples-supply-chain-secret-hoard-lasers; and Adam Lashinsky, "Apple: The Genius Behind Steve," *Fortune*, Nov. 24, 2008, http://fortune.com/2008/11/24/apple-the-genius-behind-steve/ (Cook quote).

45. In 2004, Foxconn employed five thousand engineers in Shenzhen alone. Duhigg and Bradsher, "How the U.S. Lost Out on iPhone Work"; Ngai, Chan, and Selden, *Dying for an iPhone*; Lüthje et al., *From Silicon Valley to Shenzhen*, 191.

46. Lüthje, Luo, and Zhang, *Beyond the Iron Rice Bowl*, 188–89; Ngai, Chan, and Selden, *Dying for an iPhone*.

47. Ngai, *Migrant Labor*, 102–03; Xue, "Local Strategies of Labor Control," 88–89.

48. Lüthje, Luo, and Zhang, *Beyond the Iron Rice Bowl*, 197; http://www.yueyuen .com/index.php/en/about-us-6/equipments (accessed Dec. 20, 2016); Dean, "The Forbidden City of Terry Gou"; lecture by Pun Ngai, Joseph S. Murphy Institute, City University of New York, Feb. 23, 2016.

49. Barboza, "In Roaring China, Sweaters Are West of Socks City"; Lu Zhang, *Inside China's Automobile Factories: The Politics of Labor and Worker Resistance* (New York: Cambridge University Press, 2015), 8, 23, 60; interview with Qian Xiaoyan (First Secretary, Embassy of the People's Republic of China in the U.S.A.), New York, Apr. 16, 2015; Ngai, *Migrant Labor*, 115–19. For Vietnamese government policy, see Nguyen Thi Tue Anh, Luu Minh Duc, and Trinh Doc Chieu, "The Evolution of Vietnamese Industry," 14–24.

50. Ngai, *Migrant Labor*, 66, 72, 78; Ngai, *Made in China*, 2–3, 55–56, 65–73; Ren, ed., *China on Strike*, 96.

51. Ngai, *Migrant Labor*, 86, 101; Emily Feng, "Skyscrapers' Rise in China Marks Fall of Immigrant Enclaves," *New York Times*, July 19, 2016; Ross, *Fast Boat to China*, 164–65; Richard Appelbaum and Nelson Lichtenstein, "A New World of Retail Supremacy: Supply Chains and Workers' Chains in the Age of Wal-Mart," *International Labor and Working-Class History* 70 (2006), 109.

52. Ngai, Chan, and Selden, *Dying for an iPhone*; Ren, ed., *China on Strike*, 97.

53. Ngai, *Made in China*, 32; Ren, ed., *China on Strike*, 5–9, 27.

54. Ngai, *Migrant Labor*, 120–23, 128–29; Ngai et al., "Apple, Foxconn, and Chinese Workers' Struggles," 174; Duhigg and Bradsher, "How the U.S. Lost Out on iPhone Work"; *Wall Street Journal*, Dec. 18, 2012. See also Lüthje et al., *From Silicon Valley to Shenzhen*, 184–87.

55. Charles Duhigg and David Barboza, "The iEconomy; In China, the Human Costs That Are Built Into an iPad," *New York Times*, Jan. 26, 2012; Ngai, Chan, and Selden, *Dying for an iPhone*; Ren, ed., *China on Strike*, 7, 184; Xue, "Local Strategies of Labor Control," 89, 92. For comparison, see William Dodd, *A Narrative of the Experience and Sufferings of William Dodd, A Factory Cripple, Written by Himself,* reprinted in James R. Simmons, Jr., ed., *Factory Lives: Four Nineteenth-Century Working-Class Autobiographies* (Peterborough, ON: Broadview Editions, 2007).

56. Chen, *China's Workers Under Assault*, 10, 12, 23, 46–81; Xue, "Local Strategies of Labor Control," 91–92; Ngai et al., "Apple, Foxconn, and Chinese Workers' Struggles," 172–74; Karl Marx, *Capital: A Critique of Political Economy*, vol. 1 ([1867] New York: International Publishers, 1967), 424;

Jee Young Kim, "How Does Enterprise Ownership Matter? Labour Conditions in Fashion and Footwear Factories in Southern Vietnam," in Chan, ed., *Labour in Vietnam*, 288; Ngai, *Made in China*, 80, 97.

57. "The poetry and brief life of a Foxconn worker: Xu Lizhi (1990–2014)" (accessed Aug. 4, 2016), libcom.org, https://libcom.org/blog/xulizhi-foxconn-suicide-poetry.

58. Serious as these problems are, large plants generally have better health and safety equipment and records than smaller parts suppliers with fewer resources and less subject to international scrutiny. Under pressure from Nike, conditions in the factory in Vietnam were improved and more use was made of less toxic water-based solvents. *New York Times*, Nov. 8, 1997, Apr. 28, 2000; Chen, *China's Workers Under Assault*, 82–97; Duhigg and Barboza, "The iEconomy"; Ngai, Chan, and Selden, *Dying for an iPhone*; Lüthje et al., *From Silicon Valley to Shenzhen*, 187.

59. Some Chinese factories consciously mix workers from different regions on production lines, to undercut worker solidarity. Others, usually smaller, recruit workers from particular regions or even villages, so that hometown bonds extend into the workplace and dormitories. Ngai, Chan, and Selden, *Dying for an iPhone*; Ngai, *Migrant Labor*, 129–30; Lüthje et al., *From Silicon Valley to Shenzhen*, 190; Xue, "Local Strategies of Labor Control," 93, 97–98.

60. *Bloomberg Businessweek*, Sept. 13, 2010; Duhigg and Bradsher, "How the U.S. Lost Out on iPhone Work"; Lüthje, Luo, and Zhang, *Beyond the Iron Rice Bowl*, 187; Ren, ed., *China on Strike*, 201–03; Ngai, *Migrant Labor*, 119, 130; Fallows, "Mr. China Comes to America," 62. See also *Factory City*.

61. Unlike in China, the reduction of poverty in Vietnam has not been accompanied by a large increase in inequality. World Bank, [China] "Overview," http://www.worldbank.org/en/country/china/overview#3; World Bank, "China" [Data], http://data.worldbank.org/country/china; and World Bank, [Vietnam] "Overview," http://www.worldbank.org/en/country/vietnam/overview, all accessed Dec. 2, 2016. Chinese strike data derived from *Chinese Labour Bulletin* "Strike Map," http://maps.clb.org.hk/strikes/en; U.S. data from United States Department of Labor, Bureau of Labor Statistics, "Work stoppages involving 1,000 or more workers, 1947–2015," http://www.bls.gov/news.release/wkstp.t01.htm (both accessed Aug. 16, 2016).

62. For overviews of strikes in China, see Ren, ed., *China on Strike*; Lee, *Against the Law*; James Griffiths, "China on Strike," CNN.com, Mar. 29, 2016, http://www.cnn.com/2016/03/28/asia/china-strike-worker-protest-trade-

union/; and *China Labour Bulletin*'s extraordinary interactive "Strike Map." See also *New York Daily News*, Jan. 11, 2012; Ngai, Chan, and Selden, *Dying for an iPhone*; Duhigg and Barboza, "The iEconomy."

63. The Vietnamese government is generally more supportive of worker strikes against foreign companies than the Chinese government and has less often used repressive power against them. Benedict J. Tria Kerkvliet, "Workers' Protests in Contemporary Vietnam" and Anita Chan, "Strikes in Vietnam and China in Taiwanese-owned Factories: Diverging Industrial Relations Patterns," in Chan, ed., *Labour in Vietnam;* "10,000 Strike at Vietnamese Shoe Factory, *USA Today*, Nov. 29, 2007, http://usatoday30.usatoday.com/news/world/2007-11-29 -vietnam-shoe-strike_N.htm; "Workers Strike at Nike Contract Factory," *USA Today*, Apr. 1, 2008, http://usatoday30.usatoday.com/money/economy/2008 -04-01-1640969273_x.htm; "Shoe Workers Strike in the Thousands," *Thanh Nien Daily*, http://www.thanhniennews.com/society/shoe-workers-strike -in-the-thousands-16949.html; "Vietnamese workers extract concessions in unprecedented strike," *DW*, Feb. 4, 2015, http://www.dw.com/en/vietnamese -workers-extract-concessions-in-unprecedented-strike/a-18358432 (all accessed Aug. 8, 2016); International Trade Union Confederation, *2012 Annual Survey of Violations of Trade Union Rights—Vietnam*; Kaxton Siu and Anita Chan, "Strike Wave in Vietnam, 2006–2011," *Journal of Contemporary Asia*, 45:1 (2015), 71–91; *New York Times*, May 14, 2014; *Wall Street Journal*, May 16, 2014, June 19, 2014.

64. Both the shrinking rural population and the gender imbalance stem in part from China's one-child policy. Ngai, Chan, and Selden, *Dying for an iPhone*; "Urban and rural population of China from 2004 to 2014," Statista (accessed Aug. 16, 2016), http://www.statista.com/statistics/278566/urban-and-rural -population-of-china/; Ren, ed., *China on Strike*, 21–23; Ngai, *Migrant Labor*, 35, 114.

65. Bruce Einhorn and Tim Culpan, "Foxconn: How to Beat the High Cost of Happy Workers," *Bloomberg Businessweek*, May 5, 2011, http://www.bloomberg .com/news/articles/2011-05-05/foxconn-how-to-beat-the-high-cost-of-happy -workers; Ngai, *Migrant Labor*, 114–15; Xue, "Local Strategies of Labor Control," 96; Chen, *China's Workers Under Assault*, 9.

66. Zhang, *Inside China's Automobile Factories*, 57–59; Ngai, *Migrant Labor*, 117–18; Ngai, Chan, and Selden, *Dying for an iPhone*.

67. For films dealing with Chinese factories and migrant workers, see Elena Pollacchi, "Wang Bing's Cinema: Shared Spaces of Labor," *WorkingUSA* 17 (Mar. 2014); Xiaodan Zhang, "A Path to Modernization: A Review of Doc-

umentaries on Migration and Migrant Labor in China," *International Labor and Working-Class History* 77 (Spring 2010).

68. For the factory as a sales tool, see Gillian Darley, *Factory* (London: Reaktion Books, 2003), 157–89. In China, EUPA seems something of an exception, allowing filmmakers and photographers to document its factory. For examples of tightly controlled tours, see James Fallows, "Mr. China Comes to America," and Dawn Chmielewski, "Where Apple-Products Are Born: A Rare Glimpse Inside Foxconn's Factory Gates," Apr. 6, 2015, http://www.recode.net/2015/4/6/11561130/where-apple-products-are-born-a-rare-glimpse-inside-foxconns-factory.

69. *Bloomberg Businessweek*, Sept. 13, 2010; Xing Rung, *New China Architecture* (Singapore: Periplus Editions, 2006); Layla Dawson, *China's New Dawn: An Architectural Transformation* (Munich: Prestel Verlag, 2005).

70. Neil Gough, "China's Fading Factories," *New York Times,* Jan. 20, 2016; Feng, "Skyscrapers' Rise in China Marks Fall of Immigrant Enclaves"; Mark Magnier, "China's Manufacturing Strategy," *Wall Street Journal*, June 8, 2016.

71. For example, compare two collections of images by the pioneer American photographer Lewis W. Hine: Hine, *Men at Work: Photographic Studies of Modern Men and Machines* ([1932] New York: Dover, 1977), and Jonathan L. Doherty, ed., *Women at Work: 153 Photographs by Lewis W. Hine* (New York: Dover, 1981). Of course, gender patterns have varied over time and place, with more women working in heavy industry in communist countries than capitalist ones and gender imbalances diminishing over time.

72. Countless examples can be seen by doing a Google search for images of Chinese factories.

73. For Burtynsky, see http://www.edwardburtynsky.com/site_contents/Photographs/China.html (accessed Dec. 2, 2016); for Gursky, see, for example, Marie Luise Syring, *Andreas Gursky: Photographs from 1984 to the Present* (New York: TeNeues, 2000).

Conclusion

1. Kenneth E. Hendrickson III, ed., *The Encyclopedia of the Industrial Revolution in World History,* vol. III, 3rd ed. (Lanham, MD: Rowman & Littlefield, 2014), 568; R. S. Fitton, *The Arkwrights: Spinners of Fortune* ([1989] Matlock, UK: Derwent Valley Mills Educational Trust, 2012), 228–29; Timothy J. Minchin, *Empty Mills: The Fight Against Imports and the Decline of the U.S.*

Textile Industry (Lanham, MD: Rowman & Littlefield Publishers, 2013), 31; Tamara K. Hareven and Randolph Lanenbach, *Amoskeag: Life and Work in an American Factory City* (New York: Pantheon Books, 1978), 10–11; Gray Fitzsimons, "Cambria Iron Company," Historic American Engineering Record, National Park Service, Department of the Interior, Washington, D.C., 1989; William Serrin, *Homestead: The Glory and Tragedy of an American Steel Town* (New York: Random House, 1992).

2. Lindsay-Jean Hard, "The Rouge: Yesterday, Today & Tomorrow," Urban and Regional Planning Economic Development Handbook, University of Michigan, Taubman College of Architecture and Urban Planning, Dec. 4, 2005, http://www .umich.edu/~econdev/riverrouge/; Perry Stern, "Best Selling Vehicles in America— September Edition," Sept. 2, 2016, http://www.msn.com/en-us/autos/autos -passenger/best-selling-vehicles-in-america-%E2%80%94-september-edition/ ss-AAiquE5#image=21.

3. Laurence Gross, *The Course of Industrial Decline: The Boott Cotton Mills of Lowell, Mass., 1835–1955* (Baltimore, MD: Johns Hopkins University Press, 1993), 44–45, 102–03, 229, 238–40.

4. Jefferson Cowie and Joseph Heathcott, "The Meanings of Deindustrialization," in Cowie and Heathcott, eds., *Beyond the Ruins: The Meanings of Deindustrialization* (Ithaca, NY: Cornell University Press, 2003), 4. There is a large literature on deindustrialization. In addition to this volume, see the cluster of articles on "Crumbling Cultures: Deindustrialization, Class, and Memory," ed. Tim Strangleman, James Rhodes, and Sherry Linkon, in *International Labor and Working-Class History* 84 (Oct. 2013).

5. Paul Wiseman, "Why Robots, Not Trade, Are Behind So Many Factory Job Losses," *AP: The Big Story*, Nov. 2, 2016, http://bigstory.ap.org/articl e/265cd8fb02fb44a69cf0eaa2063e11d9/mexico-taking-us-factory-jobs- blame-robots-instead; Mandy Zuo, "Rise of the Robots: 60,000 Workers Culled from Just One Factory as China's Struggling Electronics Hub Turns to Artificial Intelligence," *South China Morning Post*, May 22, 2016, http:// www.scmp.com/news/china/economy/article/1949918/rise-robots-60000- workers-culled-just-one-factory-chinas. See also *Wall Street Journal*, Aug. 17, 2016.

6. Rich Appelbaum and Nelson Lichtenstein, "An Accident in History," *New Labor Forum* 23 (3) (2014), 58–65; Ellen Barry, "Rural Reality Meets Bangalore Dreams," *New York Times*, Sept. 25, 2016.

7. Kevin Hamlin, Ilya Gridneff, and William Davison, "Ethiopia Becomes China's China in Global Search for Cheap Labor," *Bloomberg*, July 22, 2014, https:// www.bloomberg.com/news/articles/2014-07-22/ethiopia-becomes-china-s

-china-in-search-for-cheap-labor; Lily Kuo, "Ivanka Trump's Shoe Collection May Be Moving from 'Made in China' to 'Made in Ethiopia,'" *Quartz Africa*, Oct. 8, 2016, http://qz.com/803626/ivanka-trumps-shoe-collection-may-be-moving-from-made-in-china-to-made-in-ethiopia/; Chris Summers, "Inside a Trump Chinese Shoe Factory," *Daily Mail.com*, Oct. 6, 2016, http://www.dailymail.co.uk/news/article-3824617/Trump-factory-jobs-sent-China-never-come-back.html.

8. For variations of the factory under different social systems, see Michael Burawoy, *The Politics of Production: Factory Regimes Under Capitalism and Socialism* (London: Verso, 1985), and Dipesh Chakrabarty, *Rethinking Working Class History: Bengal, 1890–1940* (Princeton, NJ: Princeton University Press, 1989).

9. The documentary film *After the Factory* (Topografie Association, 2012), comparing efforts in Lodz, Poland, and Detroit at postindustrial reinvention, suggests the possibilities and limitations of such strategies.

10. 4-traders: "Hon Hai Precision Industry Co., Ltd.," http://www.4-traders.com/HON-HAI-PRECISION-INDUSTR-6492357/company/, and "Pegatron Corporation," http://www.4-traders.com/PEGATRON-CORPORATION-6500975/company/, both accessed July 5, 2016, and "Yue Yuen Industrial (Holdings) Ltd.," accessed Jan. 1, 2017; "Fast Facts About Vanguard" (accessed Jan. 3, 2017), https://about.vanguard.com/who-we-are/fast-facts/; Calvert Social Investment Fund, "Annual Report," Sept. 30, 2016, 4, 7.

Illustration Credits

Index

Page numbers in *italic* refer to illustrations.